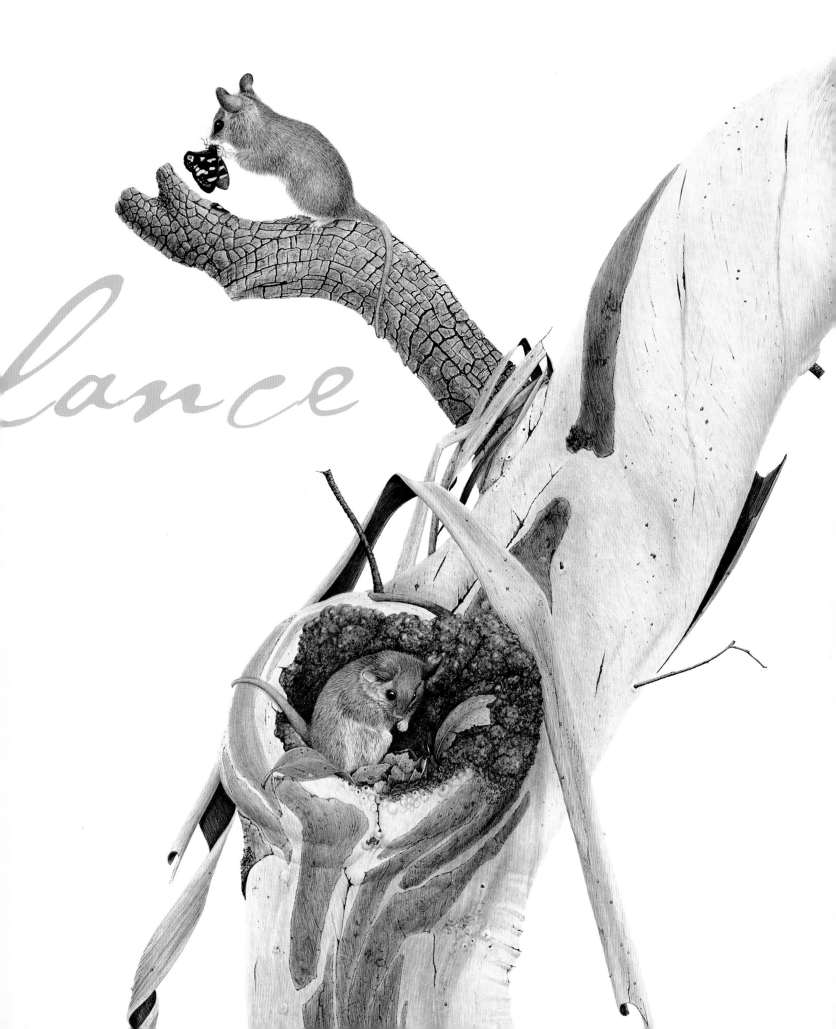

contents

Australia's marsupials represent an independent experiment in mammal evolution. Over nearly 200 million years, and beginning as obscure rat-sized creatures, they have diversified into wolf- and leopard-like types, rhino-sized herbivores and the well known kangaroos, Koala and wombats. Yet as Chris Dickman reminds us, for more than half of their evolutionary existence, the marsupials were creatures of the northern hemisphere. Until around 60 million years ago, Australia was home to not one.

The story of what happened to the descendants of the marsupial pioneers that crossed the Antarctic from South America to enter Australia is the focus of this wonderful book. Perhaps the extraordinarily warm conditions that prevailed briefly 55 million years ago (so warm that rainforests then grew in Greenland) allowed them to make the journey. Whatever the cause, when they arrived they encountered conditions unlike anywhere else: Australia's old, weathered soils were limiting to plant and animal life. That may be the reason that kangaroos hop – after all hopping is the most efficient means of locomotion ever devised – why males of the tiny antechinus die after mating, and why the Koala's brain is so small.

One of the most astonishing discoveries of recent years concerns the mammals that lived in Australia more than 60 million years ago before the marsupials arrived. Patient palaeontological excavation over many years has revealed that a diversity of superficially hedgehog-like placental mammals thrived in Australia during the age of dinosaurs. The last of these creatures lived around 55 million years ago, at the time of the marsupial incursion, leading one to speculate whether, uniquely, in Australia the marsupials drove a group of placentals to extinction.

Australia is a land of wonders and its marsupials are arguably its crowning glory. In this book, two passionate students of these astonishing mammals bring to life decades of patient observation and experience. Rosemary Woodford Ganf is one of the world's most accomplished wildlife artists, while Chris Dickman is one of its leading mammalogists. *A Fragile Balance* really is a unique resource about a fascinating evolutionary history.

Tim Flannery
September 2007

AUTHOR'S PREFACE

If you were a student in the 1960s and wished to find out about Australian marsupials, your options were quite limited. Apart from the obvious distractions of that era, few people were then studying marsupials, and compendia on marsupial biology did not exist. You could either hope to make your own direct observations of captive animals in zoological parks or in the wild, or seek original studies on marsupials that had been published in the diverse but often arcane scientific literature. Descriptive accounts of many species could be obtained in general books on Australian marsupials – including Waterhouse's *Marsupialia or pouched animals*, published in 1843, Lydekker's 1896 *Hand-book to the Marsupialia and Monotremata*, and Ellis Troughton's wonderful *Furred animals of Australia*, first published in 1941 and reprinted nine times with updates until 1967 – but accessible overviews of marsupial biology were conspicuously missing. Indeed, there was a feeling in the 1960s that marsupials were just 'second class' mammals and that people who wanted to study them were a bit odd. However, it was legitimate to study marsupials for the insights that they could bring to our understanding of the origins and biology of the 'higher' and more advanced placental mammals, including, of course, ourselves.

Fortunately, these widespread sentiments did not stop dedicated workers in several Australian laboratories from pursuing seminal studies on the ecology, reproduction and physiology of different marsupials. Most notable among these were the extensive field surveys by the Wildlife Survey Section, later the Division of Wildlife Research, of CSIRO (now CSIRO Sustainable Ecosystems), and the detailed investigations of marsupial nutrition and physiology pursued at the University of Western Australia. Individual researchers at other universities, museums and state government agencies were also beginning to uncover some of the mysteries of the marsupial way of life, in particular in reproduction, paleontology and systematics.

By the early 1970s research on marsupial biology had made many advances, so that a critical review of the field was timely. Hugh Tyndale-Biscoe's landmark 1973 volume, modestly titled *Life of marsupials*, heralded what might be called the 'modern era' of marsupial studies. On the one hand it expounded the now-prevailing view that marsupials are of intrinsic interest and should be considered as successful alternative mammals rather than as inferior runners-up to their distant placental relatives. On the other, *Life of marsupials* summarized the burgeoning research on marsupials and provided exciting ideas on how the field might develop in the future. The effect was swift and dramatic. Within four years two edited volumes entitled *The biology of marsupials* were in print, and during the 1980s edited volumes appeared on carnivorous marsupials, possums and gliders, possums and opossums, and kangaroos and rat-kangaroos.

Specialist volumes have been published since on high profile species such as the Koala, wombats, kangaroos, tree-kangaroos and the Mountain Pygmy-possum. Several field guides advise on how to identify marsupials in the field from their appearance, their faecal remains and other calling cards. In 2005, the publishers of the venerable *Walker's mammals of the world* produced a stand-alone volume on marsupials and Hugh Tyndale-Biscoe completed a new and masterly *Life of marsupials*. The following year Cambridge University Press published a text entitled just *Marsupials*.

In such an expert and crowded field, what can another book on marsupials hope to achieve? In the first instance, this book aims to provide up-to-date information to a broad audience and not just to specialist students of marsupial biology. It is aided immeasurably in this respect by the wonderful illustrations of Rosemary Woodford Ganf, which depict the extraordinary diversity of marsupials and many of their intriguing behaviors. It also aims to cover all species in Australia, and to use an evolutionary framework to interpret the many unique biological traits that characterize the marsupials of this region. For the first time, emphasis is placed on how marsupials interact with each other and with other living things, and how they modify their environment to make it more or less suitable for other species. Their impacts are often startling. For the first time too, the tangled interactions between marsupials and people are examined. What are the values of marsupials from historical, cultural, ecological and economic perspectives, and how can we reconcile the contradictory status of some species as both pests and targets for conservation? By portraying marsupials as integral components of the natural environment and of human endeavor, we hope to break new ground.

Christopher Dickman
September 2007

My career as a wildlife artist began in the Queen Elizabeth National Park at the foot of the Rwenzori Mountains, the fabled Mountains of the Moon, in western Uganda.

I had left art school at 19, determined to pursue what was considered to be an unconventional career path. My father, a veterinarian, was working with a terrestrial ecology unit in the national park. I had a unique opportunity to work alongside him and his colleagues researching both the natural history and anatomy of the wildlife. Their careful observations and recordings instilled in me the attention to detail I would need for my chosen career. Both my father and mother, a talented craftswoman herself, never wavered in their support for my pursuit of the discipline and self confidence necessary to be an artist. I am deeply indebted to them both for their loving support and confidence in me, which continues to this day.

Those early exciting years working in close proximity to potentially dangerous animals such as buffalo, lion and leopard, were not without incident. One evening I had driven alone to a hyena den. I had promised to return by dark so, at sunset, surrounded by about 30 inquisitive hyenas, I attempted to start my car. The lifeless click that came from the ignition made me realise the battery terminal was the problem. I also realised that to fix it I would need to lift the bonnet and hit the terminal with a stone. My vigorous beating of the car door panel sent the hyenas back to their den just long enough for me to execute the makeshift mechanics. Mercifully, the car started and I drove home with a considerable adrenalin rush.

Almost as exciting and scary was being asked to dance by Idi Amin. I accepted the invitation, as rumours of his brutality to his enemies made me keen to appear friendly. He asked what I was doing in Uganda and when I replied that I was a wildlife artist, he promptly asked me to paint his portrait! I accepted for the same reason as before but, fortunately, he never got back to me.

When I arrived in Australia in 1974, both the people and the wildlife seemed remarkably friendly. After a successful solo exhibition at Australian Galleries in Melbourne in 1976, I was asked by a publisher to choose an Australian wildlife subject and illustrate it in limited edition form. I was somewhat daunted by the apparent scarcity and invisibility of Australian wildlife when I was used to the abundance and visibility of African wildlife. I chose the marsupials when I realized that no-one had hand-illustrated all the then-described Australian species. I began the illustrations in 1977, and completed the third and final volume in 1989, about 8 years of actual work. I added the recently rediscovered Gilbert's Potoroo to Volume 3 in 2004.

The science behind the knowledge of marsupials has been dynamic in its progress since I began the work. The late John Calaby told me that it would be impossible for me to ever claim that I had painted every Australian marsupial while the scientific 'splitters and lumpers' continued their work. Recent advances in taxonomy have, indeed, yielded new species which I will leave to a future illustrator. However, modern digital sampling techniques now used in illustration may make my hand-illustrated effort the first and last of its kind.

The limited-edition work reached a small, mainly Australian, audience. With this new book, using the same illustration collection alongside a more accessible narrative text, my publisher and I hope to find a wider international audience. I could not have imagined a more perfect text than the one Christopher Dickman has provided. Like me, he has a strong natural history background, and he draws us into his text with anecdotes of his personal experiences in the pursuit of his science. I salute him, both as a fine field naturalist and as a very readable author able to reveal the love and dedication he has for his work.

A few years ago I experienced a kind of artistic epiphany while watching a gaunt female dingo trotting along the Strzelecki Track. I felt a great respect and empathy for her in her determined struggle to survive and provide for her pups in such harsh conditions. It was 46°C (115°F) and I was in an air-conditioned 4-wheel drive. She was backlit by the setting sun and her image burned into my memory. Two years later she became my first 3-dimensional work as a limited-edition bronze cast.

The driver of the vehicle was escorting me and a British travel journalist to the Coongie Lakes. He shared his knowledge, respect and passion for the outback and, on that and subsequent trips, he helped me realise what a privilege it is to watch Australian wildlife. We have formed a strong friendship and I now know that I no longer need to compare today's experiences with those in Africa many years before. Over the past 3 years I have made the furniture for a luxury tented safari camp he now operates in the Gawler Ranges on the Eyre Peninsula. The lessons of making the furniture from 'wild' mallee wood, added to the skills and knowledge gained from my illustration work, have led to my present primary artform, lifesized wildlife sculptures made from the 'found' wood that surrounds me.

So, like the extraordinary animals that I now live amongst, I have evolved and adapted my work to my surroundings and, for the first time in my 37 year career, I feel accepted by the art world as a 'real' artist. Christopher may have been viewed as a 'bit odd' for wanting to study marsupials in the 1960s; I may be viewed in the same way still for making them the subject of my art. But I can't think of a more wonderful life.

Rosemary Woodford Ganf
September 2007

Spotted-tailed Quoll
Dasyurus maculatus Male
Near Cradle Mountain, Tasmania

INTRODUCTION

a growing understanding

The first marsupial that I encountered in the bush was a Common Ringtail Possum (*Pseudocheirus peregrinus*). Transfixed by the light from my hand torch, the animal sat still on a horizontal branch about level with my head, without obvious signs of fear, and let me approach. Even to the uninitiated, this was a marsupial. A small, pocket-like opening on the belly indicated that the animal was a female, and a gentle pulsing from the pouch suggested that at least one joey was squirming about in there on a teat. I knew that this soft-looking creature specialized in eating leaves rather than flesh, and reached out my hand to grasp her and get a closer look. In an instant, the possum bared her lips, revealed a pair of bayonet-shaped lower teeth and plunged them repeatedly into my thumb. She was on the ground and running before I felt the pain.

Common Ringtail Possum
Pseudocheirus peregrinus Female
MOUNT LOFTY RANGES, SOUTH AUSTRALIA

That was early in 1978. I had left England only a few days earlier and arrived in Australia to begin a doctoral research program on some of the continent's magnificent but elusive marsupials. My encounter with the possum reminded me of just how much I still had to learn.

As a boy I had been fascinated by myths and stories about pouched mammals. Was the majestic Thylacine (*Thylacinus cynocephalus*) truly extinct or could it still lurk somewhere in the forgotten forests of Tasmania? How could the Koala (*Phascolarctos cinereus*) – or anything else – survive on a diet of toxic *Eucalyptus* leaves? How could it sleep so much and not fall out of the trees? Could kangaroos really box, climb trees and outrun fast dogs? I had heard that there were marsupials resembling mice, rats, cats, squirrels and even small ground-dwelling bears. There were also rumors of a 'vampire-marsupial' that would emerge under the cover of night and suck the blood from hapless chickens, ducks and other domestic fowl. And then there were the marsupial giants of the last Ice Age: bone-cracking carnivores to rival modern day lions and lumbering herbivores that would tip the scales at some 2800 kilograms (6170 pounds). Could some of these brutes have given rise to the enduring folklore about bunyips? I was enthralled and, at an early age, resolved to travel to the Great South Land to see some of the continent's enigmatic wildlife at first hand.

While a student at the University of Leeds in northern England in the mid-1970s, I was disappointed, but not really surprised, to find that marsupial studies were not much in demand. However, a science degree provided excellent training in critical thinking, and it seemed to a young sceptic that some of the established mythology about marsupials had been built upon shaky foundations that could be checked, and dispelled if necessary, by careful study. One puzzle stood out. At the time, about 30 years ago, marsupials were considered by many to be relict mammals, able to persist only in parts of the world where more advanced placental mammals had not invaded. This view seemed to explain quite neatly why many species of marsupials became extinct in Australia after European arrival – they were simply not up to the competition handed out to them by the settlers, and the rats, cats, foxes, sheep and cattle that they brought with them. But perhaps this view was too glib. I knew that some marsupials had been moved beyond the confines of Australia and had set up feral populations in the face of stiff competition; indeed, in New Zealand the introduced Common Brushtail Possum (*Trichosurus vulpecula*) had run amok and was the country's number one pest. I had also seen feral Red-necked Wallabies (*Macropus rufogriseus*) in the English Peak District – not a place for wimpy or inferior mammals.

Beyond the myths, just about every aspect of marsupial biology is intrinsically fascinating. Let us consider, for example, the reproductive biology of these mammals. Pregnancy lasts from just 11 to 42 days, depending on the species,

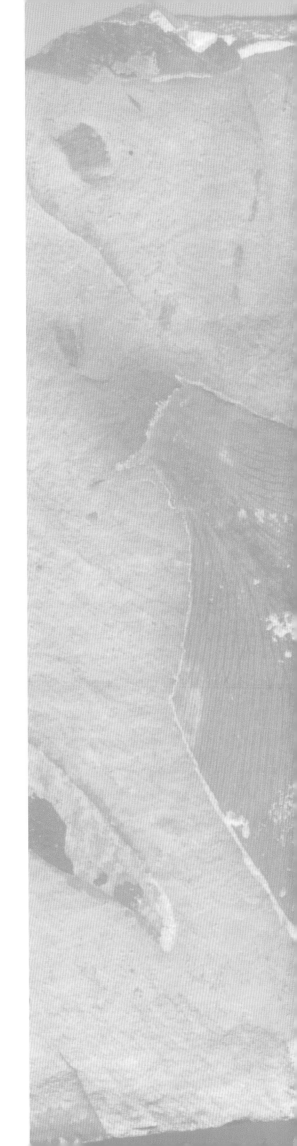

and the young at birth are not much larger than a grain of rice. Yet, these tiny creatures make their own way independently to the mother's pouch, where they attach to a nipple to obtain nourishment for the next stage in their lives. In kangaroos that nurse young of different ages, the mother has the extraordinary ability to produce two different kinds of milk from different nipples; each is attuned exquisitely to the requirements of the joey that suckles them. Among the smaller marsupials, antechinuses are renowned for their frenetic approach to reproduction. Both sexes mate promiscuously for about two weeks each winter, and then all males die. Females in these and other marsupial species appear able to bias the sex ratio of their litters, producing more males or more females depending on their age and environmental conditions. From a biogeographical point of view, there are further intriguing conundrums in explaining why marsupials occur in just the Americas and the Australian region, and how one little 'Australian' species came to be marooned on the slopes of the Andes.

In recent years our understanding and appreciation of marsupials has increased dramatically. On the one hand, we know that the marsupials diverged from other mammals much before 125 million years ago, making them profoundly different from their distant placental relatives. On the other, marsupials are providing deep insights into behavioral, ecological and evolutionary processes, and are being used increasingly as models in biomedical research. Some are considered to be pests; many others are at risk of endangerment or even extinction. There is a fragile balance for these marsupials between persistence annd oblivion.

In the 30 years since my first encounter with a marsupial, my scars have healed but my fascination with all things marsupial has only increased. The following pages aim to convey that fascination and foster an appreciation for marsupials, not so much as cute and cuddly icons, but as important and integral components of the Australian environment.

Brown Antechinus
Antechinus stuartii Male
EASTERN NEW SOUTH WALES

Red-necked Wallaby *Macropus rufogriseus*
Female with pouch joey
CRADLE MOUNTAIN, TASMANIA

Bilby

The marsupial omnivore, the Bilby, may look like a curiosity with its large, rabbit-like ears and strong claws for burrowing, but its physiological and behavioral attributes allowed it to exist in a wide range of landscapes and rainfall regimes, even the harsh dry deserts of inland Australia.

Hedley Finlayson, an early pioneer in arid zone ecology offered the following when describing this beautiful and unusual animal in 1935:

Remarkable amongst the smaller forms is the talgoo, one of the so-called rabbit bandicoots, which has carried a number of structural peculiarities to grotesque lengths yet manages to reconcile them all in a surprisingly harmonious, and even beautiful, whole.

The coat is one of the most beautiful amongst the marsupials: fine, silky, slate-blue, and quite like chinchilla. But the general aspect of the animal recalls a miniature aardvark and it resembles that African animal in being a most powerful burrower.

My first encounter with an Aardvark in the wild, while spotlighting along the lonely Trans Kalahari Highway in the back blocks of Botswana several years ago, brought home to me this striking resemblance in form of the two species: a wonderful example of convergent evolution.

Bilby *Macrotis lagotis* Male
Near Birdsville, south-western Queensland

A Bilby's burrow, which it may excavate down for 2 metres (6.5 feet), provides a secure refuge during daylight hours. Individuals generally only emerge from the burrow several hours after sunset to begin their nightly foraging activities. They eat a wide range of invertebrates, including ants, grasshoppers, termites and spiders, but also take small reptiles and mammals, and will consume seeds and bulbs when these animals are less abundant. The Bilby doesn't need access to free water. It survives the environmental rigors of the arid zone by efficiently using and conserving the small amounts of water present in the tissue of its food.

Following what can sometimes be a prolonged courtship and copulatory phase lasting many hours, female Bilbies have a remarkably short gestation period of 12–14 days (the second shortest known for any mammal) at the end of which, two (usually) 'baked beans with legs' emerge from the cloaca and make a short but tortuous journey to the pouch. The marvel of nature that permits a 12-day pregnancy quite understandably never ceases to amaze me but it astounds my female friends and associates.

Once inside the cosy, moist confines of the pouch, each newborn attaches to one of the eight nipples present. Both are nourished by the female for a further 80 days before emerging as 200 gram (7 ounce) 'cute balls of fluff'. The

young become independent a few weeks later – innocent, inquisitive, naïve and very vulnerable.

This vulnerability is reflected in the dramatic decline in range and numbers of the species over the past 200 years. It has disappeared from more than 90% of its former range. The decline has been attributed to a wide range of human activities that have disrupted, removed or modified habitat of the species.

Initial impacts came from losses of habitat as Europeans 'opened up' the country with agricultural expansion in the early 1800s. A second period of impacts during the mid to late 1800s appears to be indirectly related the deliberate introduction of successful exotic species such as the European Rabbit and the Red Fox. Populations of domestic Cats also established themselves in the wild as pastoralism spread inland.

A third phase of impact is threatening many arid zone species including the Bilby. Oddly, it is the result of a government scheme based on a conservation premise: the need to conserve the most precious resource in a dry continent – water. The legitimate water conservation efforts aim to limit the waste of water flowing from the extensive subterranean reserves bound up in the porous sandstone deposits of the Great Artesian Basin. For over 100 years this artificial flow has watered domestic livestock. The subsidized conservation scheme has the potential to transform formerly ecologically arid and semi-arid landscapes into ones that advantage certain species, particularly exotic predators. The additional pressure this scheme might apply to some native species, such as the Bilby, might ultimately nudge them over the extinction cliff edge – gone forever.

The Bilby is a non-specialist omnivore, has no strict habitat preference, has a high reproductive output, shows little or no seasonality in breeding, is semi-fossorial, is strictly nocturnal, and doesn't require free water. In short, if it were not now an endangered species, it could be a pest.

If we lose a species such as the Bilby, we are doing something drastically wrong.

Peter McRae
QUEENSLAND PARKS AND WILDLIFE SERVICE, CHARLEVILLE

WHAT IS A MARSUPIAL?

an incredible mother

The great vertebrate class Mammalia contains living representatives within three major subgroups. The largest of these, the cohort Placentalia, contains some 5100 species, and includes many mammals that are familiar to us in farm yards, as beasts of burden, as pets in the home, as carriers of disease, or as pests of our agricultural and economic enterprises. Humans too are members of this cohort. These animals are united by their possession of an advanced placentation system that nourishes young for extended periods while they grow in the uterus of the mother. The young of placental mammals are born at a relatively advanced stage of development; some, such as foals, can stand and even walk within minutes of birth. Placental mammals occur in all parts of the world. Whales, dolphins and manatees have conquered the oceans, and some – the bats – have taken to the air.

The smallest group of living mammals is placed within its own subclass, the Prototheria, within the order Monotremata, and comprises the Platypus (*Ornithorhynchus anatinus*) and four species of echidnas. Despite just squeaking in as mammals by possessing hair, mammary glands, middle ear-bones and the ability to maintain a constant body temperature, these bizarre animals are distinguished from their hairy relatives because they do not produce live young; they lay eggs. Living members of this group are confined to Australia and New Guinea.

The third group of living mammals, with about 330 species, is the cohort Marsupialia. Mammals in this group stand apart primarily because they produce miniscule young at birth, and then suckle them for extended periods at nipples on the mother's belly. In marsupials, growth in the mother's uterus is brief, and the young are born at an early stage of development relative to their placental counterparts. Marsupials are found in Australasia, defined here to include Australia, New Guinea, and islands east to the Solomon Group and west to the Makassar and Lombok straits, or Wallace's Line (see Figure 1 on page 10); and are also represented strongly in the Americas.

Whiptail Wallaby *Macropus parryi* Female and juvenile
CLARENCE RIVER, NEW SOUTH WALES

Marsupials: the six per cent solution

Marsupials comprise just over 6% of the world's mammal species, and represent a conspicuously successful alternative experiment in mammalian evolution. Living marsupials show astonishing variation in size, shape, form, behavior and the ways in which they exploit the environment. Thus, for example, the tiny Long-tailed Planigale (*Planigale ingrami*) weighs just 4 grams (0.14 ounces), some 20,000 times less than an 80 kilogram (176 pound) Red Kangaroo (*Macropus rufus*). The former species produces litters of up to eight young and spends its time foraging for insects on the soil surface or in underground cracks; the kangaroo gives birth to a single joey and grazes on fresh green grass and herbs. Some marsupials have taken to living on fruits or leaves that they pluck from under the canopy of tall trees, some specialize in eating truffles and other underground fungi, while others are ruthless hunters that prey on smaller mammals. Despite these great

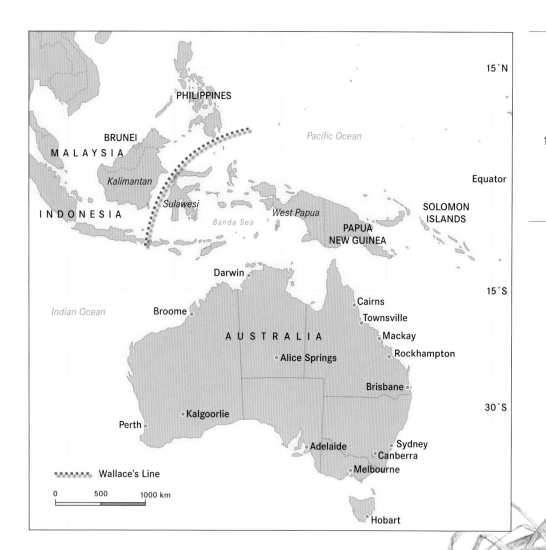

PHILIPPINES

BRUNEI
MALAYSIA

Kalimantan

Sulawesi

INDONESIA

Banda Sea

West Papua

SOLOMON
ISLANDS

PAPUA
NEW GUINEA

Indian Ocean

Darwin

Broome

Cairns
Townsville
Mackay
Rockhampton

AUSTRALIA

Alice Springs

Brisbane

Kalgoorlie

Perth

Adelaide

Sydney
Canberra
Melbourne

Hobart

········· Wallace's Line

0   500   1000 km

Figure 1 Marsupials are found in all parts of Australasia, defined here as the region south and east of Wallace's Line. The distinct and very different faunas of Australasia and south-east Asia were first recognized in the mid-19th century by the great biogeographer Alfred Russell Wallace, after whom the demarcating line is named. (Based on *Drawing the Wallace Line*, by Penny Van Oosterzee)

differences in biology, we will see that all marsupials are united by similarities in their skulls and skeletons, and, especially, their mode of reproduction. But before this we need to bring order to the discussion, and understand how marsupials are classified and named.

The science of classifying organisms began in 1735 with the publication of the *Systema Naturae* by Carolus Linnaeus, a Swedish doctor of medicine. Linnaeus continued to expand this work, and in the 1758 edition of *Systema Naturae* he introduced a two-part naming system for species that has remained largely unchanged to the present. The system recognizes species as the basic unit of interest, and groups similar *species* together within a higher level category called the *genus*. Using the example of the Red Kangaroo, above, the genus *Macropus* contains 14 species of kangaroos and wallabies that share obvious similarities, but the species name *rufus* refers only to the Red Kangaroo.

Long-tailed Planigale *Planigale ingrami* Male
NORTHERN TERRITORY

Importantly, similar genera are themselves grouped into families, families into orders and orders into classes. At the top of the pile is the kingdom, to which all animals belong. The system is hierarchical, with similarity between species decreasing at progressively higher levels in the structure.

TABLE 1 Classification and distribution of living and recently extinct marsupials

Family	Common names	Distribution	Genera	Species
'Ameridelphia' Order Didelphimorphia				
Didelphidae	Lutrine, Water and American opossums; four-eyed, mouse or murine, short-tailed, Patagonian, black-shouldered, woolly and bushy-tailed opossums	North, South and Central America	17	87
Order Paucituberculata				
Caenolestidae	Shrew opossums	South America	3	6
'Australidelphia' Order Microbiotheria				
Microbiotheriidae	Monito del Monte	South America	1	1
Order Dasyuromorphia				
Thylacinidae	Thylacine	Australia	1	1
Dasyuridae	Quolls, Tasmanian Devil, dunnarts, antechinuses, planigales, ningauis, phascogales and many others	Australia and New Guinea	13 (8)	56 (13)
Myrmecobiidae	Numbat	Australia	1	1
Order Notoryctemorphia				
Notoryctidae	Marsupial moles	Australia	1	2
Order Peramelemorphia				
Peramelidae	Bandicoots, echymiperas	Australia and New Guinea, Seram Island	3 (3)	8 (10)
Chaeropodidae	Pig-footed Bandicoot	Australia	1	1
Thylacomyidae	Bilbies	Australia	1	2
Order Diprotodontia				
Phalangeridae	Cuscuses, brushtails, scaly-tail possums	Australia and New Guinea	4 (2)	8 (19)
Burramyidae	Pygmy-possums	Australia and New Guinea	2	5
Acrobatidae	Feathertail Glider	Australia and New Guinea	1(1)	1 (1)
Petauridae	Gliders and trioks	Australia and New Guinea	3	6 (5)
Pseudocheiridae	Ringtail possums	Australia and New Guinea	6	7 (10)
Tarsipedidae	Honey Possum	Australia	1	1
Phascolarctidae	Koala	Australia	1	1
Vombatidae	Wombats	Australia	2	3
Hypsiprymnodontidae	Musky Rat-kangaroo	Australia	1	1
Potoroidae	Rat-kangaroos	Australia	4	10
Macropodidae	Kangaroos and wallabies	Australia and New Guinea	9 (2)	45 (20)

Specialists sometimes differ in their approach to classification and may even interpret the same dataset in different ways. This table, outlining the approach of this book, is based on the classification and taxonomy in a major review published in 2005: *Mammal species of the world: A taxonomic and geographic reference*, edited by Don Wilson and DeeAnn Reeder. For 'Australidelphians', unbracketed numbers show genera and species in Australia, and numbers in brackets show additional genera and species that occur elsewhere in New Guinea and other parts of Australasia.

Fat-tailed Dunnart   *Sminthopsis crassicaudata*
Male (top) and female
BETOOTA, SOUTH-WESTERN QUEENSLAND

Recognizing and grouping species just by their apearance can be problematic, and does not necessarily reflect their evolutionary relationships. Linnaeus himself grouped armadillos, insectivores, pigs and an American marsupial together in the 1758 edition of *Systema Naturae* based on their (very) superficial similarities. These mammals are placed in different orders today. Recognizing similarity from shared ancestry means looking beyond the obvious at more fundamental, or conserved, traits such as reproduction, cranial or dental anatomy. Increasingly, it also means describing and comparing species at a molecular level. Molecular studies generally target large molecules such as proteins or DNA and compare sequences of amino acids or nucleotides along parts of the molecules. The more similar the sequences, the more closely related are the two species being compared.

If morphological, molecular and other characters are used together it is often possible to gain a strong consensus about the degree of relatedness of different species and how they fit within the hierarchical structure advocated by Linnaeus. And so it is with the marsupials. After much study and debate, living marsupials are usually grouped into seven orders containing 21 families and some 92 genera. Three of these orders and families, as well as 21 genera, are confined to the Americas, with the bulk of marsupial diversity residing in Australasia. The current classification is shown in Table 1 on page 11. As well as the living forms, another five orders and 37 families of marsupials and marsupial-like taxa are known from the fossil record.

By convention, the genus + species names together are called the *scientific name* and are usually written in italic; if used repeatedly or in situations where there is no confusion, the genus name is often abbreviated to just its first letter. Thus, *Macropus rufus* would be written *M. rufus*. In the text of this book I have adopted a further convention of giving the scientific name of a species when first mentioned within a chapter together with its accepted common name, and then using the common name thereafter. As proper names, species' common names are also capitalized. This is pedantic, of course, but it serves another purpose: the common names of many marsupials are descriptive but not always accurate. For example, Common Dunnarts (*Sminthopsis murina*) are frustratingly scarce and difficult to catch, so labelling them 'common' is misleading. Likewise, Fat-tailed Dunnarts (*Sminthopsis crassicaudata*) have skinny tails when they are hungry, Yellow-footed Antechinuses (*Antechinus flavipes*) have brown feet in parts of their range, and female Red Kangaroos are actually bluish grey. Capitalized common names thus serve to identify the species and not its attributes. If species have widely used Aboriginal names, these too are given.

In the Americas all marsupials are known as 'opossums' but in Australia the similar term 'possum' is used as part of the common name for some members of the order Diprotodontia.

Distinctive features

Reproduction: the facts Of all the features that make marsupials unique, the reproductive anatomy and mode of reproduction stand out. In females, there are two lateral vaginal canals that serve to transport sperm after copulation and a single central canal that opens during birth. The uteri are also separate, rather than fused into a single uterus as in most placental mammals. In males, the testes are almost always housed externally in a scrotal sac (they are held internally only in marsupial moles), and are positioned above the penis. In placental mammals, by contrast, the testes lie below the penis and are sometimes held internally. In both sexes, urine and faeces are expelled through a common opening called the cloaca. This is also the birth opening, and males extrude the penis through the cloaca during copulation. These differences are highlighted in Figure 2 on pages 14–15. Possession of this common opening distinguishes marsupials from placental mammals but not from the Platypus and echidnas, which share a similar anatomical arrangement.

Little Long-tailed Dunnart
Sminthopsis dolichura Male
WESTERN COAST OF SOUTH AUSTRALIA

Marsupials characteristically also have short gestation periods. Unless the growth of the embryo has been arrested during development by the presence of an older, suckling young, birth follows only 11–42 days after ovulation. At birth the young are tiny and poorly developed, weighing collectively less than 1% of the mass of the mother. In small species such as the planigales, which have litters of 4–12, or ningauis, which have litters of 5–8, individual newborns weigh as little as 5 milligrams. Even in larger species that produce single young (a joey), such as the Red and Grey kangaroos, newborns do not exceed 1 gram.

As a consequence of being so small at birth, young marsupials are entirely dependent on milk from their mother to grow to independence. By the time young are weaned, the mother's investment in lactation can be truly extraordinary. In smaller species such as dunnarts and antechinuses, females may be producing milk for a third of their lives, and the mass of the litters at weaning can be more than three times the mass of the mother herself. Not surprisingly, females often look ragged at the end of lactation and suffer high rates of mortality at this time. The investment of marsupial mothers in their young has long been appreciated. When the Spanish explorer, Vicente Yáñez Pinzón, arrived back in Spain in 1500 from his travels to the New World, he presented an opossum with pouch young to King Ferdinand and Queen Isabella. The Queen reputedly checked the animal's pouch herself by poking her fingers into it, and was moved to describe the much-travelled marsupial as an 'incredible mother'.

In the large kangaroos, single young are carried in the pouch for at least 230 days, and then begin to venture out progressively until weaned some 5–8 months later. Remarkably, as soon as the first born begins to stray from the pouch, its mother will often give birth to another young that has been held for several months in an inactive state in her body. She will then suckle the two young of different ages until the older joey is weaned. The mother does this by simultaneously producing two different kinds of milk from her mammary glands, and delivering them to separate nipples: dilute milk containing mostly carbohydrates and a little protein nourishes the newborn young, and fat-laden milk sustains its older sibling. This feat is impressive, not to mention unique among mammals. But so too is the ability of marsupials to adjust the composition of their milk as each young grows. At the start of lactation the percentage of solids in milk is just 10–15% but it rises to 30% in the Common Brushtail Possum (*Trichosurus vulpecula*) just before weaning and to almost 50% in the Common Wombat (*Vombatus ursinus*). By contrast, the concentration of milk solids remains the same in most placental mammals throughout the period of lactation.

The phenomenon of diapause, or arrested development, is similar to the process of delayed implantation in some placental mammals. It occurs in all kangaroos studied so far, except the Western Grey Kangaroo (*Macropus fuliginosus*), and in the Honey Possum (*Tarsipes rostratus*), and potoroids, acrobatids and some burramyids. Except in the Swamp Wallaby (*Wallabia bicolor*), in which the gestation period is longer than the oestrous cycle,

Marsupial reproductive system

Female

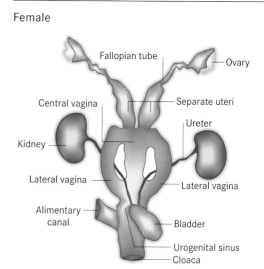

Monotreme reproductive system

Female

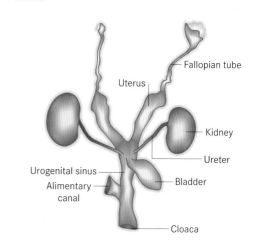

Placental reproductive system

Female

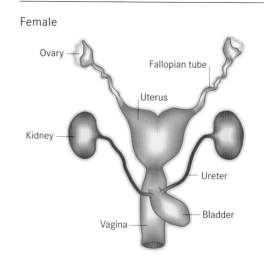

Male

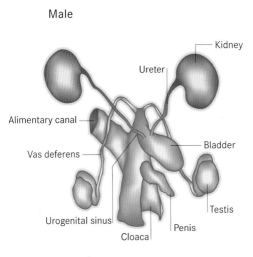

- Kidney
- Ureter
- Alimentary canal
- Vas deferens
- Bladder
- Urogenital sinus
- Testis
- Cloaca
- Penis

Male

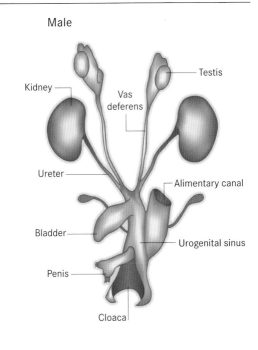

- Testis
- Kidney
- Vas deferens
- Ureter
- Alimentary canal
- Bladder
- Urogenital sinus
- Penis
- Cloaca

Male

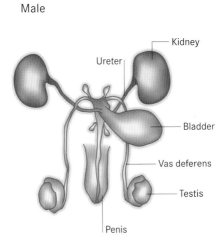

- Kidney
- Ureter
- Bladder
- Vas deferens
- Testis
- Penis

FIGURE 2 Reproductive systems of living mammals. The drawings represent the 'usual' situation and not the many specializations found among mammals, although the single functioning ovary of the monotreme is a feature specific to the Platypus. (Redrawn from figures in *The Platypus: A unique mammal*, by Tom Grant)

Swamp Wallaby *Wallabia bicolor* Female
CLARENCE RIVER, NEW SOUTH WALES

pregnancy in these marsupials coincides roughly with the length of the oestrous cycle but does not affect it. Thus, at about the same time that the female gives birth, she also becomes receptive and mates. The resultant embryo develops to an early stage, or blastocyst, of just 70-100 cells, and then enters a dormant state termed embryonic diapause. In kangaroos the blastocyst will not grow while the mother is suckling but as soon as the pouch young begins to take solid food and lessens its dependence on milk, the blastocyst resumes development. Birth usually takes place four weeks after the older young has permanently left the pouch.

Both the dormancy period of the blastocyst and the factors that reactivate it vary among marsupials. For example, in seasonally breeding populations of the Tasmanian Red-necked Wallaby (*Macropus rufogriseus*), diapause can last 8-12 months depending on the survival of the existing pouch young. In the Honey Possum, by contrast, suckling does not completely suppress the growth of the blastocyst as it does in kangaroos, and the termination of suckling does not always kick start the development of the embryo. Environmental factors such as food or time of year instead seem to stimulate the resumption of development, and diapause during gestation in the Honey Possum can vary, as a consequence, from 3 to 5 months. Diapause probably arose because it prevented a second young from being born when the pouch was already occupied. Another advantage is that it allows females to rapidly replace lost young when conditions improve, even in the absence of males.

Newborn marsupials are tiny, undeveloped and vulnerable at birth. I have studied them on many occasions in the course of my field work, but the sight of animated pinkies – no larger than a baked bean – flexing gently on their mother's nipples still takes my breath away. The hind-limbs of these miniscule creatures are just undifferentiated buds, the lungs lack air sacs, and the ventricles of the heart do not close until 3 days after birth. The eyes are black pin pricks, the ears are closed and overall development of the organs may not be complete until 30-100 days of age. The question of how such poorly developed young reach the pouch had been an enduring mystery. Early naturalists assumed that the young must grow directly out of the nipples or, if born via the birth canal, become glued to the pouch area by sticky secretions in the milk. Some even believed that

the young were sneezed into the pouch following nasal copulation! Direct observations and detailed neurological studies in recent years have dispelled these imaginative notions but the passage of young from cloaca to pouch is still fantastic.

Despite their generally poor state of development, newborn marsupials have relatively large heads and differentiated fore-limbs, and move toward the pouch using swimming movements of the arms. In several species of kangaroos and possums the mother sits upright with her legs extended in front of her during and after birth; if she changes her orientation at this critical time the young continue to move upward even if the position of the pouch has changed. Early experiments on the Quokka (*Setonix brachyurus*), a small member of the kangaroo family, confirmed that this is an anti-gravitational response. This response diminishes when the young reach the outer lip of the pouch. At this point they appear to be guided down towards the mother's nipples by smell. Once reached, the mouth of the young then envelops the nipple and effectively fuses with it for the next several weeks. The motor and sensory abilities of young that were conceived just a month earlier are quite remarkable. In some of the small marsupials, such as antechinuses, females stand on all four legs with their hips raised at birth, and the young climb rapidly down to the teats. Many small marsupials produce 2-3 times more young at birth than can be accommodated on the available nipples. However, tiny neonates do not cost the mother much in terms of resources and it is presumably advantageous for mothers to overproduce young at this stage to ensure that all her nipples become occupied.

Reproduction: the myths We can smile now at the notion of female marsupials sneezing their young into the pouch, but two other myths about marsupial reproduction remain current. The first is that all marsupials have pouches. The word 'marsupial' is derived from the Latin 'marsupium', meaning pouch or purse. This derivation makes sense when you look at the pouch area of archetypal marsupials such as kangaroos and wallabies. In these animals the pouch resembles a deep pocket on the belly of the female, and this structure fully encloses and protects the nipples. After attachment to the nipples, the young are suckled in a dark, moist, protective environment. However, in many other marsupials, such as the Numbat (*Myrmecobius fasciatus*) and all the insectivorous and carnivorous species, there

is no conspicuous pouch and the nipples are surrounded instead by rings of muscle that contract to provide temporary cover when young are present. This works well when the young are small but can be cumbersome when they have grown. In antechinuses, for example, the mother carries up to 12 young and retains them in her pouch for at least five weeks until they begin to grow hair. At this stage they protrude from the mother's belly and get bumped along the ground as she walks. The mother then adopts a curious straight-legged gait to prevent abrading her brood. Clearly, despite the name and popular perception, not all marsupials have pouches. Even among the ones that do, the pouch is variable and may open forward, to the rear, centrally, and be shallow or deep and fully enclosed.

The second reproductive myth concerns the placenta. Because marsupials produce such tiny young after a very brief gestation, it has been assumed that there is no opportunity for the growing embryos to be nourished by a maternal placenta. Of course, all placental mammals start life in intimate contact with this organ and this supposed difference between the two great groups of mammals has often been used as a point of demarcation between them. However, marsupials do in fact develop a primitive but functional yolksac placenta. For the short time that the embryo is in the uterus, it is nourished by transfer of nutrients from inside the uterus across the yolksac which is in close contact with the uterine wall. In the Koala (*Phascolarctos cinereus*), wombats and especially the peramelemorphians, a second type of placenta also develops that allows much more intimate exchange between the embryo and the circulation of the mother. This structure, the chorio-allantoic placenta, closely resembles that of the 'true' placental mammals in both structure and function. It differs primarily in functioning for the shorter gestation period that is characteristic of the marsupials.

Hard structures Given the large reproductive differences between marsupials and other mammals, it is not surprising that there are differences also in aspects of their development, their soft tissues, and the structure of their teeth, skulls and skeletons. The distinctive nature of the hard body parts is particularly important in identifying marsupials in fossil deposits, and in inferring the size, diet and general lifestyle of their owners.

The hardest of hard parts are the teeth. Marsupials show a pattern of tooth eruption and replacement that is unique among mammals, with just the third upper and lower premolars being replaced by secondary teeth. Most are dentally well endowed, with 40–50 teeth. The basic dental formula is I 5/4 (five upper and four lower incisors in each half of the jaw), C 1/1 (one upper and one lower canine), PM 3/3 (three upper and three lower premolars) and M 4/4 (four upper and four lower molars), as in most

of the American didelphid marsupials. However, there are many variations on this theme. The Honey Possum never has more than 22 teeth, wombats have only 24, while some individual Numbats have up to 52. Placental mammals also show a lot of variation in numbers of teeth, but have a basic formula of PM 4/4 and M3/3 and different arrangements of cusps on the molars. Adult monotremes have no teeth.

The numbers and degree of development of the teeth are related to diet and evolutionary history. In the omnivorous, insectivorous and carnivorous marsupials the presence of multiple incisors allows prey to be seized. In the plant-eaters, by contrast, there are 1–3 upper incisors but only one lower incisor in each half of the jaw. In these

Common Wombat
Vombatus ursinus Female
Near Salt Creek, South Australia

animals, the sharp, forward-pointing lower incisors oppose their smaller, upper counterparts, and allow delicate selection and snipping of plant leaves. However, other marsupials with this arrangement of incisors include possums and gliders which eat nectar, sap and other plant exudates, as well as the Honey Possum which specializes on nectar and pollen. Some extinct marsupials such as propleopine kangaroos probably also included meat in the diet, and the formidable marsupial 'lion' (*Thylacoleo carnifex*) certainly did so.

Other teeth also show great variation in form and function. The canines are well-developed in all except the diprotodontian marsupials – kangaroos, wombats, possums and their relatives – in which the lower and often upper canines are lost. In more carnivorous marsupials the cheek teeth (molars and premolars) tend to be narrow and aligned for cutting, whereas in more herbivorous forms the molars are flattened and provide a broad, ridged surface that allows efficient crushing and grinding of plant material. In some kangaroos, an extraordinary adaptation known as molar progression has also been discovered. As the early-erupting cheek teeth at the front of the jaw become worn down, they are shed; fresh teeth erupt at the back of the tooth-row and the whole row moves forward. This phenomenon is most obvious in species that eat abrasive foods such as grass or tree leaves. Wombats (both Common and Hairy-nosed (*Lasiorhinus* spp.)) and Koalas meet the problem of tooth wear in different ways. The teeth of wombats are open-rooted, and grow continuously through life to provide a constant grinding surface. In contrast, Koalas compensate for tooth wear by eating and chewing more, and increasing the digestibility of food by regurgitating and chewing it a second time.

Marsupial skulls can be distinguished from those of placental mammals by several features in addition to the unique pattern of tooth replacement. Some of these are not obvious to the untutored eye as they reflect differences in the shape or arrangement of bones that are common to the skulls of all mammals. Other differences are more conspicuous. The nasal bones of marsupials are large and widen posteriorly; those of placental mammals are thin and expand towards the tip of the snout. Marsupials usually have at least two pairs of openings, or holes, in the palate; placentals have just one pair. A further difference is the shape of the lower jaw. Except in the Honey Possum and the Koala, the lower jaw bone

is directed inward to form an internal angular process – the so-called 'marsupial shelf'. This inward inflection allows for insertion of the internal pterygoid muscles, which produce sideways movements of the jaw during feeding. Not surprisingly, the shelf is most obvious in terrestrial browsers and grazers that need to generate large bite forces to process their fibrous plant food.

Behind the skull, the arrangement of bones in the marsupial skeleton follows the basic mammalian pattern but with two distinctive features. The first is the presence of forward-projecting epipubic bones on the pelvic girdle of marsupials. These are conspicuous bony elements in most species, although they are present only as cartilaginous remnants in the Thylacine (*Thylacinus cynocephalus*) and as small, barely ossified knobs in the marsupial moles (*Notoryctes* spp.). Epipubic bones have been found on the fossilized remains of two early placental mammals and in a now-extinct group known as multituberculates. They are also present in modern monotremes but are otherwise characteristic of marsupials. These bones have long been considered as important sites of attachment for abdominal muscles that support the testes and pouch. However, recent research suggests that they form part of a complex muscular linkage between the pelvis and hind-limbs that stiffens the trunk of the body during movement.

The second distinctive skeletal feature of marsupials relates to the great structural variability of the feet and toes. Many species retain five digits on the hind-feet and may use the first digit as an opposable big toe to assist in climbing. This is the arrangement in proficient tree climbers such as the Common Brushtail Possum and Feathertail Glider (*Acrobates pygmaeus*). In many terrestrial marsupials, such as some dasyurids and bandicoots, the first toe is greatly reduced; in kangaroos it is entirely absent. The second and third toes of peramelemorphians and diprotodontians are slender and bound closely together by an envelope of skin, with only the claws protruding. This condition, termed syndactyly, provides a fur-grooming comb for the owner. In the peramelemorphians, kangaroos and rat-kangaroos, the fourth toe is enlarged and elongated, and forms the major axis of the foot. These adaptations reduce the lateral mobility of the foot but allow increased forward speed, most notably in the hopping motion of kangaroos. Modifications to the front foot are less dramatic. However, the third and fourth fingers of the

Feathertail Glider *Acrobates pygmaeus*
Male (upper left) and female
NORTH-EASTERN NEW SOUTH WALES

marsupial moles are greatly expanded and equipped with stout claws for digging, while the fourth finger of the spectacular Striped Possum (*Dactylopsila trivirgata*) is slender, elongated, and used for winkling larvae from bark and rotten wood. The second and third fingers of several other arboreal marsupials oppose each other, allowing them to firmly grasp tree trunks and branches.

Soft structures and behavior

Compared to ecologically similar placental mammals, the facial area of the marsupial skull is large but the cranial cavity is small. In association, the cerebral hemispheres of the marsupial brain tend to have simple folds and lack the central connection of nervous tissue that is a characteristic of placental brains. Does this mean that marsupials are not very smart? Certainly, early research seemed to suggest so. Superficially at least, marsupials do not have highly evolved social systems, modes of communication nor advanced behavior such as play. However, recent studies have uncovered considerable complexity in marsupial behavior as well as good learning ability. Individuals can learn new tasks and pass the information on to their young or other relatives; some kinds of behavior, such as fear of new predators, can even be learned by animals watching each other.

I can vouch for the learning ability of marsupials from watching them and trying to capture them over many years. Some learn quickly if given a new source of food and will visit traps so readily that they become 'trap-happy'. Others, such as bandicoots, like the food but not the experience of being captured, and seem able to work out how to get the food scot-free. In one project on Southern Brown Bandicoots (*Isoodon obesulus*) near Perth, I found that peanut butter and jam sandwiches were bandicoot magnets, and achieved high rates of capture in wire cage traps using this bait. However, some individuals soon learnt that the bait could be extracted from the back of the traps by pulling it through the wire mesh. I stopped this by covering the end of the traps with hessian sacks. This worked temporarily until several enterprising individuals learnt that they could tug the sacks off and get the food without being caught. I had to up the ante again by tying the sacks on. After just one more trapping session a single animal learnt how to push the traps over so the trap door would not close,

and passed this information on to its neighbors. The only way for me to catch animals then was to peg the traps down. Undeterred, the larger bandicoots worked out that they could still get free food by entering traps, standing just before the trap door trigger, and leaning forward to pluck the bait from the back of the trap. I suppose I could have tied the bait to the back of the traps after first affixing the hessian sacks and then pegging the traps down, but at bandicoots 4 : Chris 0 it was time to start work on a more gullible study species. I came away with great respect for the learning ability of marsupials.

Other insights into the complexity of marsupial behavior are coming from observational studies in the field and from detailed neurological and physiological experiments in the laboratory. For example, marsupials have been recently discovered to have trichromatic, or color, vision, a trait that is shared only with some primates. As most marsupials are nocturnal, what function does color vision serve? We do not know. Most marsupials are well-endowed with specialized skin glands, and use olfaction as a major channel of communication. In addition, all marsupials studied to date possess a vomeronasal organ which probably allows interpretation of pheromonal signals between individuals. In light of this research, intriguing suggestions are beginning to emerge that marsupials make particularly extensive use of chemical communication and may interact in ways that we do not appreciate. If this is so, it is likely that the true complexity of marsupial behavior – and the extent of marsupial intelligence – remains to be discovered. It is just that we have not yet been smart enough to recognize it!

Desert Bandicoot *Perameles eremiana* Female
SOUTH OF MOUNT CROMBIE, SOUTH AUSTRALIA

Koala

In the 21st century, the Koala serves as one of Australia's international icons. In the last 20 years, the Pope, the President of the USA, and countless international visitors and Australians have been photographed with a Koala. The people are smiling; the Koala looks indifferent. But the politics of Koala management is anything but smiles or indifference.

On Kangaroo Island in South Australia, the 18 individuals imported from Victoria in the 1920s had become more than 27,000 Koalas by 2004. The population of this foliage-eating marsupial is seriously overbrowsing the local eucalypts on which it depends.

Koala *Phascolarctos cinereus* Male
CLELAND CONSERVATION PARK, SOUTH AUSTRALIA

To avert an impending disaster, a committee of scientists, wildlife managers and welfare specialists recommended in 1996 that the population be culled. The political response was to turn down that recommendation and introduce a major program of sterilization and translocation. A contraceptive agent is now being tested. In sorry contrast, some local populations in New South Wales (NSW), where the Koala is a vulnerable species under state legislation, are suffering silent extinction.

The entire NSW Koala population is not in any immediate danger of extinction – it is too widespread and local populations can coexist with people. However, we need to know how each Koala population survives the local combination of threats to its existence, how local communities plan to conserve their local Koala population, how the state and federal governments handle Koala conservation, and how much active research is conducted into why Koala populations thrive or fail. This is today's world of Koala politics but over the course of European settlement in Australia the story has been very different.

The colony of NSW was 10 years old (1798) before the Koala was discovered by Europeans on the southern outskirts of Sydney. In 1863, naturalist John Gould wrote, 'it is nowhere very abundant' and 'It is very recluse in its habits, and, without the aid of the natives, its presence among the thick foliage of the *Eucalypti* can rarely be detected'. Ominously, Gould also noted, 'Like too many other of the larger Australian mammals, this species is certain to become gradually more scarce, and to be ultimately extirpated'. The cross-disciplinary exercise of ecological history has shown a more complex picture. By the end of the 19th century, Koala skins were a major item in Australia's export fur trade and the Eden region in south-eastern NSW boasted two Koala skinning factories. Koalas are scarce in the region today but exploitation did not bring about the dramatic decline. The primary cause of the loss was intensive clearing of the rich, fertile lands that supported Koala habitat.

In Queensland, Koalas were shot for the fur trade in their millions in the early decades of the 20th century but public pressure and the political response banned the trade in 1927. The Koala is not listed as vulnerable in Queensland, except in the South Eastern Queensland Bioregion where it is recognized as vulnerable in the face of the rapidly increasing human population. In Victoria, the primary management problem is a number of overabundant local populations. The IUCN Australian mammal working group concluded in 2005 that the Koala did not fit the criteria to be considered nationally vulnerable, and the federal government review of its status in 2006 did not add it to the national list of threatened species. So, what is at issue for Koala conservation?

While field zoologists have been straining their necks looking up in the trees for Koalas or scanning the ground for their distinctively shaped dry faecal pellets, which last long after the Koala has moved on, the laboratory scientists have been diligently examining the biochemistry of the leaves eaten by Koalas. The story from the test tubes is that anti-nutritional factors, or plant secondary metabolites, play a role in the nutritional quality of the leaves that they select. The rich soils that support Koala food trees were those selected for settlement and crops, so Koalas and farmers (and later growing regional urban centres) competed for the same land. Steeper land on low nutrient soils was suitable for neither. As the trees were cleared for agriculture, Koala habitat was lost or fragmented. Mortality is increased by roaming dogs (a big problem in coastal NSW), by cars on country roads and by the ever-present threat of bushfire and drought, exacerbated by climate change. Loss can be accelerated if the fertility of the population is decreased by the disease chlamydiosis, which may appear in stressed populations.

Koala conservation thus depends on land-use planning skills that are wildlife friendly, that recognize the importance of high quality habitat and that acknowledge that our view of the Koala is at least as much culturally driven as it is by the careful study of its biology.

Daniel Lunney
NEW SOUTH WALES DEPARTMENT OF ENVIRONMENT
AND CLIMATE CHANGE, SYDNEY

early models, early days

Marsupials are now confined mostly to the southern hemisphere, reaching their greatest numbers in Australia and comprising a significant part of the mammalian fauna of South and Central America. One species, the adaptable Virginia Opossum (*Didelphis virginiana*), is common in the United States and has extended its range to south-western Canada in the last 60 years. Fossil remains of marsupials are known from other parts of the world, and their broad distribution suggests that these ancestral forms were outstanding travellers and colonizers. Where did the ancient marsupials arise, and why are the modern species now restricted to just two of the southern continents?

Ideas put forward by biogeographers such as Philip Darlington in the 1950s and 1960s held that the northern hemisphere was a major centre for the evolution of new faunal groups and that, as new species arose, they pushed the earlier forms to progressively further-flung parts of the Earth. Australia's isolation in the southern oceans was thought to make it a difficult place for new and advanced forms to disperse to, thus leaving the continent free for the fortunate marsupials that had managed to get there at an earlier time. In this view, it was no surprise that the Antipodes should be full of odd, relict species that could not survive elsewhere. Support for this way of thinking came from other studies proposing that early marsupials had arrived in Australia from the north, island-hopping from Peninsular Malaysia across the islands of the great Indonesian archipelago by chance or rafting ability.

In the 1970s, old and formerly discredited ideas about continental drift were revived, allowing a radically different interpretation of marsupial origins to be constructed. To appreciate this, we need to take a brief detour to look at the Earth's geological history (see Table 2 on page 26 and Figure 3 on page 27).

At the beginning of the Triassic Period some 250 million years ago, the world's landmasses were coalesced into a single supercontinent, termed Pangaea. Pangaea began to break up during the Triassic, with the vast northern land-mass, Laurasia, separating from its southern counterpart, Gondwana. These huge continents themselves also began to fracture, with Africa breaking free from the eastern Gondwanan landmass about 170 million years ago, and India, Madagascar and New Zealand separating in turn over the next 100 million years. Australia and eastern Antarctica went their separate ways some 38–45 million years ago. South America and western Antarctica were the last components of the ancient Gondwanan landmass to dissociate, sundering perhaps 30 million years ago. Laurasia, comprising modern-day Asia, Europe and North America, remained united as the break-up of Gondwana took place.

These titanic events occurred because the continental land masses sit on plates in the Earth's crust and are propelled by convection currents that lie deep beneath the surface. These currents continue to shape the geography

Common Spotted Cuscus
Spilocuscus maculatus Male
MCILWRAITH RANGE, QUEENSLAND

TABLE 2 The geological time scale and major events in the evolution of marsupials

				Time intervals	Notable events for Australia
PHANEROZOIC	Cainozoic	Quatenrary		Holocene 10,000 ya–present	Europeans visit from 1606, settle from 1788
					Dingo introduced 3500–4000 ya
					Thylacine and Tasmanian Devil subsequently disappear from Australian mainland
				Pleistocene 1.75 mya–10,000	Last Glacial Maximum 18,000–20,000 ya
					Loss of megafauna 45,000–55,000 ya
					Humans arrive at least 45,000 ya
		Tertiary		Pliocene 5.3–1.75 mya	Grasslands and arid habitats develop; grazing marsupials are diverse
					Large bodied marsupials – megafauna – make their first appearance
			Miocene	Late Miocene 11–5.3 mya	'Dim age' for marsupial fossils in Australia but forest-dwellers diminish
				Middle Miocene 16.411 mya	Icehouse conditions; forest and forest-dwelling marsupials dwindle
				Early Miocene 23.5–16.4 mya	Greenhouse conditions in Australia; marsupial diversity very high
				Oligocene 33.7–23.5 mya	Marsupials appear in Australian fossil record
				Eocene 53–33.7 mya	Australia separates from Antarctica; marsupials are present but conditions for fossil preservation are poor
				Paleocene 53–65 mya	High marsupial diversity in South America; oldest Australian marsupials appear in late Paleocene
					Asteroid collision at Paleocene–Cretaceous boundary; demise of dinosaurs
	Mesozoic	Cretaceous		Late Cretaceous 97–65 mya	Marsupials in the northern landmass, Laurasia; some disperse to South America towards the end of this period
				Early Cretaceous 135–97 mya	First marsupial and placental fossils appear
				Jurassic 203–135 mya	The great southern landmass, Gondwana, breaks apart; isolated islands form in New Guinea region
					Marsupial and placental mammals diverge
				Triassic 250–203 mya	First mammals appear in late Triassic as the single supercontinent, Pangaea, begins to sunder

ya   years ago
mya   million years ago

of the Earth's land surface today and, in the Australasian region, have been pushing the Australian plate northward since its rift with Antarctica at the stately clip of about 6 centimetres (2.4 inches) a year. Some 30 million years ago, the northward-drifting Australian plate collided with the Pacific plate, resulting in buckling and massive uplifting of land along the active edges. The uplifted land now forms part of present-day New Guinea. Dramatic and large-scale mountain building over the last 10–15 million years has given the island its highest peak of 4884 metres (16,020 feet) and the distinction of housing some of the world's most impressive equatorial glaciers (other such glaciers occur in East Africa and Ecuador). Although parts of New Guinea existed as isolated islands as long ago as 160 million years, the shape of the present landmass is clearly much more recent. It was connected by land to Australia at the time of uplift 30 million years ago but lower-lying areas have been submerged on several occasions since. The most recent land connection formed during the last ice age when sea levels fell by as much as 160 metres (525 feet). New Guinea has been an island since rising sea levels breached the land bridge to form Torres Strait 15,000 years ago.

If the Earth's land surface has been reshaping itself for the last 200 million years, we no longer need to postulate how marsupials or other faunal groups got from one fixed point to another by island-hopping or by making improbably long oceanic voyages. Instead, we need to ask where the ancestral faunas were when major geological events were unfolding and how these events have shaped the distributions that we see today. Stepping back to the early Triassic, fossil evidence indicates that mammal-like reptiles were widespread and that recognizable 'true' mammals were yet to appear. There has been much discussion about how to distinguish a mammal that has just crossed the line from its mammal-like but still reptilian ancestors, and inevitably any such line must be somewhat arbitrary. However, there is some consensus that an early mammal can be identified if it has differentiated teeth with occlusion between the upper and lower molars, and if the jaw hinge rests on the dentary bone of the lower jaw and the squamosal bone of the skull. These were novel features in the early mammalian lineage, and probably permitted strong bite forces and access to a wide range of foods.

The earliest undoubted mammals have been identified from teeth and skull remains in late Triassic deposits aged 200–210 million years in localities in many parts of the world. Several different mammals have been described from this period, with the structure of their teeth suggesting that they could eat different kinds of prey ranging from plants to insects and small vertebrates. As Pangaea was beginning to break apart at this stage in mammalian evolution, some of these early mammals probably became separated soon after they arose. The earliest known marsupials turn up in the fossil record much more recently but unfortunately the affinities they may share with the late Triassic forms are not clear. The oldest fossil marsupial, described as *Sinodelphys szalayi*, is quite spectacular. It was recovered from rocks aged 125 million years in the Yixian Formation in China, and details of its skull, skeleton and even fur have been superbly preserved. Structures of the fore- and hind-feet suggest that the living animal was an agile climber that would

Continental movements

FIGURE 3 Australia severed its last links to Antarctica about 38 million years ago (top map) before moving north into warmer latitudes. The two continents had remained joined as the rest of Gondwana fragmented over the previous 200 million years (middle maps). In the early Triassic (bottom map) the world's landmasses were joined together as the supercontinent Pangaea. (Redrawn from figures in *Prehistoric mammals of Australia and New Guinea*, by John Long et al.)

have been at home in trees and on the ground. At least 20 genera of slightly younger marsupial or marsupial-like fossils have been recovered from North America, and lend force to the argument that marsupials arose on the Laurasian land mass. Intriguingly, although the marsupial fossil record dates back just 125 million years, some genetic analyses suggest that marsupial and placental mammals are so divergent that they may have gone their separate ways as early as 180 million years ago. If so, some important fossil discoveries are still to be made.

Dispersal and diversification

Before the break-up of the northern supercontinent, early marsupials were present in many parts of Laurasia and had also dispersed south across land bridges into South America. Although they have since disappeared, marsupials were present in central Asia and northern Africa in the Oligocene (23–34 million years ago) and in southern Asia in the middle Miocene as recently as 11–16.4 million years ago; they persisted in Europe for at least 30 million years. There is an early record of a marsupial from Madagascar but this fossil evidence has been disputed. Given the early separation of Madagascar from the rest of Gondwana, marsupial presence on this large island would represent a surprising dispersal event. Marsupial hegemony in Laurasia did not last; apparently few species survived to the end of the Miocene (6 million years ago). We do not know what caused the demise of marsupials over such a vast area but several authors have speculated that numbers dwindled as placental mammals gained the ascendancy.

As marsupials declined in the northern hemisphere, they continued to flourish in both South America and Australia. Marsupials were certainly present in South America at least 60 million years ago, with ancestral animals probably arriving from the north as much as 5–10 million years earlier. The oldest undoubted fossils have been uncovered at a site in modern Peru, although there are exciting but contested claims of still older remains from southern Bolivia. Whatever the precise ages of these deposits, fossil-bearing rocks dated at 60 million years and slightly younger have yielded large numbers of marsupial fossils, and show that South America at this time was an important nursery for marsupial evolution. For much of this long period, South America was effectively isolated from North America owing to the sundering of the land connection between them, which allowed the marsupial fauna to radiate into a spectacular array of different species.

This situation lasted until about 3 million years ago, when the Central American land bridge formed and allowed waves of invasion of placental mammals from the north in what has been called the 'Great American Biotic Interchange'. Many unique South American marsupials disappeared at about this time. Among the host of invaders from the north were carnivores such as bears, raccoons and cats; their arrival coincided with losses of marsupial carnivores such as the impressive thylacosmilids, or sabre-tooths. It is tempting to suggest that the invading hordes were responsible for the rout of South America's marsupials. With the exception of the sabre-tooths, however, many of the continent's other carnivorous marsupials were heading for oblivion before the placental invaders arrived, so another cause must be sought for their demise. The most plausible suggestion is that they were pushed by a home-grown enemy, the flightless phorusrhacids, or 'terror birds', which appear increasingly in the fossil record from around 20 million years ago. With some species standing at almost 3 metres (9.8 feet) and possessing a massive head and ferocious beak, these formidable birds were the largest predators in South America and would probably have been the equal, at least, of any of the extant marsupials.

The Great American Biotic Interchange ended the dominance of marsupials in South America, although a still significant fauna of about 94 species remains. At least 13 species occur on the Central American isthmus and in south-eastern Mexico, including four species that are found nowhere else. Unlike their extinct relatives, the living species are all small (less than 5 kilograms; 11 pounds). Most are omnivorous and only partly terrestrial, with many spending their lives above ground in the trees. Only one marsupial, the Virginia Opossum, has exploited the opportunity to colonize large areas of North America above Mexico and it probably began its northward advance in the last million years.

As the fortunes of marsupials waxed and waned in most parts of the world after their exodus from Asia and North America, immigrants to eastern parts of the southern supercontinent, Gondwana, fared quite differently. Here, in the Australian part of Gondwana, the early marsupials were spectacularly successful. Arriving some

60–65 million years ago when marsupial diversity in South America was high, the ancestors of the modern marsupial fauna gave rise to a bewildering range of odd beasts. These included the bizarre yalkaparidontians or 'thingodonts', dog-sized wynyardiids, calf-sized ilariids (the generic name *Ilaria* is derived from an Aboriginal word meaning 'strange'), subterranean moles, large predators and many other forms. In total, the Australian marsupial radiation yielded seven orders and at least 31 families. Extinctions over the long span of time since colonization of the continent have whittled the present-day marsupial fauna down to just four orders and 16 families. The last living members of two families, Chaeropodidae and Thylacinidae, expired in the 20th century. Nonetheless, even this impoverished fauna is impressively diverse. In 1959 it moved the great Australian mammalogist Ellis Troughton to comment that '… Australia [had] fostered the greatest phylogenetic deployment of a single mammalian Order that the World can ever know'.

Eastern Barred Bandicoot *Perameles gunnii*
Female with young aged approximately 2 months
NEAR LAUNCESTON, TASMANIA

Where did Australia's marsupials come from? At a superficial level the answer is straightforward: South America. The oldest marsupial fossils discovered so far in Australia come from deposits at Murgon in south-eastern Queensland, and date to the Paleocene–Eocene boundary, 50–55 million years ago. Although known only from isolated teeth and other fragmentary materials, some of these ancient marsupials show affinity with South American marsupials of similar age. At this time, South America, Antarctica and Australia were still joined as a single land mass. Reconstructions of the early southern climate indicate that it was warm and humid, and findings of fossilized leaves and other plant tissues show further that extensive tracts of southern beech forest had flourished. These conditions were certainly suitable for early marsupials – fossils of at least eight now-extinct species have been discovered since 1982 at accessible sites near the northern Antarctic Peninsula, which extends towards the tip of South America. These marsupials were part of the South American fauna, and their middle Eocene age confirms their presence in Antarctica when this land mass was still part of the Gondwanan land bridge.

Acknowledging South America as the source of Australia's marsupial fauna is just the beginning. We need to ask which marsupial group, or groups, succeeded in colonizing Australia, and whether there was just one dispersal event or several. We might also ask whether any marsupials made the reverse journey back to South America. Great strides have been made in recent years in disentangling the relationships between the Australasian and American faunas but many uncertainties remain. The account below outlines some of the ideas that have been advanced.

In 1982, Frederick Szalay at the City University of New York shattered the comfortable notion of the time that the Australian and American marsupial radiations were discrete and separate. He argued that an obscure Andean marsupial, the Monito del Monte (*Dromiciops gliroides*), was in fact more closely related to Australian marsupials than to any of its brethren in South America. Szalay described detailed similarities in the anatomy and structure of the ankle bones of the monito and Australian

marsupials, and even suggested that *Dromiciops* might be the ancestor of the Australian fauna. This startling proposal has received support from two quite different sources. Firstly, laboratory studies have revealed more genetic similarities between *Dromiciops* and Australian marsupials than between it and its South American counterparts. Secondly, fossil discoveries have suggested that some of the 55 million year old dental material uncovered at Murgon comes from marsupials related to the Monito del Monte. These finds confirm that the marsupial family to which the monito belongs, the Microbiotheriidae, had an intercontinental distribution and thus was well placed to have provided the early stock that gave rise to Australia's marsupials.

On closer inspection, however, the story becomes more complex. Detailed molecular and morphological comparisons suggest that *Dromiciops* is not related equally to all Australian marsupials but shows much stronger affinities to some groups than to others. If this is so, it implies that the Australian marsupial fauna was founded by a much more diverse original stock from South America than by *Dromiciops*-like creatures alone.

What evidence of other marsupial ancestors do we have? In the last few years genetic analyses have revealed an unexpectedly close relationship between Australian bandicoots and South American marsupials, and a more distant relationship between the bandicoots and their Australian relatives. Although bandicoots have never been recorded from the rich faunas of South America, these relationships suggest that bandicoots may have been among the immigrants to Australia, perhaps accompanying the relatives of *Dromiciops* on their trek across Antarctica. Intriguingly, ancient fossil remains of a bandicoot-like animal have been identified from the deposits at Murgon. As no earlier fossils are known from the continent, it seems quite plausible that this group arose elsewhere in Gondwana before entering Australia.

The Murgon deposits have yielded other fascinating insights into the early marsupial invasion of Australia. In 1993 a team from the University of New South Wales led by Mike Archer described a marsupial that they called *Thylacotinga bartholomaii*. Although represented only by isolated molars, the shape, size and pattern of cusps and other structures of these teeth were characteristic of South American marsupials of similar age, and leave little doubt that this species was an early immigrant from Antarctica or South America. Six years later Archer and his colleagues described an even more enigmatic marsupial, *Djarthia murgonensis*. The teeth and jaw fragments of this small and primitive species exhibit characteristics of both Australian and South American marsupials. These features make it difficult to specify the origin of this animal but certainly do not preclude it from being a potential ancestor of the Australian marsupial fauna. 'Djarthia' is the Koolaburra Aboriginal word for 'elder sister', and well describes the possible relationship of this marsupial to the rest of the Australian radiation. The species name *murgonensis* honors the town of Murgon, site of so many important fossil discoveries.

As a result of these findings, it seems likely that Australia was seeded by a diverse range of marsupials in the 20–25 million years before the continent was separated from Antarctica. Only some survived. Despite being now confined to South America, the Monito del Monte is recognized as being part of the Australian radiation, or 'Australidelphia' (Table 1), and a descendant of ancestral microbiotheriids that were widespread across the once-continuous Gondwanan landscape. All other American marsupials fall within the 'Ameridelphia'. Because of the early appearance and widespread distribution of microbiotheriids, some researchers have suggested that they may have actually arisen in Australia and made the reverse crossing back to South America. Although this is unlikely because of Australia's isolation at the eastern end of Gondwana, clearer resolution of the continent's early marsupial history will have to await the discovery of further fossil material.

New Zealand is usually omitted from discussions of mammalian evolution owing to the paucity of fossil sites aged more than a million years and its unimpressive mammal tally of just three living species of bats. But, this changed in late 2006 when the fragmentary remains of one or more extinct mouse-sized mammals were discovered. Dating to the early Miocene, over 16 million years ago, these remains may represent a previously unknown mammalian lineage that was present in Gondwana when New Zealand separated about 82 million years ago. This early separation predates the arrival of early marsupials in eastern Gondwana by at least 17 million years but does not preclude the possibility of marsupials subsequently dispersing to New Zealand.

It is now tempting to speculate that the erstwhile 'land of birds' could have been another frontier for marsupials, but no fossils have yet been found.

Australasian beginnings

After the tantalizing insights of the wonderful Murgon fossils, the story of marsupial evolution in Australia goes quiet for some 30 million years. This long 'dark age', lasting to the late Oligocene 25–26 million years ago, has yielded no useful fossil material and thus frustrates attempts to track the fortunes of different marsupial groups after their arrival Down Under. When fossils begin to reappear, at sites in South Australia, Queensland and Tasmania, their numbers are modest. But soon after, in the Miocene between 15 and 23 million years ago, marsupials underwent such an astonishing flowering in diversity that they represented some 85% of the mammals known on the continent at that time (monotremes and bats made up the remaining 15%). What happened? Geological evidence confirms that the Australian climate changed from dry, icehouse conditions into a lush greenhouse period near the Oligocene-Miocene boundary, and this favored the formation of huge lakes and rivers, and the spread of extensive tracts of rainforest. Marsupials thrived under these productive conditions. The best-known and most diverse fossil sites are in the Tirari Desert in South Australia and at Riversleigh in north-western Queensland. Here, discoveries of stunningly well-preserved teeth and skulls show that the forests were stalked by different kinds of thylacines, marsupial 'lions' and smaller carnivores, variously sized possums, koalas and wombats, and early kangaroos. These sites have now dried out and look very barren to the untutored eye. But if you walk among the surface rocks or stand on a bluff overlooking the rich fossil beds, you can still sense their antiquity and importance. In 1994 the Riversleigh fossils, as well as the more recent fossil deposits at Naracoorte in South Australia, were justifiably inscribed on the World Heritage List for their outstanding natural universal values.

Good times never last, and this was certainly true of the halcyon conditions of the early Miocene. At about the mid-point of this epoch around 15 million years ago, the greenhouse climate gave way to progressively drier and colder conditions. Australia, and much of the world, entered the icehouse again. The great continental forests began to shrink and were replaced progressively by vast

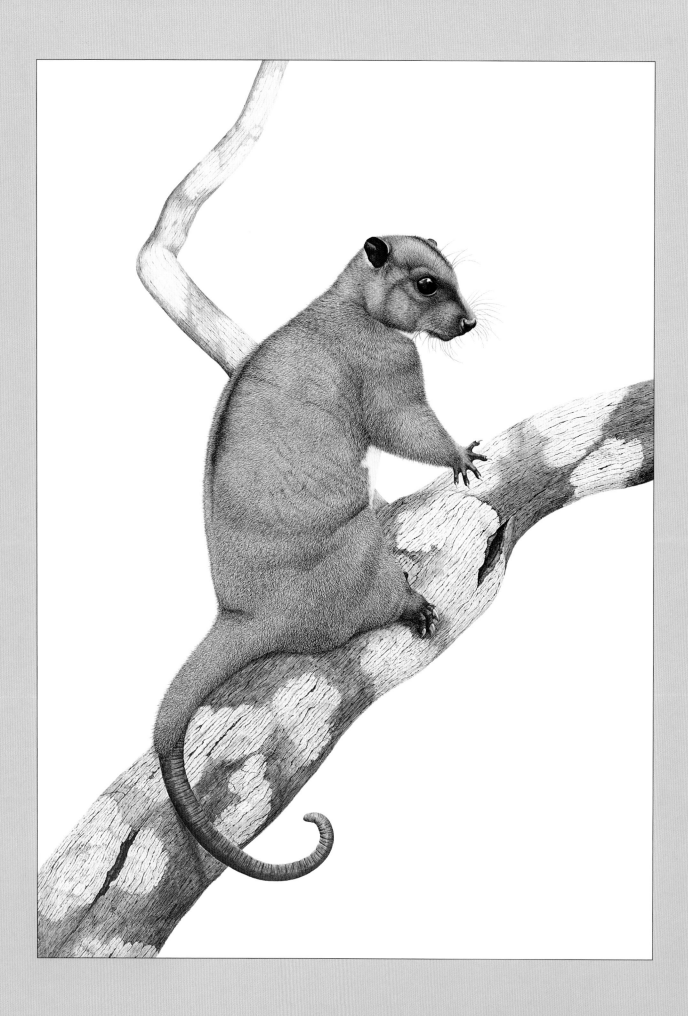

areas of grassland and savanna woodland. Despite a brief return to milder and wetter conditions about 5 million years ago, the continually deteriorating climate drove a slow and inexorable decline in the diversity of Australia's tree-dwelling marsupials. Many species of possums, bandicoots and dasyurids disappeared, as did several entire families such as the enigmatic wynyardiids, miralinids and yingabalanarids. On the other side of the coin, the expanding grasslands set the scene for an explosive radiation of now-familiar grazing marsupials such as kangaroos and wombats. Aridity increased, with hardy dryland plants appearing 2-4 million years ago, and 'true' desert features such as sand dunes about a million years later.

Cool, dry times make poor conditions for preservation, thus few good sites are known for marsupial fossils during the 'dim age' of about 5-12 million years ago. The evidence that is available has revealed one intriguing trend. As marsupial diversity declined from the mid-Miocene, the average size of the survivors went up. This is initially surprising but can be explained by at least two processes. Firstly, as body size increases, the ratio of surface area to body volume declines. In other words, big animals have relatively less skin than small animals and, as a consequence, have a relatively smaller area over which heat and water can be lost. This is clearly advantageous in cool, dry conditions. Secondly, dry conditions favor plants with tough, water-conserving leaves and stems. Such plants are difficult for small herbivores to eat and process but can be handled readily by larger species. Large herbivores have increased gut capacity, and thus can store poor-quality food for long enough to ensure that it is digested. If large herbivores are present, these in turn provide the food to support large carnivores.

At the beginning of the Pleistocene, 1.75 million years ago, Australia's marsupial fauna was dominated by large and often strange looking beasts that have collectively been termed 'megafauna'. Among this bestiary were marsupial 'lions' or thylacoleonids, averaging over 100 kilograms (220 pounds); bizarre, long-snouted giants called palorchestids; rhinoceros-sized 'wombats' or diprotodontines, that tipped the scales at 2800 kilo-

Southern Common Cuscus *Phalanger mimicus* Female
CLAUDIE RIVER, QUEENSLAND

grams (6170 pounds); giant short-faced kangaroos or sthenurines; and large carnivorous kangaroos or propleopines. Not surprisingly, marsupials that would be familiar today – although 'super-sized' as a consequence of their low quality diet – were also present. Much of the megafauna persisted until the arrival of humans in Australia some 45,000-50,000 years ago but then disappeared. We return to the question of what caused this extinction event in the next chapter, Recent History. By the end of the Pleistocene 10,000 years ago, some of the larger surviving species had undergone an extraordinary 'dwarfing' process that reduced their body mass by some 40-50% but still left some large extant marsupials: the Common Wallaroo (*Macropus robustus*), the Red Kangaroo (*M. rufus*) and the Western Grey (*M. fuliginosus*) and Eastern Grey (*M. giganteus*) kangaroos.

As marsupials were adapting to changing conditions in Australia, representatives of at least three orders dispersed north to colonize New Guinea and many of the surrounding islands in the region. New Guinea's fossil record is sparse; the oldest deposit dates back just 3.1 million years and another dozen sites are yielding more recent material. However, evidence from these sources, as well as phylogenetic comparisons of living marsupials and studies of plate tectonics, can be used to reconstruct the likely sequence of colonization.

In the first instance, geological studies suggest that a land bridge connected Australia to its northern neighbor for the first several million years after the uplift of New Guinea, with the bridge being breached by marine flooding towards the end of Australia's fossil 'dark age'. The first colonists probably trekked north while the land bridge was still intact. Calculations made using 'molecular clock' techniques (which assume fairly regular rates of change in DNA and other molecules) provide some support for an early Miocene dispersal event but suggest that it took place around 20 million years ago when a formidable sea barrier was in place. As New Guinea's earliest colonists were the ancestors of present-day cuscuses and bandicoots, which do not like to get their feet wet, a sea crossing seems unlikely; fine tuning is needed to establish the exact date of 'first contact' in New Guinea. Re-establishment of land bridges allowed more episodes of overland dispersal 8-12 million and 2.7-4.7 million years ago, with the ancestors of New Guinea's remaining forest-dwelling marsupials arriving in these later waves. Rapid fluctuations in sea levels, beginning some 2.5 million years ago, led to successive land bridges forming and sundering. As Australia's forests had receded by this time with the cooling and drying of the climate, species adapted to grassland and open savanna habitats made up the most recent dispersal events. The main exception was the reverse dispersal of two cuscus species, the Southern Common Cuscus (*Phalanger mimicus*) and Common Spotted Cuscus (*Spilocuscus maculatus*) into Australia. They presumably used slivers of gallery forest along watercourses to make their move.

Despite the recency of New Guinea's marsupial invasion, around 15 species are known only as fossils, at least six of which disappeared within the last 50,000 years. Unlike the giants that roamed Australia during the Pleistocene, the largest of New Guinea's extinct marsupials were middle-weights of 100–200 kilograms (222–440 pounds) that appear to have been forest-dwelling browsers and grazers. The causes of these extinctions are not clear. People arrived in New Guinea at least 40,000 years ago and may have had some impact; equally, climatic fluctuations have been extreme over the same period, as exemplified by the coverage of glaciers which has ranged from almost zero to more than 5000 square kilometres (1930 square miles). Without a reliable chronology of events, we may never know what caused the larger species to disappear. However, three recent extinctions are more readily interpreted. The Thylacine and two species of the small wallabies known as pademelons (*Thylogale* spp.) disappeared within the last 2000–3000 years, coincident with the arrival of wild dogs (*Canis lupus*). Dogs probably competed with the Thylacine and, with humans, may have increased hunting pressure on the wallabies to unsustainable levels.

It is easy to marvel at the extraordinary marsupials of times past and lament their demise but less easy to reconstruct the species or the communities they formed. Can you imagine gazing upon Australia's Miocene marsupials, when the diversity of forest-dwelling species approximated or even exceeded that of the rich mammal faunas in the rainforests of the Amazon or Borneo today? While a time machine would be needed to see the real thing, we may still glimpse times past in the pockets of wet forest in north Queensland and the highlands of New Guinea. The ancestors of today's rainforest marsupials retreated to these refugia as conditions dried in the mid-Miocene, allowing them to escape the worst of the deteriorating climate. Many of their descendants appear little changed; some living possums, forest wallabies and bandicoots bear striking similarity to now-extinct taxa from the Miocene environment of central Australia. Perhaps, to (mis)quote Conan Doyle, there are lost worlds everywhere.

Red-legged Pademelon
Thylogale stigmatica Male
CAPE YORK, QUEENSLAND

Leadbeater's Possum

I can still remember my first day of work on Leadbeater's Possum, even though it was over 23 years ago. It was raining. Not heavily, just a slow but constant drizzle.

In retrospect, I should have expected it to be raining. The montane ash forests of the Central Highlands of Victoria where Leadbeater's Possum live are among the wettest environments in southern Australia; some areas receive over 2 metres (6.5 feet) of rain each year. This makes for truly spectacular wet forests; mature Mountain Ash (*Eucalyptus regnans*) trees can top 100 metres (330 feet) making them the tallest flowering plants in the world. The wet conditions combined with highly productive soils also make for incredibly dense vegetation in the understorey, and close to and on the ground.

It's a tough place to work (even a 50 metre (164 feet) walk is a major physical journey) but it's nevertheless a stunningly beautiful one. And it's an ideal place for a small, feisty and enigmatic animal to live. Yet only comparatively recently has Leadbeater's Possum been known to inhabit these forests. The species was actually described in the 1860s from a handful of specimens captured in most un-forest-like tea tree thickets 150 kilometres (95 miles) to the south. When these areas were extensively cleared and Leadbeater's Possum not seen again, it was presumed that the species had joined the long list of extinct Australian mammals. Then in 1961 the species was rediscovered near Marysville in the Central Highlands of Victoria.

The possum's name comes from John Leadbeater, the taxidermist at the National Museum of Victoria who prepared specimens of the species for scientific description. An attempt was made in the 1970s to give the animal a new and more appealing name: the Fairy Possum. The new name never caught on – and probably just as well as it does not do the species justice. Leadbeater's Possum is, in many ways, hardly fairy-like. It is actually highly aggressive and lives in groups run by a dominant matriarch. Colony members band together to fight off intruders, maintain strict territory boundaries and drive away enemies like forest owls. The alarm call of a colony member will quickly bring others rocketing through the dense understorey to its aid. And you have to admire the spunk of these 120 gram (4 ounce) animals that are brave enough to attack an 85 kilogram (187 pound) biologist mimicking its alarm call. I can still recall an amazing night in the forest with my father who stood absolutely gob-smacked while possums jumped on my arms and head biting my face and nose while I was mimicking alarm calls. Perhaps the response was so aggressive because we were standing close to one of the colony's nest trees, a massive dead tree roughly 2.5 metres (8 feet) in diameter and a whopping 20 metres (65 feet) in circumference. Nest trees are a critical part of the existence of Leadbeater's Possum. Individuals spend up to 75% of their lives inside tree hollows. But not just any tree will do.

Leadbeater's Possum
Gymnobelideus leadbeateri Male
NEAR MARYSVILLE, VICTORIA

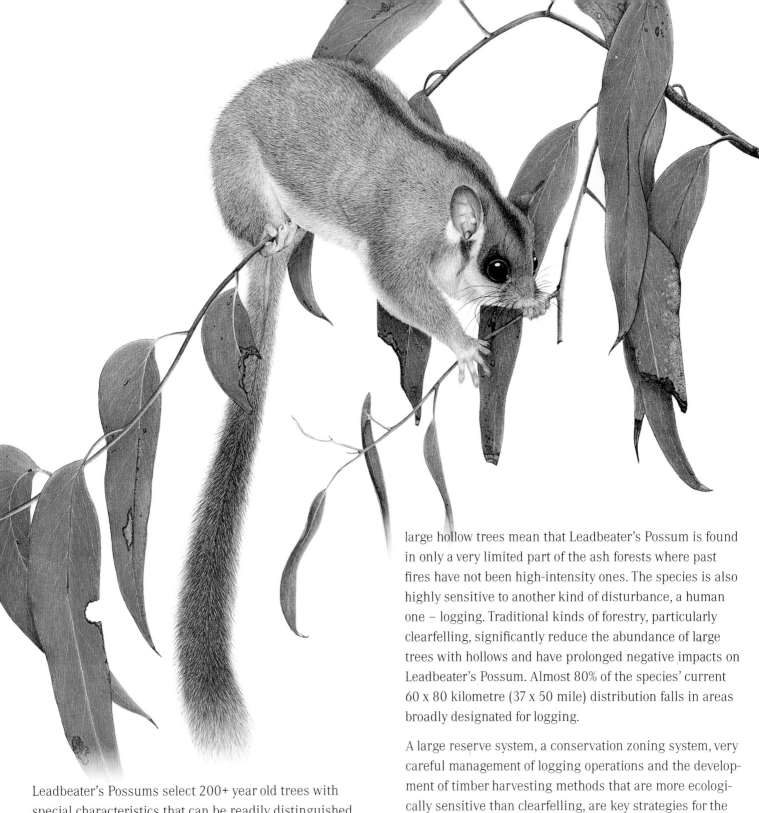

Leadbeater's Possums select 200+ year old trees with special characteristics that can be readily distinguished from other large trees in a stand of Mountain Ash. They are usually short, dead and highly decayed, and often have long narrow cracks or fissures where animals can squeeze through into the central hollow chamber where the animals make a nest of tightly woven bark strips. Colonies have not one but many nest trees in their territories and frequently swap between them, probably to keep the owls guessing or to try and escape the many, many fur-inhabiting fleas and mites that parasitize them. These specialized requirements for many special kinds of

large hollow trees mean that Leadbeater's Possum is found in only a very limited part of the ash forests where past fires have not been high-intensity ones. The species is also highly sensitive to another kind of disturbance, a human one – logging. Traditional kinds of forestry, particularly clearfelling, significantly reduce the abundance of large trees with hollows and have prolonged negative impacts on Leadbeater's Possum. Almost 80% of the species' current 60 x 80 kilometre (37 x 50 mile) distribution falls in areas broadly designated for logging.

A large reserve system, a conservation zoning system, very careful management of logging operations and the development of timber harvesting methods that are more ecologically sensitive than clearfelling, are key strategies for the conservation of Leadbeater's Possum. The importance of these strategies cannot be overestimated.

The species has already been brought back from 'extinction' once and, following its momentous rediscovery, it would be nothing short of a complete tragedy to lose it again. This time it would be permanent.

David Lindenmayer
CENTRE FOR RESOURCE AND ENVIRONMENTAL STUDIES,
AUSTRALIAN NATIONAL UNIVERSITY, CANBERRA

fall of the giants

Let us now move forward in time and consider the events of the late Pleistocene. This was a period of great change in many parts of the world, and also the period in Australia that saw the demise of some of the largest and most distinctive marsupials that have ever lived. What happened? There were dramatic shifts in climate during the Pleistocene, accompanying changes in vegetation, and the first appearance of human beings in the island continent. Could these events have brought giant marsupials to their knees? If we can understand the events of the recent past, we can better appreciate the legacy that they left for modern marsupials and how their present-day communities are structured.

The last 50,000 years

If we could return to the Australian environment of 50,000 years ago, the marsupial fauna of the time would be quite different from what we know today. Although many of the smaller species would look familiar, the landscape was dominated by the giants of the megafauna. Among this prehistoric zoo were some 15 species of kangaroos that reached body weights of 100 kilograms (220 pounds) or more, giant wombats, rat-kangaroos and an outsize species of koala that probably tipped the scales at about 16 kilograms (35 pounds). Other marsupials were larger still. Three species in the now-extinct family Diprotodontidae probably exceeded 100 kilograms (220 pounds) when fully grown, with *Diprotodon optatum* achieving a maximum mass of about 2800 kilograms (6170 pounds). This species has the distinction of being the largest known marsupial and is colloquially known as the 'marsupial rhinoceros'. Two members of another family that is now extinct, the Palorchestidae, reached body weights of about 100 and 500 kilograms (220 and 1100 pounds) respectively. These bizarre, long-snouted creatures probably bore a passing resemblance to modern-day tapirs. Remarkably complete skeletons of many of these giant beasts have been found, and it is clear from the structure of their teeth and skulls – and in some cases even the fossilized contents of their last meal – that they were grazers or browsers.

One large carnivore stalked the plant eaters. This formidable predator, *Thylacoleo carnifex*, the so-called 'marsupial lion', averaged perhaps 100–130 kilograms (220–290 pounds), and was equipped with large, powerful incisors and enormous (4–5 centimetre; 1.5–2 inch) carnassial, or cutting premolar, teeth in each jaw. The marsupial lion was memorably described as 'one of the fellest ... of predatory beasts' by Sir Richard Owen, who was perhaps the foremost palaeontologist in 19th century England. Surprisingly, Owen's conclusion about the predatory nature of *T. carnifex* was disputed for several decades.

Desert Rat-kangaroo
Caloprymnus campestris Male
LAKE EYRE BASIN, SOUTH AUSTRALIA

Some contemporaries and later workers argued that the species' unusual teeth allowed it to cut fibrous plant material such as stems or tough fruits. However, cuts discovered on the fossilized bones of several giant herbivores have been attributed to *T. carnifex* and suggest that the carnassial teeth were used to butcher the carcases of prey and rend the flesh into bite-sized chunks. And, recent analyses by Steve Wroe and colleagues at the University of Sydney have confirmed that *T. carnifex* had the most powerful bite for its size of any known carnivore. This 'lion' was clearly no vegetarian.

In total, some 50 species of marsupials that roamed Australia 50,000 years ago were no longer present at the start of the Holocene era 10,000 years ago. Another half dozen species disappeared from New Guinea over the same period. Intriguingly, with the exception of a single, small species of *Antechinus*, only large species weighing 40 kilograms (88 pounds) or more were affected. The debate about 'whodunit' has raged since 1838 when Sir Richard Owen first described *Diprotodon* and continues

to stir much passion within the scientific community today. And we are still not sure what caused the extinction of these marsupials.

There are two main hypotheses that explain the disappearance of the megafauna. At their most basic, the first proposes that the climate dried and cooled so severely that the large beasts could not survive the new conditions, and the second proposes that they were exterminated in a killing spree by newly arrived human colonists. Other ideas, such as one proposing that the megafauna succumbed to epidemic disease, are interesting but lack empirical support.

Shifts in Australia's climate over the last 50,000 years and longer have been identified using diverse sources such as concentrations of carbon dioxide and methane in cores taken from the Antarctic ice sheet (higher concentrations indicate warmer, 'greenhouse', conditions), changes in the ratios of two

Cinnamon Antechinus
Antechinus leo Male
Cape York Peninsula, Queensland

isotopes of oxygen and of hydrogen from the ice sheet and from deep sea cores (indicative of changing temperatures), and dating of the emergence of coral reefs and land from the sea. These signals show with remarkable consistency that the Australian environment has oscillated between prolonged periods of cool, dry conditions and shorter periods of warmth. The cool periods are sometimes termed 'ice ages' because of the pronounced effects of glaciation in the northern hemisphere, although significant ice sheets in the Australasian region have always been limited to parts of Tasmania, the south-eastern highlands of the Australian mainland, and the mountains of New Guinea and New Zealand. Over the last glacial cycle, beginning 110,000 years ago, moderate temperatures prevailed for about 70,000 years, punctuated only by a short, cold period some 60,000–70,000 years ago. Temperatures then began to plunge about 38,000 years ago, dropping as much as 8°C at southerly latitudes to reach an ice age low 18,000–20,000 years later, before warming to present conditions some 10,000 years ago. This period of severe climatic conditions has been termed the Last Glacial Maximum.

The falling temperatures and increasing aridity that presaged the plunge to the ice age low would have reduced the availability of free water and also triggered an expansion in dry open grassland and woodland at the expense of wet forest. Some authors argue that these conditions would have provided insufficient food and water for marsupials with large appetites; hence, the megafauna faded from hunger and thirst while their smaller relatives survived. Further evidence that has been used to support the climate hypothesis is that some large species survived the deteriorating conditions by 'down-sizing'. For example, the Eastern Grey Kangaroo (*Macropus giganteus*), at a modest 50 kilograms (110 pounds), appears to be a direct descendant of the appropriately named *Macropus titan*, which was twice its body mass. Several other large species of kangaroos, koalas, devils and perhaps wombats also persisted by undergoing dwarfing. If food and water had become scarce in the Last Glacial Maximum, it would be reasonable to expect strong selection for small body size and the attendant lower maintenance costs.

As compelling as this sounds, the climate hypothesis has been criticized on several grounds. There is some evidence that climatic cycles of the kind seen over the last 50,000 years had in fact occurred over much

of the last 2 million years, so it is not clear why the Last Glacial Maximum should have been so lethal. In addition, if water was limiting, it may have been so only in northern Australia. Sedimentary deposits in valleys, flood-out areas and dry lakes suggest that water was abundantly available in much of southern Australia. Most importantly, recent studies suggest that the megafauna may have died out before climatic conditions became severe, perhaps somewhere between 45,000 and 55,000 years ago. These dates have been disputed but, with the other inconsistencies just noted, they give considerable pause for thought.

What about the impact of people? As with the megafauna, there is acrimonious debate about the timing of key events, most notably the question of when people first arrived in Australia. Evidence from several dating techniques provides some certainty that people were in Australia at least 45,000 years ago, and hints that occupation may have begun as much as 10,000–15,000 years earlier. These older dates have been disputed on several grounds, particularly with respect to whether fossil remains have lain undisturbed since deposition and technical issues associated with the dating techniques. If we take the conservative approach of accepting a human arrival date of 45,000–50,000 years, the first people may have overlapped with megafauna for a very brief period; certainly it would have been less than 5000 years. Could they have exterminated all the large marsupials in such short order?

Using the terminology of warfare, 'Blitzkrieg' and 'attrition' have both been proposed as possible mechanisms. The Blitzkrieg, or lightning war, hypothesis predicts that the first colonists over-hunted the megafauna in a systematic and concerted manner and presumably used the meat as the mainstay of their diet. Annihilation could have taken a millennium or so. By contrast, the attrition hypothesis predicts that megafauna formed a lesser part of the diet of the first colonists and that people pushed them to the brink by imposing a small additional source of mortality. For long-lived and slow-breeding species, any additional mortality could push death rates above birth rates and thus drive populations on a slow but inexorably downward spiral. This process might take several thousands of years depending on hunting pressure and efficiency. A third hypothesis predicts that it was not Aboriginal hunting that caused megafaunal collapse but rather an extensive and continent-wide burning regime.

Fires might kill animals that could not escape the flames but their most telling effect would have been to replace the cover of shrubs and trees with grasses over large areas, thus depleting the food supply for large leaf-eating animals.

Given the probably small temporal overlap between people and megafauna, the attrition hypothesis is hard to support. The time needed for people to achieve a slow extermination was simply not available. Another problem for this hypothesis, and for Blitzkrieg, is that there are precious few, if any, 'kill sites' known where butchered megafaunal remains have been found; and we have no evidence that Aboriginal hunting technology of the time was sufficient to effectively hunt giant marsupials. What about fire? Isotope analyses of fossilized eggshell fragments from Emus (*Dromaius novaehollandiae*) show an abrupt transition in the diet of these large birds 50,000 years ago from leafy shrubs, trees and grasses to tough scrubby material, and this shift has been interpreted as arising from a fire-induced change in vegetation. However, direct evidence of fire, such as the amount of charcoal in layers of sediment, fails to confirm a significant increase in the frequency or extent of fire at any time over the last 100,000 years, except over the last 10,000 years. Taken together, these findings provide little support for the proposition that humans were responsible for the dramatic collapse of Australia's megafauna.

Where to from here? In the above account I used the dates that are agreed by most workers for when the megafauna made their last stand and when people first arrived, but also noted that these dates are contentious. In fact, there are at least a dozen sites in Australia and two more in New Guinea that have reported dates for megafauna in the order of 12,000–42,000 years ago. Many of these sites also contain tools and other human artefacts, ostensibly suggesting the possibility of a long period of human–megafaunal coexistence. However, analyses of the stratigraphy of these sites suggest that most have been disturbed by floods and other events that could have resulted in old fossil material being redeposited in younger sediments. Such disturbance is evident, for example, if skeletal remains are found to be separated from each other and disarticulated. The problem is that it is often the sediments containing the fossils that are dated, rather than the fossils themselves. If the sediments are recent, the fossils they contain may be interpreted erroneously to be recent too. One

or two sites do have stronger claims that megafauna survived with people for several thousand years but resolution of this thorny problem probably awaits direct dating of the fossil bones themselves.

The most recent explorations accept that humans coexisted with megafauna for a short period but come back to the fact that animals weighing above 40 kilograms (88 pounds) were most severely affected. Hugh Tyndale-Biscoe advanced the idea that large species would likely be susceptible to environmental change owing to their slow reproduction. *Diprotodon*, for example, would have been weaned at 1000 days and would have achieved sexual maturity at 4–8 years. In his book *Australia's mammal extinctions*, Chris Johnson of James Cook University used data on reproduction and body size in living marsupials to predict the fecundity of their extinct relatives, and confirmed that species with females that produced less than one young a year were most likely to die out. With some exceptions, such as tree-kangaroos, these slow reproducers were the large ones. Johnson suspected that slow breeders might be particularly vulnerable to hunting pressure, especially if they were terrestrial and occupied open habitats where they would be easy to spot. Using simulation modelling to assess whether hunting losses could have population-level effects on megafauna, he found that even light off-take by hunters could drive populations to extinction. This result was similar irrespective of whether hunting was indiscriminate or aimed primarily at juveniles. Most surprisingly, populations were predicted to crash after just a few hundred years of light hunting. In essence, the modelling suggested that megafauna were uniquely susceptible to very small increases in mortality owing to their slow reproduction and hence inability to replace hunting losses. In this analysis, over-hunting by the earliest human colonists most certainly did the deed.

This conclusion neatly combines an attrition-like mechanism with a Blitzkrieg result and also resolves many of the other inconsistencies that have bedevilled previous debate. For example, it does not require the first people to have used sophisticated hunting technology nor to have dined on megafauna at every opportunity. Few kill sites would be expected if hunting was light, sporadic

Bennett's Tree-kangaroo *Dendrolagus bennettianus*
Female with pouch joey
DAINTREE, QUEENSLAND

The event that most obviously correlates with the disappearance of these species is the introduction of the Dingo (*Canis lupus dingo*) some 3500–4000 years ago. This canid was brought to the Australian mainland by Asian seafarers but never reached Tasmania. As a large and social carnivore, the Dingo could have been expected to compete with the Thylacine and perhaps the Devil for food, and to have preyed upon the Tasmanian Native-hen directly. The dominance of carrion in the diet of the Devil may have reduced competitive overlap with the Dingo for live prey, hence allowing it to persist for longer. These kinds of interactions are quite plausible and rely on the probably reasonable but untested assumption that the Dingo was a superior predator to its rivals.

Whatever the case, other developments were unfolding at about the same time as the arrival of the Dingo, and these probably also contributed to the extinctions of the three native species. Most importantly, human populations on the Australian mainland began to rise and expand into new habitats about 4000 years ago, and there

and all over within a few hundred years. Even dwarfing can be interpreted. As small marsupials breed faster than large ones, down-sizing should reduce sensitivity to hunting and thus allow long term persistence. Time will tell if this is the last word in the debate but for now the weight of evidence supports the human overkill hypothesis.

The last 5000 years

The Australian climate has been relatively stable and benign over the last 8000 years and, until the arrival of Europeans 220 years ago, has been correspondingly a period of minimal faunal turnover. However, three native vertebrates became extinct on the mainland of Australia within this period but remained extant in Tasmania. The Tasmanian Native-hen (*Gallinula mortierii*) disappeared from southern Victoria some time within the last 4670 years, while the Thylacine (*Thylacinus cynocephalus*) became extinct on the mainland perhaps 3000 years ago. The Tasmanian Devil (*Sarcophilus harrisii*) survived until about 430 years ago, making its last known stand on the mainland in the far south-west of Western Australia.

is some evidence that new hunting technologies, such as improved blades and stone spear tips, may have helped to fuel these increases. Rock paintings and cultural artefacts confirm that the Thylacine, Tasmanian Devil and Native-hen were all hunted at about this time. Remains from contemporary archaeological sites show that people were also killing a broader range of prey species than they had before, and this probably had the added effect of reducing the prey available to the meat-eating marsupials. Equivalent developments did not extend to Aboriginal society in Tasmania, thus human impact on the three native species there changed little, if at all. Indeed, all three species were still present in 1642 when the Dutch explorer Abel Tasman first charted the waters of Van Diemen's Land, and in 1803 when the island was colonized and claimed by the British as a penal colony.

Eastern Grey Kangaroo *Macropus giganteus*
Male with juvenile, and female with juvenile (rear)
VICTORIA

Tasmanian Devil

The Devil growled, turned towards my face and snapped her jaws shut a little too close to my jugular. Was this the snarling, whirling Tasmanian Devil of popular image? No, this was Cleopatra, a usually gentle old female Devil that lived on the Freycinet Peninsula on Tasmania's east coast and I had known her since she was a pup. She had become trapped several nights in a row and was in a very grumpy mood. A small, delicately built female with the finest silky fur, Cleopatra was distinctive among the 150 Tasmanian Devils that lived on the peninsula in being so comfortable about being trapped and handled that she freely expressed her mood. This is rare in wild Devils, though most will lose their fear of being handled after repeated captures.

Tasmanian Devils are larger than life characters, full of bluster and belligerence. They really do spin around quickly (but only 180°), as portrayed in the cartoon, to chase interlopers at a carcase. Devils are also highly intelligent (I put them in the same

Tasmanian Devil *Sarcophilus harrisii* Male
NEAR CRADLE MOUNTAIN, TASMANIA

class as Border Collies) which, along with their nature, adapts them well to be highly effective in their dominant ecological role of top predator and specialized scavenger. They take on prey two to three times their size and are the only mammalian carnivore in Australian ecosystems to have special adaptations for scavenging. Their robust teeth are resistant to breaking, and their massive skulls with strong jaw musculature enable them to eat all of the thick skin and bone in a carcase except for wombat skulls. They forage over long distances at night, moving with a distinctive loping gait, and ambush or wear down macropod prey. Devils do not expend unnecessary energy and use their excellent sense of smell to find any animal that is already dead, then dominate the carcase against other scavengers. Devil dining on a carcase is a noisy affair. The cracking of bones and growling, which can rise to a caterwauling screech, ensure that Devils frequently have company at dinner. The record is 22 around a cow but more typically two to three Devils can fit around a wallaby.

Devils are very social creatures and apparently know each other very well. Female pups establish a home range that broadly overlaps that of their mother, leading over time to concentrations of related females sharing the same space. Two female young from Cleopatra's first litter established home ranges in the spine of hills between the Freshwater and Saltwater lagoons on Freycinet Peninsula where their mother lived. I trapped one daughter through her first short (21 day) pregnancy (signalled by the pale, flaking skin on her face, and harrowed expression and demeanor) and again a few days after she gave birth. She has a personality similar to her mother. Males are more likely to disperse away from the place they were born and up to four different males can sire the usual litter of four pups (female Devils have four teats in the pouch and newborn are the size of long-grained rice, the smallest of any mammal of this size). The few days of mating are the only time of year when females allow a male to dominate. Females are very choosy, selecting the experienced older and larger males, which are also the most dominant. They do not respect a male or allow him to mate unless he can put on a display of cave-man tactics, overcoming her and dragging her to the den where he keeps her captive for 12–72 hours.

The largest remaining marsupial carnivore now that the Thylacine has gone, Devils once roamed across mainland Australia. A small number of them travelled across the dry land bridge to Tasmania during the ice ages. They were isolated from mainland Australia, and their subsequent removal there, when sea levels rose 13,000 years ago. In the 1980s to mid-1990s Devils were a common carnivore in Tasmania, particularly in the open, dry eucalypt forests and woodlands of the east and in coastal areas. They were even heavily persecuted on some farms for their depredations on poultry and young sheep; but it is the island founder effect, from the low genetic diversity of the initial small Tasmanian population, that may be their undoing.

Since the mid-1990s, Tasmanian Devils have been declining dramatically from an infectious facial tumor, which arose through a genetic change. This cancer behaves like a parasite, spreading from Devil to Devil through biting and the direct transfer of tumor cells which then grow as a tissue graft. The genetic similarity of Devils means that they don't recognize the tumor as pathogenic. The majority of biting, and thus probably transmission, occurs during the mating season making the disease 'frequency dependent'. Typical infectious diseases die out at low population density. Frequency dependent diseases have the potential to cause extinction. The impacts of the disease could be far reaching because it has coincided with the introduction of feral Red Foxes to Tasmania. As the largest Fox-free Australian island, Tasmania had been a refuge for many species of marsupial that have become extinct on the Australian mainland. The decline of Devils could facilitate the spread of Foxes and the extinction of many more wildlife species. The first actions to mitigate the extinction risk have been to address potential ecosystem impacts, increase Fox eradication efforts and establish Devil 'insurance' populations, designed to retain genetic diversity for up to 50 years until either Devils become extinct and can be reintroduced or natural resistance evolves. Disease suppression is being trialled and longer term measures, such as a possible vaccine, are being investigated.

Menna Jones
School of Zoology, University of Tasmania, Hobart

spectacular variety across the continent

If we accept the historical reconstruction, marsupials have been present in Australia for at least 60 million years. The 160 or so species present at the time of European arrival are the survivors of a magnificent evolutionary experiment and, while they share many commonalities, they also occupy different geographical ranges, achieve different densities, eat different foods and exploit the environment in a wide variety of ways. Are there general patterns in the distribution and abundance of modern marsupials? If there are, can we discern the processes that shape and influence them?

Ecological studies that deal with such questions about marsupials have been confined mostly to recent years. In 1954, for example, two Australian ecologists wrote a highly influential text called *The distribution and abundance of animals*, and marsupials rated not a mention. Admittedly, authors Herbert G Andrewartha and L Charles Birch were primarily concerned with the ecology of insects but the omission of Australia's flagship mammal group from their work was striking. Just 20 years later a review by Eleanor Russell was able to cite over 90 studies on marsupial ecology; now the literature is voluminous. To try to bring order to this enormous information base, it is most convenient to place marsupials into manageable groups of similar species, and then ask what processes influence where the groups occur. We can still review individual species as exemplars of their groups but do not need to do this for each species on a case by case basis. There are many ways of allocating species to groups, the most obvious being on the basis of their phylogenetic relatedness. However, in the account that follows, I have chosen to use an alternative approach based on the ecological guild. This recognizes ecological similarities between species and is appropriate given the primacy of ecological forces in dictating where marsupials now occur.

The guild concept

American ecologist Richard Root was among the first to explore the idea of the ecological guild, defining it in 1967 as 'a group of species that exploits the same class of environmental resources in a similar way'. The guild concept was not taken up immediately because of the practical difficulty of defining group limits and because of uncertainty about which environmental resources were most important for guild membership. Research in the 1980s showed that our intuitive groupings of species usually correspond well with the results of more objective clustering methods and recent use of the guild concept has provided much insight into how species-rich communities are organized and structured. There is also broad recognition that diets and food requirements provide a good basis for assigning guild membership. Because species that eat similar foods often share similar ancestry, some coincidence

Black-footed Rock-wallaby
Petrogale lateralis (front) Male
PEARSON ISLAND, WESTERN AUSTRALIA

Purple-necked Rock-wallaby
Petrogale purpureicollis Male
Recognized in 1924 for its brightly colored head and shoulders, this macropod was confirmed as a species distinct from the closely related Black-footed Rock-wallaby only in 2001.
WESTERN QUEENSLAND

can usually be expected between guilds constructed on ecological and phylogenetic criteria. Using diets as the basis here, we can define three broad marsupial guilds: carnivores, omnivores and herbivores. Each guild can be divided in turn into two or more 'subguilds', and this is a convenient level at which to work. With just a few anomalies, the diets of most species can be readily characterized – they are strikingly different between subguilds, and correspond with differences in body size, feeding behavior and habitat use.

Diets and foraging behavior

Carnivores This guild can be divided into marsupials that chiefly consume vertebrate flesh (the meat-eaters) and those that eat mostly insects and other invertebrates (the insect-eaters). This is a large guild, comprising perhaps 60 species that occur in most parts of Australia.

Meat-eaters This group includes four Australian species of quolls (*Dasyurus* spp.), the Tasmanian Devil (*Sarcophilus harrisii*) and the recently extinct Thylacine (*Thylacinus cynocephalus*). Quolls are the junior members of this subguild, with the small but feisty Northern Quoll (*Dasyurus hallucatus*) achieving a body weight of just several hundred grams and the Spotted-tailed Quoll (*Dasyurus maculatus*) a maximum of 7 kilograms (15 pounds). The smaller quolls season their diet by eating a lot of invertebrates and even fruit but the larger members of the group tend to be obligate flesh-eaters that hunt their prey by stalking and pouncing. Spotted-tailed Quolls seem to prefer prey up to about their own body weight, but they also hunt larger quarry such as Swamp Wallabies (*Wallabia bicolor*) that may be twice their size. These quolls are astonishingly agile and have even been observed raiding tree hollows to extract Greater Gliders (*Petauroides volans*). All marsupials should be handled with respect but particular care is needed in the case of quolls. In 1988, describing one of the two known species of quoll from New Guinea (*Dasyurus spartacus*), Steve Van Dyck of the Queensland Museum captured the essence of the genus. 'The name *spartacus* embodies many features shared by this dasyurid with the notorious Thracian gladiator – strength, a tenacious fighting spirit, and the capacity to draw blood.'

The largest of the meat-eaters was the Thylacine. This magnificent marsupial probably achieved a body weight of 35 kilograms (77 pounds), which it sustained by doggedly pursuing and killing wallabies and kangaroos. Old news reel footage of the last known Thylacine, taken at the Hobart Zoo in 1933, reveals the impressive depth of the jaw, sharp flesh-puncturing teeth and huge gape that made the species such an effective predator. In between the quolls and Thylacine in size, at 7–9 kilograms (15–20 pounds), is the Tasmanian Devil. This species is an effective hunter that can run down mammals up to the size of possums and small wallabies. It also has massive jaw muscles and bone-crunching molars that allow it to scavenge and totally consume the carcases of much larger mammals. The faecal pellets of Tasmanian Devils are often white or light grey from partly digested bone, providing good evidence of a scavenged feast some time before.

What is it like to handle such formidable carnivores? Having spent many years working on the Devil's smallest relatives, the ningauis and other tiny dasyurids, this question had often crossed my mind. In late 2005 I got my chance to find out. By good fortune I was able to tag along with Nick Mooney, a wildlife expert at the Tasmanian Department of Primary Industries and Water, during routine monitoring of Devils at one of his study sites. Expecting to lose a considerable quantity of blood or perhaps a stray digit while extracting a Devil from a trap, I was astonished at how calm the captured animals were. Not only did they raise little protest at being rudely roused from their slumbers, they serenely allowed us to weigh and mark them, check their sex and condition, and even inspect their jaws and teeth. Clearly, these courtesies are not extended to possums and other prey!

Insect-eaters Members of this subguild are smaller than any of the meat-eaters, with most weighing under 100 grams (3.5 ounces) and six species weighing 10 grams (0.35 ounces) or less. Most are dasyurids. The tiniest species prefer small insects; larger insect-eaters select progressively bigger prey. Smell is often used to detect buried prey but all dasyurids have keen hearing, good close-up vision and spectacular agility that enable them to tackle formidable prey. I have watched antechinuses in furious pursuit of large spiders, beetles and centipedes, and even jump from logs and other vantage points to take moths on the wing. Other researchers have observed tiny planigales getting a ride on the back of large grasshoppers while they attempt to bite through the insect's neck to subdue them.

Julia Creek Dunnart
Sminthopsis douglasi Female
JULIA CREEK, QUEENSLAND

Members of this subguild usually eat a wide variety of invertebrates but they do show some dislikes. Ants are seldom eaten by choice, slugs are avoided and hairy caterpillars may be tried once, but once only. If you roll a hairy caterpillar between your fingers the 'hairs' break off in your skin and irritate for hours; it must be a not-to-be-repeated experience for a dasyurid when they get these barbs in the mouth, up the nose and on the tongue! Worms, snails and slaters and other crunchy morsels are seldom eaten but make seasonal appearances in the diet of some species if other types of prey are hard to find. Most hunting occurs at night; only animals that are very hungry will make an appearance by day, and even then they usually stick close to leaf litter or other protective cover. The Dusky and Swamp antechinuses (respectively *Antechinus swainsonii* and *A. minimus*) are exceptional in foraging as much by day as by night; only the Numbat (*Myrmecobius fasciatus*) is routinely active by day, perhaps using the striking bars on its coat to avoid detection by predators.

When prey such as spiders and centipedes are being eaten, it is clear that we are using the term 'insect-eater' quite loosely. This is even more so for the larger members of this group, which supplement their diet with small vertebrates and occasionally flowers, berries and other plant products. For example, both species of mulgara, the Brush-tailed Mulgara (*Dasycercus blythi*) and Crest-tailed Mulgara (*D. cristicauda*) consume other small marsupials, rodents and small lizards, to the extent that they may comprise a third of the diet. Mulgaras and related species such as the Kowari (*Dasyuroides byrnei*) weigh up to 140 grams (4.8 ounces), and readily tackle prey, such as rats, that may be of similar size. The killing is carried out swiftly, usually by well-targeted bites to the neck of the prey. I once saw a Brush-tailed Mulgara enter the nest of a Crimson Chat (*Epthianura tricolor*) that had been concealed in the middle of a spinifex (*Triodia* sp.) hummock and kill three nestlings in a matter of just seconds. The Atherton Antechinus (*Antechinus godmani*), weighing up to 125 grams (4.4 ounces), also has a

Kowari *Dasyuroides byrnei* Male
Shown eating Tessellated Gecko *Diplodactylus tessellatus*
Captive-bred from stock originating on
Coorabulka Station, south-western Queensland

ferocious reputation and probably eats other small mammals. However, not all large insect-eaters show such flesh-eating tendencies. The Brush-tailed Phascogale (*Phascogale tapoatafa*), a highly arboreal dasyurid weighing up to 310 grams (11 ounces), seldom eats vertebrates as part of its natural diet. And the largest member of the subguild, the 700-gram (25-ounce) Numbat, is a specialist termite eater.

With some trepidation, I have included Australia's most enigmatic marsupials – the Northern and Southern marsupial moles (*Notoryctes caurinus* and *N. typhlops* respectively) – here as insect-eaters too. Examination of the gut contents of a small number of moles has revealed the remains of various insect species, including their eggs, larvae and pupae. Ants have often been found, perhaps as a result of accidental ingestion. However, accounts from Aboriginal people suggest that moles also eat plant material and underground fungi. Provisional placement of the two species of moles in the insect-eater subguild seems reasonable until quantitative information becomes available.

The carnivorous marsupials in general have short, sharp incisor teeth, well-developed canines, and cheek teeth that allow food to be cut and chopped rather than ground. Only the Numbat and marsupial moles depart significantly from this general plan. Uniquely among land mammals, the Numbat always has more than seven cheek teeth in the lower jaw, with the numbers of these teeth varying between individuals and frequently even on each side of the jaw in the same animal. The teeth do not wear markedly with age, perhaps suggesting that they are little used in the feeding process; termites are soft-bodied and may simply be crushed by the long and powerful tongue against the roof of the mouth. The teeth of the moles are simple, often blunt and widely separated. The molars are rooted lightly in the bones of the jaw, giving the impression that they would be ineffective at coping with hard prey. The guts of all carnivorous marsupials are relatively short and simple. This reflects the fact that animal-based diets are high in protein, minerals, vitamins and water, and relatively easy to digest.

Omnivores Marsupials in this guild eat a very broad range of foods but most nonetheless tend to rely on one or two food groups more than others. This provides the basis for defining the four subguilds: insect-eating

omnivores, fruit-eating omnivores, fungus-eating omnivores and generalist omnivores. In the first three subguilds one food group (insects, fruits or fungi) makes up half or more of the diet on a volumetric basis and a variety of other foods comprises the rest of the intake. The generalist omnivores are able to take virtually any foods available from their local environment. Omnivores tend to be at the small end of the marsupial size range: the smallest weigh less than 10 grams (0.35 ounces) and none exceed 5 kilograms (11 pounds). In Australia, about 37 species of marsupials can be categorized as omnivorous.

Insect-eating omnivores Members of this group are drawn from three families, and include the Feathertail Glider (*Acrobates pygmaeus*: Acrobatidae), pygmy-possums (*Burramys parvus* and *Cercartetus* spp. Burramyidae) and, from the Petauridae, the Striped Possum (*Dactylopsila trivirgata*), Leadbeater's Possum (*Gymnobelideus leadbeateri*) and wrist-winged gliders (*Petaurus* spp.). These species eat variable amounts of nectar, pollen, gum, sap and other plant exudates in addition to invertebrates. Nectar-feeding is particularly efficient in the Feathertail Glider and pygmy-possums; fine brushes on the tips of their tongues allow them to soak up liquids like a sponge. The larger gliders feed somewhat more robustly, using their bayonet-like lower incisors to gash the bark of trees to obtain sap. In fact, the largest member of the subguild, the Yellow-bellied Glider (*Petaurus australis*), leaves unambiguous evidence of its presence in the deep and distinctive V-shaped notches it cuts in some two dozen species of food trees. The cuts are sometimes visited by smaller gliders that mop up residual flows of sap and thus are excellent sites to stake-out if you wish to see these charming marsupials at close quarters. Northern Queensland naturalist Rupert Russell reported seeing Feathertail Gliders, Sugar Gliders (*Petaurus breviceps*) and Common Brushtail Possums (*Trichosurus vulpecula*) licking sap in turn at a cut in the trunk of a Mountain Stringybark tree (*Eucalyptus resinifera*), and I have seen Sugar Gliders feeding head down on sap oozing from an incised Manna Gum (*Eucalyptus viminalis*) tree in the Australian Capital Territory.

Two remaining insect-eating omnivores have notable diets and feeding habits. The Striped Possum spices its diet with leaves, fruits and the honey of native bees, and uses its uniquely elongated fourth finger to winkle grubs from fissures in logs or tree trunks. Also unusual is the Mountain Pygmy-possum (*Burramys parvus*). This alpine and subalpine endemic feasts on fat-laden Bogong Moths (*Agrotis infusa*) over summer but adds berries, flowers, leaves and, uniquely, seeds, to its diet as well.

Fruit-eating omnivores The only confirmed member of this small group, the Musky Rat-kangaroo (*Hypsiprym-nodon moschatus*) of north Queensland, picks up much of the fruit it eats from the forest floor after it has fallen. This unusual species

sometimes climbs in low vegetation to pluck fruit as it ripens, and also eats seasonally available invertebrates and fungi.

Two other species can be placed provisionally within this subguild, the Southern Common Cuscus (*Phalanger mimicus*) and Common Spotted Cuscus (*Spilocuscus maculatus*). In Australia both species are confined to

Musky Rat-kangaroo
Hypsiprymnodon moschatus Male
Atherton Tableland

Herbert River Ringtail Possum *Pseudochirulus herbertensis* Female
Atherton Tableland, Queensland

forest habitat at the northern tip of Cape York Peninsula, where they have been observed clambering in the tops of several species of rainforest trees searching for fruits, buds, flowers and leaves. It is not clear how much fruit they eat or whether their diets vary with season. However, in New Guinea where cuscuses are more diverse, there is a trend for small cuscus species (0.9–3 kilograms; 2–6.6 pounds) to be more frugivorous than their larger relatives (up to 10 kilograms; 22 pounds). Averaging just 2–3 kilograms (4.4–6.6 pounds), the Australian species are at the smaller end of the size range, and could thus be expected to include a lot of fruit in the diet.

Fungus-eating omnivores Except for perhaps the Rufous Bettong (*Aepyprymnus rufescens*) (which eats modest amounts of fungus and is included here for convenience), all rat-kangaroos of the family Potoroidae can be placed into this subguild. Six species have been studied in some detail: the Brush-tailed Bettong (*Bettongia penicillata*), Tasmanian Bettong (*Bettongia gaimardi*), Northern Bettong (*Bettongia tropica*), Gilbert's Potoroo (*Potorous gilbertii*), Long-nosed Potoroo (*Potorous tridactylus*) and Long-footed Potoroo (*Potorous longipes*). Their diets comprise from 50% to over 90% fungal material, with invertebrates, plant leaves, fruits, seeds, roots and tubers making up most of the remainder. Observations of the remaining species of rat-kangaroos suggest that their diets are similar. Most of the fungi that are eaten are underground truffle-like species, often with hard casings around the edible flesh. Bettongs spend much time snuffling about in leaf litter, apparently locating the fungi using their keen sense of smell. On Barrow Island, off the north-western coast of Western Australia, Burrowing Bettongs (*Bettongia lesueur*) forage so intently for buried food that you can sit and watch them for long periods sniffing at the ground just centimetres from your feet. All the rat-kangaroos are terrestrial and equipped with short, powerful fore-limbs and stout claws that allow efficient digging. They also have long, blade-like premolar teeth that have been interpreted by some authors to be adaptations for cracking the tough husks that cover the fruiting bodies of their favored fungi.

Generalist omnivores This group, containing the bandicoots and bilbies, is generalist in the sense that its members can eat virtually all kinds of food, from insects and other invertebrates to eggs, small vertebrates, tubers, fruits, seeds and fungi. However, individuals will eat different quantities of each of these food groups, shifting their diets between seasons and localities depending on what is available. They are also quick learners when it comes to exploiting new foods. When working on a population of Southern Brown Bandicoots (*Isoodon obesulus*) south of Perth in the 1980s, I found that slices of bread smeared with plum jam and peanut butter were irresistible to any passing animals. They would also investigate tropical fruits such as bananas and dates, cooked egg and smelly fish oils, and readily ate tinned cat or dog food. Smell is very keenly developed in bandicoots, with all species able to detect and dig up buried prey. In the same population of Southern Brown Bandicoots near Perth, I found that animals could find and uncover mealworms that I had buried at depths of 30 centimetres but at the same time ignore 'control' sites that I had dug and refilled with just loose soil. Animals occasionally found and ate frogs buried at depths of 30–35 centimetres on the edge of a swamp. The unearthly screams of the expiring amphibians made night-time fieldwork a memorable, if unnerving, experience.

Although bandicoots take some prey from the soil surface, their compact bodies, cone-shaped heads and powerful fore-limbs with flattened claws showcase their digging prowess. Their teeth are sharp and well-designed for piercing the bodies of prey animals and for chopping up the softer parts of plants. The gut is relatively short and simple in bandicoots, but the presence of a caecum is indicative of animals that include some plant matter in the diet. The caecum of the recently extinct Pig-footed Bandicoot (*Chaeropus ecaudatus*) is more developed than in other bandicoots, and this accords with anecdotal observations and some direct dietry evidence that this species ate more grass and seeds than its living relatives.

Herbivores This large guild contains over 60 species of Australian marsupials. Ecologically, we can recognize two subguilds: browsers or leaf-eaters, and grazers. Members of these groups show similarities in their diets and way of life but are quite divergent in aspects of their dental morphology and digestive physiology. They are also drawn from several different families.

Leaf-eaters These marsupials eat the shoots and leaves of shrubs and trees, very occasionally supplementing their diet with buds, flowers and other plant products. Membership of this guild is very heterogeneous, and includes the Koala (*Phascolarctos cinereus*), Greater

Tasmanian Bettong *Bettongia gaimardi* Male
EASTERN TASMANIA

Glider, three species of brushtail possums (*Trichosurus* spp.), Scaly-tailed Possum (*Wyulda squamicaudata*), seven species of ringtail possums, Swamp Wallaby, and the two Australian tree-kangaroos, Bennett's (*Dendrolagus bennettianus*) and Lumholtz's (*D. lumholtzi*). Although not included here, the larger species of cuscus and several species of New Guinea possums are browsers, as were many of the now-extinct megafauna from the Pleistocene in Australia.

Given the diversity of species within this subguild, it is not surprising that some are generalist browsers that eat the leaves of a wide range of plant species (e.g. the brushtail possums and Swamp Wallaby), and others are more specialized. In the latter category, the Koala is notoriously choosy about eating the leaves of *Eucalyptus* trees, although it has been observed to dine occasionally on the leaves of *Melaleuca* and even *Casuarina* trees. The Greater Glider is even more conservative; it eats the leaves of just one or two species of *Eucalyptus* in any one locality, although some two dozen species are used over the species' entire range.

Scaly-tailed Possum *Wyulda squamicaudata* Female
NORTHERN KIMBERLEY, WESTERN AUSTRALIA

Except for the ground-dwelling Swamp Wallaby, all browsers are skilled climbers and depend on trees for shelter as well as for food. Tree hollows or tangles of branches and foliage are favored for nests by most species, although Koalas are unusual in resting while propped up in the fork of a tree. Despite their arboreal preferences, most browsers have to descend to the ground from time to time to switch trees or feed on low shrubs, and can bound quickly if pressed. Koalas will cover several hundred metres between trees, while Common Brushtail Possums are even more versatile and become almost terrestrial in rock outcrops or savanna woodland where trees are scarce.

Grazers Members of this large subguild share a common diet of grass, forbs and low shrubs, and include the kangaroos, wallabies and wombats. The smallest species, such as the Monjon (*Petrogale burbidgei*) and Nabarlek (*Petrogale concinna*), weigh only 1250–1350 grams (44–47 ounces) and tend to select succulent, low-fibre plant foods when these are available. The larger kangaroos are 50–60 times heavier and, like wombats, can include much tougher and more fibrous plant material in their diet. Middle-sized grazers such as hare-wallabies (*Lagorchestes* spp.) and forest-dwelling pademelons (*Thylogale* spp.) add fruits to their diet when in season, and most species will eat the leaves of selected shrubs. If times

are tough, unusual foods such as *Acacia* leaf litter may be eaten. Most grazers are terrestrial, and feed and shelter at ground level. However, rock-wallabies display remarkable agility when clambering on steep slopes and almost sheer rock faces, and climb the lower limbs of trees on occasion to access leaves if ground-level food is scarce. Wombats, alone among the large marsupials, forage at ground level but dig extensive underground burrows.

Monjon *Petrogale burbidgei* Female
MITCHELL PLATEAU, WESTERN AUSTRALIA

R.Woodford Ganf
1987

Red-necked Pademelon *Thylogale thetis* Male
DORRIGO PLATEAU, NEW SOUTH WALES

Being vegetarian may seem to have a lot of advantages over being a carnivore or an omnivore in that green food is readily available and usually easy to find. But, appearances can be quite deceptive. Plants pose many challenges for the animals that eat them. The first problem is that the most nutritious components of plants – the carbohydrates, proteins and lipids – are locked inside leaf and stem cells behind walls of tough and indigestible cellulose. The second problem is that plants often place their most nutritious parts in places where it is difficult for herbivores to find them, such as high above ground at their growing tips or underground in deep roots or tubers. In addition, many plants repel herbivores by lacing the contents of their cells with distasteful or toxic chemicals, by deploying external defences such as spines and poisonous hairs, or by embedding shards of silica within their tissues.

The herbivores adopt several approaches to overcome these defences. The simplest and most obvious is to selectively eat the most nutritious and less well defended parts of plants, such as buds and young leaves. This works well for agile, arboreal herbivores such as possums that can climb high to reach growing shoots, and for smaller terrestrial herbivores such as rock-wallabies with slender snouts. When leaf material has been selected and gathered, all herbivores then begin the digestive process by chewing and grinding the food to fragment it and release the cell contents. This process is aided by the production of copious amounts of saliva. In kangaroos and wallabies, the cellular fragments are further digested in a highly elaborated fore-stomach that houses huge numbers of cellulose-fermenting microbes. Without these, plant cellulose could not be digested. In the Koala, possums and wombats, by contrast, microbes are housed in a greatly enlarged caecum or in expanded parts of the colon. The Koala has the distinction of having perhaps the largest caecum, for its size, of any mammal. Particles of food may remain in this specialized fermentation chamber for 7–12 days, allowing full digestion of 25–50% of the stored plant fibres and release of considerable amounts of energy.

Beyond these common solutions to overcoming plant defences, the herbivorous marsupials show an astounding range of additional adaptations. Wombats have open-rooted molars that grow continuously through life, thus allowing them to maintain the complex, ridged surfaces that are needed for grinding fibrous plant food. In some kangaroos, by contrast, the front cheek teeth are shed as they become worn and are replaced by fresh molars that erupt at the back of the tooth row. This system works well but if animals have tough diets for several seasons or survive for longer than average, they face starvation as their supply of new molars runs out. In his 1995 book *Kangaroos*, Terry Dawson reports seeing an old female kangaroo with just two very worn molars in each of her jaws. Once food is in the mouth, all herbivores produce large quantities of saliva to lubricate the passage of plant fibres. In some of the browsing possums, food is passed through the gut twice to maximize the extraction of energy and nutrients – soft faeces are excreted by day and eaten by animals while in their nest and then passed as hard pellets after full digestion at night. In contrast, Koalas increase the digestibility of their food by regurgitating and masticating it a second time. Along with the wombats, Koalas also possess a specialized gland in the stomach wall that secretes gastric juices to assist the digestion process. All these browsers produce stomach enzymes that neutralize toxic compounds in plant leaves.

Before leaving the herbivorous marsupials, we should look at one remaining species whose placement poses problems – the Honey Possum (*Tarsipes rostratus*). This diminutive (9 gram; 0.3 ounce) species is the only marsupial – one of a very small and select group among all the mammals – to subsist entirely on nectar and pollen. It has reduced numbers of teeth (up to 22, compared with 42–50 in dasyurids), a long, sharply pointed snout and an elongate, brush-tipped tongue for probing flowers, all of which suggest that this species has been specialized on its unusual diet for a very long time. As a guild with just one species is not exactly a guild, the Honey Possum is considered here as an honorary, if tenuous, herbivore.

Dietary guilds: some caveats It is convenient to pigeon-hole species into guilds and subguilds but it is worth remembering that our categories provide much better fits for some species than for others. We have seen that the Honey Possum sits uneasily in the herbivore guild and that there is insufficient evidence to confidently label the marsupial moles as insect-eaters. But there are other species that also give pause for thought. For example, erstwhile meat-eaters such as the Eastern Quoll (*Dasyurus viverrinus*) and Northern Quoll switch to omnivory when times are tough; possums and, occasionally, wallabies are not above checking out freshly killed prey on the sides of roads. In New Zealand, the Common Brushtail Possum is notorious for raiding the nests of native birds for eggs. Our knowledge of species' diets is often based on just one or two detailed studies; it is tempting to think that these are definitive and that we can even extrapolate the results to other guild members. Is this a reasonable assumption? Let us consider the bettongs. The surviving species persist in small areas of high rainfall that are a fraction of the vast areas that they occupied at the time of European settlement. Quite naturally, all dietary studies have been made on these remnant populations and from these we assume that bettongs are fungus-eating omnivores. Yet, each species once occupied arid areas where fungi are rare or do not occur. Does this mean that bettongs are not really dependent on fungus, or does it mean that the arid environment has suffered a loss of fungal diversity in recent years? Current research by Graeme Finlayson, at the University of Sydney, suggests the former possibility. Working at Scotia Reserve in western New South Wales, where both Brush-tailed and Burrowing bettongs have been recently reintroduced, Finlayson's work indicates that both species are thriving on a mostly vegetarian diet. The bottom line from these observations is that, while guilds are convenient for grouping ecologically similar species, their boundaries – like marsupial diets – should be seen as flexible rather than rigidly unchangeable.

Burrowing Bettong
Bettongia lesueur Female
BERNIER ISLAND, WESTERN AUSTRALIA

Distribution and abundance

Marsupials occur in all terrestrial habitats in Australia, as well as on many offshore islands and throughout the New Guinea region. Unlike the Yapok (*Chironectes minimus*) of South America, no Australian species has taken to freshwater habitats, although many seek refuge in the dense vegetation of riparian and lacustrine situations. Many marsupials also occupy coastal or subcoastal habitats, and two species – the Mountain Pygmy-possum and Dusky Antechinus – are found in the high country at altitudes above 2000 metres (6560 feet). It is reasonable to assume that marsupials are not distributed at random across the continent, but we face considerable challenges in identifying common determinants of distribution at the continental scale. Certainly, more marsupial species live at low than at high latitudes and in areas where the environment is structurally complex. And, as predicted by ecological theory, more marsupials occur on the mainland than in Tasmania or on smaller islands. However, distributional patterns and the processes that shape them become easier to discern if we look at marsupials on a guild basis. I have adopted this approach below.

Two other decisions need to be noted. Firstly, I have placed most emphasis on identifying the ecological forces that dictate species' current distribution. However, as we saw in the chapter, Origins and Evolution, historical events are also important in shaping species' distributions. For example, tree kangaroos and cuscuses now occur in north-eastern Queensland because their ancestors moved south from New Guinea when a land bridge provided the opportunity to disperse. Within Australia, the drying of the southern corridor across the Nullarbor has separated formerly continuous populations of possums and potoroos. The ancestral populations have since diverged, with Common Ringtail Possums (*Pseudocheirus peregrinus*) and Long-nosed Potoroos persisting in the east and Western Ringtail Possums (*Pseudocheirus occidentalis*) and Gilbert's Potoroos in the west. While acknowledging the importance of such historical and biogeographical events, and noting them in passing in the account below, I have taken the view that species can only persist where they do now because of the prevailing ecological conditions. If the conditions are unsuitable, species do not persist. How long would cuscuses have survived on Cape York if they had dispersed there and found no trees? I have also placed emphasis on patterns of distribution and abundance that prevailed in the early years of European settlement, deferring discussion of factors that influence modern distributions to the chapter, Conservation.

Carnivores

Meat-eaters   All four species of quolls have suffered large reductions in their ranges, and now occur patchily around the periphery of continental Australia and in Tasmania. They occupy a variety of timbered habitats, venturing occasionally into lightly wooded country and coastal heathland. There is little evidence that any species is restricted by narrow habitat requirements, food or soil type, although the Northern Quoll shows a marked preference for rock outcrops. The most forest-dependent species appears to be the Spotted-tailed Quoll, which is confined largely to tall open and closed forests that receive at least 600 millimetres (23 inches) of rain each year. These productive forests supply a range of medium-sized mammals that the quoll includes in its diet, as well as opportunities for using arboreal routes for stalking and pouncing on those prey. Curiously, records of this striking species turn up from time to time in odd places such as on the edge of towns and in low scrub or farmland. Most of these records are close to patches of forest, with inland occurrences probably reflecting animals dispersing along riverine corridors or forested stock routes. Eastern Quolls are less reliant on heavy forest, and occur in highest densities in Tasmania where mixed forest and pasture is available.

Uniquely among the meat-eaters, the Western Quoll (*Dasyurus geoffroii*) was once widespread across the arid inland in open grassland and shrubland habitats. It is not clear how this species was able to conquer Australia's harshest climatic region when other quolls could not, although some evidence suggests that the Western Quoll is unusually tolerant of temperature extremes. Animals can maintain a normal body temperature at 0°C (32°F) and are able to greatly limit water loss at high temperatures. Western Quolls do not dig burrows but may limit their exposure to temperature extremes by using the below-ground warrens of Burrowing Bettongs and other diggers.

The two remaining meat-eaters are (or were, in the case of the Thylacine) confined to Tasmania. The Thylacine appears to have been confined mostly to the open forests and woodlands of northern and eastern Tasmania as these habitats provided the densest populations of wallabies and other prey. The Tasmanian Devil, with its broad-ranging diet, occurs throughout the island. Intriguingly, both species were widespread on the Australian mainland until recently, with the Thylacine even extending into New Guinea. The demise of the latter species, around 3000 years ago, was probably driven by the introduction of Dingoes (*Canis lupus dingo*) about 1000 years earlier, and probably accompanied by improvements in the hunting technology of Indigenous people (see chapter, Recent History). The Tasmanian Devil managed to persist until just over 400 years ago, until it too succumbed. Like the Eastern and Spotted-tailed quolls, the Devil is abundant locally in Tasmania where its food sources remain common and wild dogs are scarce. However, it is facing a serious threat from a virus that causes debilitating facial tumors. If the Red Fox (*Vulpes vulpes*) becomes established in Tasmania following its introduction there in 1999 or 2000, all extant members of the meat-eating subguild will be at risk.

Insect-eaters   Insects and other invertebrates occupy all terrestrial habitats in Australia, and so too do the marsupials that prey upon them. Despite this common food source, however, there is little evidence that the distributions of insectivores are generally governed by similar environmental or biotic factors.

Southern Ningaui *Ningaui yvonneae*
Female with young aged approximately 60 days
BILLIATT CONSERVATION PARK, SOUTH AUSTRALIA

Perhaps the most easily explained distribution is that of the Numbat, the only dietary specialist among the insectivorous marsupials. This species presently occurs in the south-west of Western Australia and is closely associated with the distribution of its preferred prey of mound-building termites. In the late 1950s John Calaby recorded 25 species of termites in Numbat droppings, the most frequent being a distinct south-western form of the abundant termite *Coptotermes acinaciformis*. He also recorded 27 species of ants but these were in small numbers and apparently ingested incidentally. If the distribution of the *Coptotermes* is overlain with that of the Numbat, the correspondence is quite good. Calaby also noted that fallen hollow logs make critical nesting sites for the Numbat and that these are often hollowed out by the termite. Since *Coptotermes* is restricted to eucalypt forests and woodlands, it is perhaps no surprise that Numbats attain their highest numbers in these habitats.

Intriguingly, Numbats were rare or absent from one widespread forest type – jarrah (*Eucalyptus marginata*) – at the time of Calaby's studies. Jarrah wood is hard and resistant to termite attack. Harvesting operations have since altered the structure of some parts of the jarrah forest by removing trees and dumping small logs and branches on the forest floor. These materials provide nest sites for Numbats and appear to have allowed populations of termites to increase, so allowing the Numbat to extend its range locally into the modified areas.

At the time of European settlement, Numbats occupied mulga, scrub and other habitats across much of southern and inland Australia, and they presumably depended upon other, diverse communities of termites. The cause of the subsequent range reduction is uncertain but the introduced Red Fox is likely to have had a large effect. The scarcity of hollow logs in lightly timbered habitats in drier areas perhaps made Numbats more conspicuous to the new marauding predators, and the habitats would have offered poor cover for escaping them.

While the Numbat specializes by diet, the two species of marsupial mole have become morphologically specialized for an almost exclusively subterranean way of life. They are blind and lack external ears, a calloused, horny shield protects the snout, and powerful spade-like claws are present on the fore-feet. Although widely distributed in central and north-western Australia, the moles live principally in firm, sandy soils and river flats where

Giles' Planigale *Planigale gilesi* Male
ANNA CREEK STATION, SOUTH AUSTRALIA

burrowing is easy. In these habitats moles also have access to diverse and abundant supplies of ants and other insects, such as beetle larvae, which appear to form at least part of the natural diet. Moles have never been collected in calcareous earth soils, nor from red or yellow clay areas, presumably because these soils provide difficult conditions in which to burrow. Sadly, moles also seem to be getting scarce. In *Our sandhill country*, published in 1933, Alice Duncan-Kemp noted how colonies of moles, or *ka-ko-ma*, could be found in the magnificent red sand dunes of western Queensland, undermining the ground to the point that it was dangerous for both horsemen and stock to travel. In 18 years of intensive research in this region, with 10 years spent on Duncan-Kemp's old property at Mooraberrie, I have never seen a mole or anything resembling mole tracks. New survey methods based on digging trenches to uncover moles' sand-filled tunnels offer the best hope for clarifying the status of these mysterious animals in future.

The distributions of the remaining less specialized marsupials range from being very localized to almost pan-continental, and we might expect the factors controlling their distributions to vary accordingly. Perhaps the best examples of species limited by apparently simple environmental factors are the several species of planigales. These tiny marsupials are characterized by their unusual 'squashed' appearance: they have curiously flattened heads and their limbs are attached laterally to the body. Three species, the Long-tailed Planigale (*Planigale ingrami*), Giles' Planigale (*P. gilesi*) and Narrow-nosed Planigale (*P. tenuirostris*), show a close association with grey, brown and red clays, and black earth soils, all of which are characterized by deep surface cracks. Provided these soils are present, these planigales can occupy a variety of habitats from open grassland and savanna to heavily vegetated creek channels. Both of the major soil groups are situated chiefly on riverine plains and associated lowlands;

many collecting localities of the planigales that do not correspond with these soil groups lie on riverine formations. Both Giles' and Narrow-nosed planigales use soil cracks for foraging – their compressed heads and bodies squeeze into the narrowest of fissures to obtain prey. The deepest reaches of soil cracks are also secure retreats from larger insect-eating dunnarts and predators, as well as shelter from the summer heat and winter cold.

In contrast to its three relatives, the Common Planigale (*Planigale maculata*) is not restricted by soil type or land form but is confined mostly to coastal areas receiving more than 1000 millimetres (39 inches) of rainfall each year. This species uses soil cracks opportunistically but otherwise makes nests of bark and grass on the soil surface. It is presumably exposed to less extreme fluctuations of temperature than its three relatives, which occupy more arid regions, and is physiologically less able to withstand cold.

Narrow-nosed Planigale
Planigale tenuirostris Male
North-western New South Wales

Fat-tailed False Antechinus
Pseudantechinus macdonnellensis Male
Macdonnell Ranges, Northern Territory

Pilbara Ningaui *Ningaui timealeyi* Female
with young aged approximately 50 days
Pilbara, Western Australia

The distributions of several other species of insectivorous marsupials also seem related to rainfall. Like
the Common Planigale, the Brush-tailed Phascogale
prefers coastal areas receiving at least 1000 millimetres
(39 inches) of rain a year and disappears entirely in
areas where annual rainfall is less than 500 millimetres
(20 inches). The distribution of the Dusky Antechinus
is similar, at least in eastern Australia. Why these
species are restricted to wet areas is a matter of
speculation. However, as Brush-tailed Phascogales
and Dusky Antechinuses are among the largest species
in the insectivore guild and need a lot of food each
day to maintain their body sizes, it is possible that
their distributions are limited indirectly by rainfall
and more directly by the higher productivity of insects
in wetter habitats.

Still other insectivores have favored habitats that restrict
where they occur. The Fat-tailed False Antechinus
(*Pseudantechinus macdonnellensis*) is confined to rock
outcrops over most of its range, although enterprising
animals also make homes in the large termite mounds
of the Tanami and western Simpson deserts. The range
of the Long-tailed Dunnart (*Sminthopsis longicaudata*)

Wongai Ningaui
Ningaui ridei Female
Near Cook, western South Australia

also encompasses rock outcrops but small ridges on the hind-feet and its remarkable, counter-balancing tail allow this species to clamber nimbly on boulders and loose scree slopes too. By contrast, the Julia Creek Dunnart (*S. douglasi*) is confined to the flat Mitchell grass downs country of northern Queensland, using crevices in the soil of this inhospitable region for shelter. Several more species are restricted to grasslands dominated by spinifex grasses (*Triodia* spp.). The diminutive ningauis, for example, use their cone-shaped heads to bore through the closely knit and spiny leaves of the spinifex hummocks as they search for small insects. The Pilbara Ningaui (*Ningaui timealeyi*) seems to prefer large, dense hummocks but both the Southern and Wongai Ningauis (*N. yvonneae* and *N. ridei*) use hummocks of any shape or size. Most activity takes place within the secure interior of the hummocks but I have seen both of the latter species climbing gingerly on the outside of hummocks, using their semi-prehensile tails to help stay above the pin cushion below.

Spinifex is also used for shelter by several desert-dwelling dunnarts, including the Ooldea Dunnart (*Sminthopsis ooldea*) and Lesser Hairy-footed Dunnart (*S. youngsoni*). These insectivores are small enough to wriggle their way into hummocks without doing themselves too much damage. The Sandhill Dunnart (*S. psammophila*), by contrast, is too stocky to shimmy through spinifex leaf spines without getting impaled, and seems to prefer old doughnut-shaped hummocks with open centres and gaps between the leaves. Wildlife biologist Sue Churchill reports that this species has a most unusual trick up its sleeve. If a ready access point into a hummock cannot be found, animals spring from a crouched position over the spiny outer leaves and land directly in the plant's open centre. There is presumably a strong premium on judging the jump correctly each time!

So far we have looked at insectivores with distributions that can be explained readily by their association with one or two environmental factors. But how reliable is it to assume that an association of this kind is real and not simply a lucky coincidence or correlation? And what can we say about species that have distributions with no obvious associations with environmental factors? One way out of these dilemmas is to model the environment experienced by a species in known parts of its range and then ask the model to predict other areas where the species is likely to be found. If the prediction is correct, and the actual and predicted distributions coincide, we can assume with some confidence to have identified the environmental parameters that restrict the species' range. If not, the parameters in the model can be replaced until the prediction is correct.

Habitats and soils have been used successfully in some models for insectivorous marsupials but climate is a surprisingly general and reliable predictor. Take the antechinuses. Climate models indicate that the Brown Antechinus (*Antechinus stuartii*) occurs in warm coastal localities in eastern Australia that receive an annual rainfall of 1430 millimetres (56 inches), whereas the smaller Agile Antechinus (*A. agilis*) occupies much cooler areas where rainfall averages only 1071 millimetres (42 inches). Few areas

Sandhill Dunnart
Sminthopsis psammophila
Male (left) and female
Eyre Peninsula, South Australia

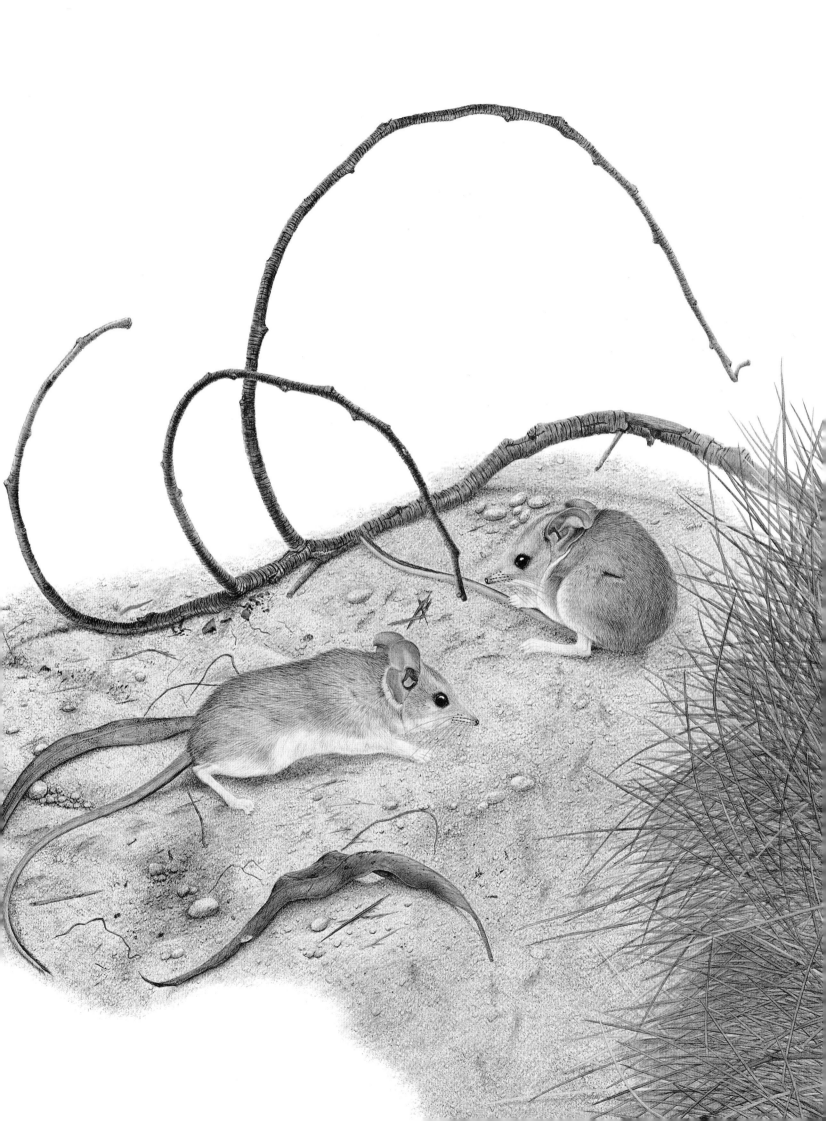

Ooldea Dunnart
Sminthopsis ooldea Male
COMMONWEALTH HILL STATION, SOUTH AUSTRALIA

are predicted to be suitable for both species and the distributions do not in fact overlap much. However, one predicted area of overlap, near Bowral in the southern tablelands, has recently been shown to harbor both species, and suggests that the climate models are robust.

Another example is even more striking. Climate modelling has shown the Hairy-footed Dunnart (*Sminthopsis hirtipes*) to occur in central and western areas of the continent that receive an average of 246 millimetres (10 inches) rain a year. Under this regime, this distinctive dunnart would not be expected in the dry interior of the Simpson Desert, where rainfall is less than 200 millimetres (8 inches) in many areas. Yet, in mid-1992, after a year and a half of above-average rainfall in the desert, Hairy-footed Dunnarts turned up in my study area for the very first time. Although dunnarts had probably been holed up in an oasis somewhere in the desert, the known range expanded eastward – some 500 kilometres (310 miles) – in apparent response to the wetter conditions. Of course, if species are shifting around the landscape in response to such climatic shifts, much

more profound fluxes in distributions can be expected when the full effects of global warming become apparent. We return to this issue later.

These examples show that predictive modelling can be a very useful means of identifying the factors that determine where species occur. But it can also get things spectacularly wrong. On a purely climatic basis, for example, both the Narrow-nosed and Giles' planigales are predicted to range well into Western Australia but neither has been recorded there despite considerable survey effort. The Yellow-footed Antechinus (*Antechinus flavipes*) is likewise predicted to occur in southern coastal New South Wales and along much of the Victorian coast, where it has never been turned up. Such failures seem to cast doubt on the modelling process but they might actually indicate that different factors operate in different parts of species' ranges. The planigales, for example, probably do not occur in the far west due to the lack of suitable cracking soils. In the case of the Yellow-footed Antechinus, field studies invoke another factor entirely: competition.

The Brown and Agile antechinuses, and the Subtropical Antechinus (*Antechinus subtropicus*), occupy all wooded habitats in south-eastern Australia, and achieve densities of up to 50 animals per hectare (2.5 acres) in some places. The Yellow-footed Antechinus typically maintains much lower densities than its relatives and, where they do occur together, there is competition for food. In south-eastern Queensland the Yellow-footed Antechinus does extend its range to the coast and even onto off-shore islands but only in places where the Subtropical Antechinus is absent. Where the ranges of the two species abut, there is some evidence that the shared distributional boundary ebbs and flows between years. This perhaps reflects a temporary advantage that one species obtains over the other as climatic or environmental conditions shift.

If abutting, or parapatric, distributions indicate the potential for competition, several other pairs of insectivores may turn out to place limits on each other's distributional boundaries. These include such similar species as the Common Dunnart (*Sminthopsis murina*) and the White-footed Dunnart (*S. leucopus*) in Victoria and southern New South Wales, and Gilbert's Dunnart (*S. gilberti*) with both the Little Long-tailed Dunnart (*S. dolichura*) and Grey-bellied Dunnart (*S. griseoventer*) in Western Australia. Obtaining confirmation of competition is not easy and often involves experimentally removing species from an area to see if putative competitors move in and take over the previously occupied space. Despite the logistic challenges, there are wonderful opportunities here for enterprising postgraduate students!

Yellow-footed Antechinus
Antechinus flavipes Two males
Shown in dry sclerophyll forest habitat
with common heath *Epacris impressa*
MOUNT LOFTY RANGES, SOUTH AUSTRALIA

So far it is clear that the distributions of individual insectivore species can be limited by food, soils, vegetation type, climate, competition and combinations of these factors. And there are undoubtedly more factors that we have not listed. Rather than continue to examine each species in turn, a more productive approach is to ask if there are general, unifying patterns that allow us to explain and predict where species occur. A good first step here is to overlay the known distributions of all species in the subguild and thus compile a map of the species' densities. I first constructed a species density map some years ago, and was stunned by what it showed: as many as 8-9 species co-occur in the arid central regions of Australia, while just 2-3 species co-occur in the wetter and more productive forests and heaths on the continent's periphery (Figure 4). This pattern runs counter to the more usual finding elsewhere in the world – that the mammalian faunas of arid regions are impoverished. A further surprising trend in Australia – in contrast to the pattern for North and South American mammals and for mammalian faunas generally – is that the fewest species of insectivores are found at low latitudes.

Several explanations can be put forward to account for these patterns. Firstly, there may be excellent opportunities for speciation in arid habitats, so that in effect the deserts act as 'hotspots' for the generation of new insectivores. Scincid lizards are also very diverse in arid Australia and perhaps have responded to similar opportunities to radiate in the splendid isolation of the central deserts. It is possible also that much evolution of the insectivorous marsupials has taken place in arid environments, so that many species are adapted to the harsh conditions. Secondly, despite their apparent simplicity, necessary resources such as food, shelter, vegetation and other habitat components may be more readily available in arid than in temperate forested regions, thus allowing more species to be supported. A third possibility is that, even if these resources are not very abundant or diverse in the deserts, harsh environmental conditions keep populations of resident species at low levels and thus prevent any from becoming dominant. This process has been termed 'rarefaction'. It has some merit as an explanation because populations of desert insectivores never achieve densities of more than five animals per hectare (2.5 acres) – an order of magnitude less than the densities of some of their forest-dwelling relatives. At such low densities competition

Species density

Insect-eaters

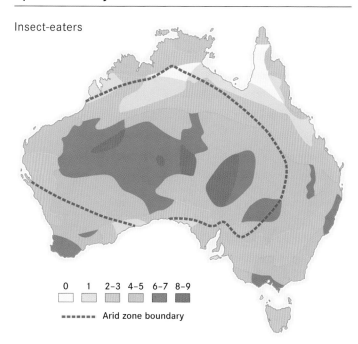

| | 0 | 1 | 2-3 | 4-5 | 6-7 | 8-9 |

- - - - - - - Arid zone boundary

FIGURE 4 Species density map of insect-eating marsupials compiled by overlaying distribution maps for all species in the subguild. Lighter colors indicate areas where insect-eaters are scarce or absent, stronger colors show areas where the distributions of up to nine species overlap. The area within the dashed red line is the arid zone.

is unlikely and no single species would be able to gain ascendancy. Finally, there is some evidence that desert insectivores are very mobile, capable of travelling more than 3 kilometres (1.9 miles) in a night and over 10 kilometres (6.2 miles) in a week. Although such movements do not compare with the grand movements of nomadic desert birds, they are much greater than those of their sedentary relatives in the forests. Movements of several kilometres make possible a lot of interchange of species across broad inland regions, thus increasing the chance of many species occurring together when local conditions are suitable to support them.

Overlaying maps of species distributions provides some insight into general factors that determine where species occur but it is really a coarse first step. If you go to any area and expect to find all the species whose ranges overlap, you will be disappointed. On average, less than half the species predicted from their distributions will be found. The reason is simply that, within broad areas, different species prefer different habitats and seldom stray beyond them. At the habitat level, hummock grasslands, arid woodlands and other types of desert vegetation, still have more species of insectivores than temperate forests, woodlands and heaths. The difference is stark: hummock grasslands support an average of 5.3 species, three times as many as the average for rainforest and heath. Detailed studies in hummock grasslands show that they provide a surprisingly broad range of opportunities for different species to seek food and shelter. As we saw earlier, the hummocks themselves act like giant pincushions and allow entry only by the smallest species such as ningauis. Larger insectivores forage around the edges of hummocks; species such as the mulgaras dig burrows in the sand under the protective spiky leaves of the hummock grass above. As hummock grasslands also have sparse shrubs, low trees, cracks in the soil, leaf litter and other components of habitat that harbor food or shelter, there are great opportunities for many species to use this vegetation type. These apparently simple habitats thus provide more diverse habitat components than forests and they are exploited by the extraordinary range of insectivorous marsupials that we see.

Omnivores

Insect-eating omnivores These marsupials are confined largely to the forests and heathlands of southern and eastern Australia. The ranges of individual species are dictated principally by their feeding and nesting requirements, which depend in turn on the distributions of preferred trees and shrubs.

Consider Leadbeater's Possum. This rare species occurs largely in high altitude Mountain Ash (*Eucalyptus regnans*) forest at some 200 localities within a restricted area of 4000 square kilometres (1545 square miles) in the central highlands of Victoria. Its diet includes *Eucalyptus* nectar and sap, insects gleaned from bark and leaves, honeydew from herbivorous lerp insects, and gum exudates from understorey wattles. Despite this broad diet, pioneering studies in the 1980s by Andrew Smith at Monash University suggested that just wattles and tree hollows are the key habitat components for this possum. Wattles are important because they produce gum. Captive animals gouge wattle branches to increase the flow of gum using specialized and stereotyped patterns of behavior. Gum is usually indigestible but the gut of Leadbeater's Possum is distinguished by expansions of the caecum and colon that seem to allow gum to be efficiently fermented. Tree hollows are important in providing shelter. As this possum lives in colonies of up to eight animals, suitably large hollows are needed to accommodate them, and these occur in tall, old trees. In combination, the exacting food and nest requirements of Leadbeater's Possum limit it to a smaller area than most other members of its subguild.

Recent discoveries of Leadbeater's Possum in subalpine woodland and lowland swamp forest in Victoria suggest that the species is more ecologically flexible than had been thought. The new populations are small, and work by Dan Harley at Monash University suggests that the swamp forest site has similar structural elements, at least, to those identified for the Mountain Ash sites: tall mountain Swamp Gum (*Eucalyptus camphora*) trees that provide hollows for nests, and a dense understorey of paperbarks (*Melaleuca* spp.) and tea-trees (*Leptospermum* spp.) for food.

Sugar Glider *Petaurus breviceps* Male
Near Tumut, New South Wales

In contrast to Leadbeater's Possum, the Sugar Glider is very widely distributed in a crescent from south-eastern South Australia along the east coast and across the 'Top End' to the Kimberley in Western Australia. It also occurs in New Guinea and has spread to occupy most habitats in Tasmania since its introduction there in 1835. Apart from its obvious gliding membrane, the Sugar Glider bears a strong resemblance to Leadbeater's Possum and shares with it a similar taste for insects and plant exudates, as well as such attributes as an elaborated caecum and colon for the processing of gum. Further evidence of the ecological similarity of the two species comes from observations that they compete for food resources in their small area of overlap in Victoria, with Leadbeater's Possum holding its own in patches of dense forest. Given these similarities, why are the ranges of these two guild members so different?

Studies of the Sugar Glider throughout its vast range indicate that it is so flexible that it can use a multitude of resources in most wooded habitats. At Willung in Victoria, Andrew Smith found that the glider spent 43% of its foraging time consuming wattle gum, 11% of its time on *Eucalyptus* sap, and 28% of its time searching for insects; honeydew, pollen and other plant foods were the objects of the rest of its foraging time. But at Jervis Bay in coastal New South Wales, Jon Howard from the University of New England observed gliders spending over 90% of their time feeding among *Banksia* and *Eucalyptus* flowers for nectar and pollen. In northern Queensland and in New Guinea, gliders visit a very wide range of tree and shrub species, using virtually whatever insect and plant resources are available.

In addition to their catholic diet, Sugar Gliders are not very choosy about where to live. Provided that tree hollows are available at least 3–4 metres (10–13 feet) above ground, the gliders will readily use cavities in a wide range of dead or living trees and can persist well in roadside corridors if old trees are available for shelter. Several hollows are usually used within a shared home range. The importance of tree hollows has been demonstrated experimentally during attempts at reintroducing gliders at Daylesford, Tower Hill and Organ Pipes National Park in Victoria. Food plants were abundant in these localities but because the forest was young and regenerating, natural tree hollows were scarce. Provision of artificial nest boxes allowed Sugar

Gliders to be reintroduced successfully but only after earlier attempts, without the erection of nest boxes, had failed.

Between the distributional extremes of Leadbeater's Possum and the Sugar Glider, the ranges of most other insect-eating omnivores can be interpreted as being related directly to their specific food and nest requirements and, in some locations, to competition. Before leaving this group, I would like to mention two species that are special because of their limited distributions and atypical feeding behavior: the Mountain Pygmy-possum and the Striped Possum.

The Mountain Pygmy-possum became a national celebrity in 1966 when it was discovered living in a ski hut at Mount Hotham in the Victorian high country. It had previously been known only from fossil deposits at the Wombeyan Caves in New South Wales and Buchan Caves in Victoria, and was of considerable scientific interest owing to the similarity of its serrated, blade-like upper premolars to those of the rat-kangaroos. This tiny possum is exceptional in being the only Australian mammal restricted to alpine and subalpine habitats above 1370 metres (4494 feet). Dedicated work by Linda Broome at the NSW Department of Environment and Climate Change suggests that the distributional range of the Mountain Pygmy-possum covers about 500 square kilometres (193 square miles) but within this area, the species occupies patches of habitat that collectively cover less than 10 square kilometres (2417 acres). The fossil record confirms that the Mountain Pygmy-possum was once more widespread and has been retreating to its present high altitude refugia as conditions have warmed since the Last Glacial Maximum some 18,000 years ago. What factors place such severe limits on its distribution?

Unlike virtually all other members of its subguild, the Mountain Pygmy-possum does not depend on trees. It lives above the winter snowline in locations where trees are stunted or absent, and relies instead on boulders and dense, low shrubs for shelter. The species is strictly terrestrial; it builds nests at ground level among and under the boulders. These structures are particularly important in winter in preventing snow from settling on the ground and allowing a buffered space to develop below the snow pack that keeps temperatures at a balmy 0–2°C (32–35.6°F). Unusually among marsupials, the Mountain Pygmy-possum survives the winter by hibernating, entering this phase in late summer and emerging up to 7 months later when snow-melt heralds the arrival of spring. During hibernation animals drop their body temperature to around 2°C (35.6°F), warming up to normal operating temperature for a day every 2–3 weeks so that they can eat. The adaptive significance of this possum's unusual diet can be appreciated at this point. The animals fatten up over summer by feasting on the flowers, seeds and fruits of plants growing in the boulder fields, and also eat large numbers of Bogong Moths (*Agrotis infusa*) that arrive at the boulder fields in late spring. In captivity animals cache seeds as temperatures fall; if they do this in the wild, such stores may provide a ready source of food to sustain possums as they arouse over the long winter period. The unusual

premolar teeth of this species allow animals to crack the tough outer seed coats to eat the nutritious endosperm inside. The Mountain Pygmy-possum thus appears to have specialized in exploiting a niche that was once widespread but now exists only at the highest altitudes.

The Striped Possum also has a restricted range in Australia, being limited to the dense rainforests and fringing open forests of far north Queensland. It was once reputed to be a fruit-eater with a penchant for bananas but more recent observations of this attractive, and spectacularly agile, black-and-white animal suggest that it has a stronger prediliction for insects. It has been observed stabbing its bayonet-like lower incisors into fissures in tree bark to uncover grubs and other

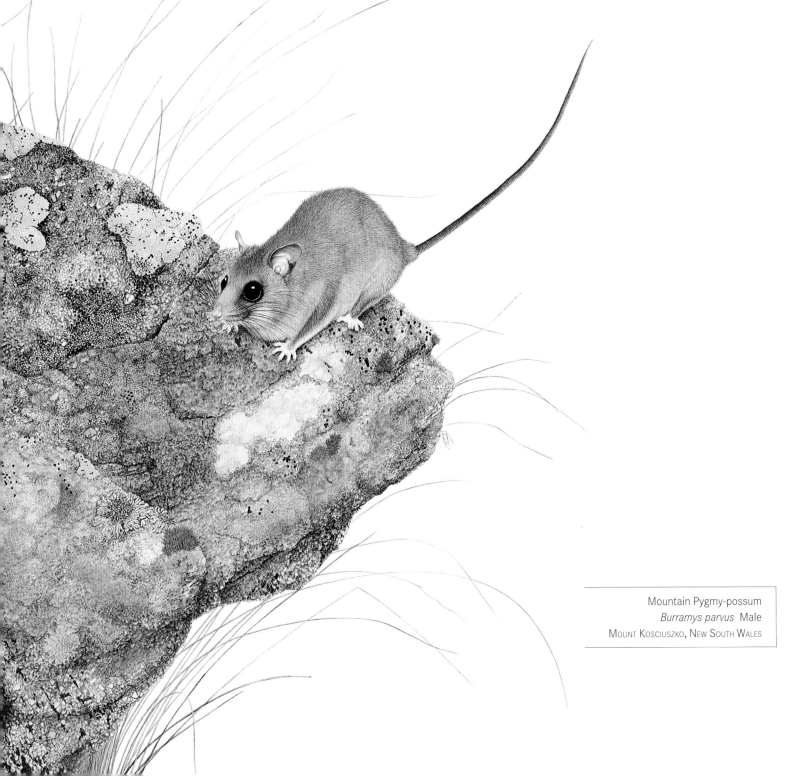

Mountain Pygmy-possum
Burramys parvus Male
Mount Kosciuszko, New South Wales

buried prey, and also uses its elongated fourth finger to remove them once exposed. Analyses of stomach contents confirm its relish for insects, revealing the remains of ants, termites and beetle larvae. Although Striped Possums include plant products and occasionally small vertebrates in their diet, none of these foods is restricted to particular forest types. There is no clear evidence either that Striped Possums have specific nesting requirements, as there are records of animals using nests in tree hollows, tangled vines and clumps of epiphytes. Early research in New Guinea suggested that access to free water is an important constraint on the distribution of this possum. North-eastern Queensland certainly receives high rainfall each year; even so, we do not know whether it is free water that is important for the Striped Possum or associated factors such as forest productivity.

Fruit-eating omnivores The only card-carrying member of this subguild, the Musky Rat-kangaroo, is endemic to the Wet Tropics of northern Queensland. The tropical rainforest that predominates in this region provides a twilight environment that allows this marsupial to forage safely by day, and shelter in the abundance of logs, fallen branches and vine thickets at night. Most importantly, the high diversity of trees in the rainforest ensures that the Musky Rat-kangaroo has access to a smorgasbord of fruits for much of the year. Animals prefer large fleshy fruits from trees such as figs, palms and quandongs, and deftly manipulate the fruit in their fore-paws to strip away the flesh. The strong dependence of the Musky Rat-kangaroo on tropical rainforest suggests that its distribution has always been restricted.

Striped Possum
Dactylopsila trivirgata Female
Atherton Tableland, Queensland

The distributions of the two guest members of the fruit-eating omnivore subguild are similarly restricted to tropical forest, although both occur further north than the Musky Rat-kangaroo. The confinement of both cuscus species to the tip of Cape York Peninsula probably reflects the recency of their arrival in Australia from New Guinea some 18,000 years ago, and subsequent lack of time or opportunity to move further south. The Southern Common Cuscus is mostly confined to primary rainforest and exploits large trees for both fruits and den sites. The Common Spotted Cuscus, by contrast, is sometimes found in mangroves and paperbark trees several hundred metres from rainforest. This species sleeps in temporary nests in the tree canopy; being liberated from any dependence on tree hollows for dens, it sometimes uses regenerating rainforest and other wooded habitats to obtain food or disperse. Despite these and other ecological differences between the two species of cuscus, both are critically dependent on dense forest and do not persist in lower rainfall areas on the western side of Cape York where rainforest gives way to more open woodland.

Fungus-eating omnivores   Members of this group have suffered such drastic reductions in range since European settlement that it is difficult now to specify which factors influence their distributions. Certainly, for the worst-affected species such as the bettongs, the forces that now dictate where they occur will be quite different from those operating two centuries ago. Nevertheless, some generalizations are possible.

Bettongs are now found only in peripheral regions of Australia, occurring generally in forested and wooded habitats with a grassy understorey. Two species, the Rufous Bettong and Tasmanian Bettong are now, and have historically, been restricted to high rainfall areas in Tasmania and the eastern mainland. Most probably, rainfall itself is a secondary influence here, with both species responding to the dense cover and abundance of foods available in high rainfall areas. The Rufous Bettong, at least, seldom drinks free water and instead obtains its needs from fungi and juicy green food. A third species, the Northern Bettong, may be tentatively placed with the first two. Now restricted to a small area on the Atherton Tableland in Queensland's Wet Tropics, the Northern Bettong may have once ranged more broadly in drier districts inland, as suggested by undated but recent fossil material. Three further bettongs, the Burrowing Bettong,

Brush-tailed Bettong and extinct Nullarbor Dwarf Bettong (*Bettongia pusilla*), were once widespread in arid regions and were clearly independent of free water and dense groundcover. The Burrowing Bettong is one of the most proficient of all digging marsupials, while the Brush-tailed Bettong builds impressive subsurface nests using materials that it carries with its tail. We know nothing about the ecology of the extinct Nullarbor taxon but the burrowing and nesting behavior of the other two species would have assisted them in avoiding the harsh climate of the inland.

Our knowledge of the Brush-tailed Bettong is better than for most other species in the subguild thanks to a series of intriguing studies carried out on remnant populations in Western Australia. One such study, by Per Christensen of the Western Australian Department of Conservation and Land Management (now Environment and Conservation), confirmed the association of this bettong with open forest associations that provide a cover of grass and herbs. After fire, the bettongs forage on fungi, insects, seeds and other nitrogen-rich foods, prompting Christensen to suggest that the bettong's apparent 'fire dependency' is actually a means of exploiting this essential food component. Importantly too, fire promotes the growth of shrubs such as poison peas (*Gastrolobium* spp.) which provide food, shelter and an indirect measure of protection from predators by their production of sodium monofluoroacetate (1080) poison. Bettongs have high tolerance to this poison but introduced Red Foxes and feral Cats (*Felis catus*) do not (we return to this subject in the chapter, Conservation). Christensen concluded that the present distribution of the Brush-tailed Bettong was correlated principally with plant associations adapted to hot fires. At Scotia Reserve, in western New South Wales, this bettong prefers sandy areas interspersed with patches of leaf litter and trees with a low canopy – habitats structurally similar to those in Western Australia and, with the exception of fungi, containing similar food resources. Recent work on the Tropical Bettong by Karl Vernes at James Cook University suggests that this species also is well adapted to fire-prone habitats, with animals eating truffles that increase in the post-fire environment.

Brush-tailed Bettong *Bettongia penicillata* Female
SOUTH-WESTERN WESTERN AUSTRALIA

Like the bettongs, the living species of potoroos are now found in moist forest or heath sites in south-eastern and south-western Australia. All occur in areas of high rainfall: annual averages exceed 750 millimetres (29 inches) for the Long-nosed Potoroo, 800 millimetres (31 inches) for Gilbert's Potoroo and 1100 millimetres (43 inches) for the Long-footed Potoroo. The three species occupy dense ground-level vegetation for shelter but show little evident preference for particular soil types. There is little doubt that use of dense surface vegetation reduces the risk of being caught by native and introduced predators, and this helps to explain why the ranges of potoroos have contracted less than those of the plains-dwelling bettongs since the arrival of European settlers. The major exception is the Broad-faced Potoroo (*Potorous platyops*). This now-extinct species once occurred coastally in South Australia and Western Australia but also extended into dry inland habitats with patchy understorey and sparse or no tree cover. Its secrets will almost certainly remain secure: a mere handful of living animals was collected between 1839 and 1875 when this, the smallest potoroo, made its last farewell.

Broad-faced Potoroo
Potorous platyops Male
SOUTH-WESTERN WESTERN AUSTRALIA

Generalist omnivores   Along with the fungus-eating omnivores, several members of this subguild have also suffered greatly from the shocks and after-shocks of European settlement. The most severely affected were the six species that had extensive distributions in arid and semi-arid regions, with three of these having disappeared entirely and the others now being confined to tiny parts of their former ranges. In comparison, five species with coastal or subcoastal distributions remain extant and occupy ranges that have declined by half or less since European arrival.

The Bilby (*Macrotis lagotis*) is the only generalist omnivore to retain a presence in arid Australia and it is instructive to review the attributes that allow it to persist there. In the first instance, given that it now occupies some of the harshest and most open habitat on the continent, it would be reasonable to expect the Bilby to tolerate high temperatures and perhaps be susceptible to the cold. Yet, the reverse is true. At temperatures just 2–3°C (35.6–37.4°F) above zero the tissue insulation of this paradoxical species is greater than that of its counterparts in temperate coastal areas; at 40°C (104°F) it is the only species of perameloid marsupial that is unable to maintain a stable body temperature. In part, this is because the Bilby has a very limited capacity to reduce heat by evaporative cooling; free water is scarce in its inland habitat, so the usual means of staying cool such as panting or licking saliva on the body surface have been abolished in the interest of water conservation. By slowing its rate of metabolism, producing concentrated urine and dry faeces, the Bilby is further able to reduce its overall water turnover to about half that of bandicoots in temperate regions. Apart from its physiological capabilities, the key that has opened the arid plains to the Bilby is its ability to dig spectacularly deep, sheltering burrows that can go down to 2 metres (6.5 feet) and exceed 3 metres (10 feet) in length. In addition to providing a buffered temperature when daytime surface

temperatures may exceed 50°C (122°F), the high humidity at the end of a burrow would assist animals greatly in reducing their water use.

The Bilby was common over much of Australia south of 18° until the early 1900s, when populations in many areas crashed. Within this vast range it occupied woodland, grassland, shrub-steppe and savanna habitats, preferring deep, loamy soils that allowed easy burrow construction. Rocky areas were avoided. As the Bilby has a diverse diet that includes invertebrates, small vertebrates, bulbs, seeds, green plant material and fungi, it seems unlikely that its distribution could have been restricted by its access to food. The distributional ecology of this species was probably therefore centred squarely on its digging proficiency and ability to find suitable soil landscapes.

Lesser Bilby *Macrotis leucura* Young male
KOONCHERA DUNE, NORTH-EASTERN SOUTH AUSTRALIA

Like its larger relative, the Lesser Bilby (*Macrotis leucura*) appears to have relied on deep burrows to escape the extremes of the inland climate and to conserve both energy and water. It constructed burrows in sandhills, and seems to have blocked the entrances to maintain humidity when in residence. Unfortunately, as Lesser Bilbies were last reported in 1931, few other observations of their behavior or physiological abilities have been documented. There is a similar paucity of observations on the Desert Bandicoot (*Perameles eremiana*), which was last collected in 1943. Apart from the extinct Pig-footed Bandicoot, which probably dug short burrows under piles of grass, other bandicoots of arid Australia most likely excavated shallow depressions under nests made from local plant materials and woody debris. On occasion, they probably made opportunistic use of the burrows of more proficient diggers such as the Burrowing Bettong or even the large woody above-ground structures of the stick-nest rats (*Leporillus apicalis* and *L. conditor*). All were probably nocturnal and able to obtain most or all of their water from food. We might conjecture further but, in reality, the opportunity to discover more about the adaptations to aridity of this unfortunate group of omnivores vanished in the 20th century along with the animals themselves.

Generalist omnivores away from the arid zone have commanded a lot more research attention simply because they are easy to catch, live near people and are still around. This does not mean that there is any real agreement on the factors that limit them; indeed, the five species that occupy coastal and subcoastal areas use a range of quite different habitats. The Southern Brown Bandicoot, for example, reaches its highest numbers in heath and responds well to patches that have a recent history of fire. However, it can also use forest, scrub and even open paddock environments, especially if there are no other bandicoot species present to challenge it. The larger Northern Brown Bandicoot (*Isoodon macrourus*) uses a similar range of habitats, but seems most dependent on groundcover. The Eastern Barred Bandicoot (*Perameles gunnii*) exhibits a clear preference for grassland but often tolerates cultivated farmland and suburban backyards. This species achieved a measure of fame in the 1980s and 1990s by making a last, doomed stand on the Australian mainland at the small town of Hamilton in western Victoria. Although the bandicoot exploited gardens, a creek reserve and even the local rubbish tip, this last remnant of the wild mainland population has been virtually eliminated by introduced predators, disease and motor vehicles. The Long-nosed

Northern Brown Bandicoot
Isoodon macrourus Male
Near Darwin, Northern Territory

Bandicoot (*Perameles nasuta*) is also familiar to people in the leafier suburbs of Brisbane and Sydney, and seems to prefer sites that offer dense cover for daytime shelter and open habitats for foraging at night. The Rufous Spiny Bandicoot (*Echymipera rufescens*), found on the tip of Cape York Peninsula, is the fussiest of the Australian bandicoots. This species evolved in the dense, wet forests of New Guinea and probably moved south along forested corridors in the late Pleistocene with the cape's two species of cuscus. It does not stray far from rainforest.

From this brief survey, three factors emerge, tentatively, as limiting the distributions of these bandicoots: nest requirements, the availability of free water and competition. In pioneering field studies in the 1970s, Greg Gordon at the University of New South Wales proposed that nests are the objects of most significance in home ranges of the Northern Brown Bandicoot. This species, and indeed all bandicoots in temperate and tropical regions, shows no inclination to burrow in the wild but constructs ground-level nests of grass and leaves. It is not that these bandicoots are incapable of digging, as they can excavate burrows about 0.5 metres (1.6 feet) deep in captivity. Critically, they burrow only if surface conditions are too hot or if surface nesting materials are not available. Thus habitats providing access to key nest and shelter components such as grass, logs and woody debris are important determinants of distribution.

In physiological studies at about the same time, Tony Hulbert and Terry Dawson at the University of New South Wales documented the high water requirements of the Long-nosed Bandicoot and suggested that, unlike the Bilby, it is unable to obtain its water needs from food alone. They suggested that its distribution is limited to temperate and tropical areas by its need for free water. It is consequently no surprise that this species is confined to a narrow belt on the east coast of Australia where annual rainfall is at least 750 millimetres (29 inches). The other four omnivores with temperate or tropical distributions also occur in high rainfall areas but appear to be less dependent on access to free drinking water. They may respond instead to the high productivity and diversity of foods available in high rainfall environments.

Evidence for competition comes from two general observations. Firstly, bandicoots usually avoid each other but are renowned for their pugnacious fighting behavior if confined. As a general rule, mammals that show such intolerance among themselves are also intolerant of related species so that, if two species occur together, the larger prevails. Secondly, the distributions of bandicoots generally do not overlap but, when they do, the species segregate into different habitats. This pattern is exactly what is expected when competition occurs. In Tasmania, for example, Southern Brown Bandicoots avoid paddocks and areas of low groundcover when Eastern Barred Bandicoots are present but exploit the grassland when they are absent. Similarly in northern Queensland the Long-nosed Bandicoot is restricted to patches of rainforest, whereas the Northern Brown Bandicoot is found only in grassland, low scrub or open forest. On Cape York Peninsula and in isolated patches of rainforest elsewhere where the former species is sparse or absent, the Northern Brown Bandicoot invades the rainforest. Turf wars of this kind usually occur if an important resource, such as food, is in short supply. Further work is needed to resolve whether competition really shapes the distributions of bandicoots and, if it does, what the animals compete for.

Herbivores

Leaf-eaters   Because they eat leaves and need to climb trees and shrubs to get them, the leaf-eaters, or folivores, are very largely restricted to Tasmania and the forested parts of the Australian mainland. However, the distributions of individual species vary enormously: the Common Brushtail Possum has perhaps the broadest distribution of any Australian marsupial, while the ranges of the small rainforest possums and tree kangaroos are among the smallest.

For the Common Brushtail Possum, the most relevant question is not what limits its distribution but why is it distributed so broadly? Although this familiar species has disappeared from many of its former haunts in the drier parts of Australia, it can still be found in abundance in eastern, south-western and northern parts of the continent. It is also common in Tasmania, and (in)famously abundant in New Zealand, where it was introduced in 1858. To thrive in such varied regions, the Common Brushtail Possum needs to call on a bagful of ecological and physiological tricks. Although classified here as a leaf-eater, this possum borders on being omnivorous; it eats leaves, shoots and fruits from eucalypts and a wide range of other trees and shrubs. It is not above raiding birds' nests for eggs or

Common Brushtail Possum
Trichosurus vulpecula Female
Tasmania

killing newly hatched nestlings, and has been recorded on occasion taking meat from pets' bowls in residential back yards. It is able to nest in a wide variety of places, from tree hollows to rock piles to holes in creek banks. In my local area in eastern Sydney, possums can sometimes be found nesting in caves in steep-sided sandstone gullies and in clefts in old sandstone quarries. Most people who live near urban or rural bushland will have had possums living under their eaves. If you are lucky enough to be able to sleep through the noise of possums stampeding over your tin roof at night, the heavy breathing, guttural gurgles and shrieks of duelling possums outside your window will usually be enough to alert you to their presence!

The Common Brushtail Possum is also a good colonizer, having a relatively high reproductive rate with young that mature and disperse early. Unusually among the possums, it is also tolerant of dry environments owing to its ability to use water very sparingly and to produce concentrated urine. The species copes with high temperatures (above 40°C; 104°F) by panting, using the process of evaporative water loss to cool the mouth and circulating blood. Conversely, at low temperatures possums shiver to maintain their body heat. As may be expected for such a widely distributed species, the Common Brushtail Possum is quite variable in appearance and physiology. The brain structure varies more within this possum than in most other marsupials that have been studied, and components of the blood and parts of the genome differ between regions. Some of these variations are easier to interpret than others. Possums are largest in the cooler, southern parts of their distribution, for example, and this is presumably because large size means more efficient retention of body heat. The tendency for mammals to be larger in progressively more polar areas was first noticed in the 19th century, and has become known as Bergmann's Rule after its discoverer. Another common pattern, termed Allen's Rule, is that limbs and tails become shorter in cold climates, again to retard heat loss. I am not sure if Common Brushtail Possums are stubbier-limbed in Tasmania than on the mainland but we might expect this if Allen's Rule is correct. Possums in cooler climates also tend to be darker, while those in inland or in tropical regions are silvery-grey or red. Here, long dark fur provides better insulation and heat retention than the lighter colors, a pattern known as Gloger's Rule. Common Brushtail Possums in Western Australia have

striking tails with black, white and grey bands of variable width. I have no idea what advantages these different patterns might convey, if any, but they were wonderfully useful in allowing me to distinguish individuals during a population study near Perth some years ago.

The success of the Common Brushtail Possum raises two questions. Firstly, what are the factors that limit other leaf-eating marsupials? Secondly, why are the other subguild members not as successful as the Common Brushtail?

Taking the first question, there is good evidence that most other subguild members are limited by some combination of specific food or nesting requirements. The Rock Ringtail Possum (*Petropseudes dahli*), for example, is found exclusively on rock outcrops in northern Australia. It feeds on the leaves, flowers and fruits of a variety of trees and shrubs but is limited in extent by the availability of protective rock clefts and fissures for nesting. Four other species of ringtail are limited to the dense rainforests of north-eastern Queensland and cannot persist in the more open forests that surround them. The Daintree River and Herbert River Ringtail possums (*Pseudochirulus cinereus* and *P. herbertensis* respectively) consume the leaves, and sometimes fruits and flowers, of a variety of rainforest trees and use either tree hollows or large epiphytic ferns for shelter. The Lemuroid Ringtail Possum (*Hemibelideus lemuroides*) prefers young, digestible leaves from a limited number of rainforest trees, whereas the spectacular Green Ringtail Possum (*Pseudochirops archeri*) selects leaves with more fibre. This last species is also unique among possums in being able to eat the leaves of figs and stinging trees (*Dendrocnide* spp.). It also eschews the use of tree hollows, instead sleeping curled up on a branch amid tangled foliage in the upper reaches of tall trees.

In contrast to the restricted ranges of these ringtail possums, the Koala has a wide distribution in eastern Australia but is limited locally by its notoriously narrow food requirements. Early work by Ian Eberhard suggested that, to overcome the challenges of processing the toxic and indigestible compounds in *Eucalyptus* leaves, southern populations had become adapted to exploiting just the Manna Gum and chemically allied species. Indeed, there is good correspondence in the distribution of the Koala with this food source. On the other hand, there is no evidence that Koalas are limited by the availability of nests or dens as they sleep in the forks of trees. Not only this, the pelt has the highest insulation value recorded for any marsupial, and a high resistance to wind.

Before tackling the second question that we raised, it is instructive to review the factors limiting another leaf-eater, the Common Ringtail Possum (*Pseudocheirus peregrinus*). This species occurs broadly throughout much of eastern Australia and has been the subject of many studies. It can achieve high densities in some areas. In an isolated remnant of bushland at North Head, in Sydney, for example, up to 20 ringtails can be seen in a single sweep of a spotlight on a calm night; most will be sitting in thickets of coastal tea-tree but some will be walking slowly and deliberately over grassy lawns. Despite its occasional predilection for human fingers and thumbs, this species depends almost exclusively on the leaves of various species of *Eucalyptus* and understorey shrubs. It faces two common problems associated with eating eucalypts. Firstly, the nutritional quality of the leaves is low. Most species provide little nitrogen (usually less than 2.5% of the dry weight of the leaves) but an abundance of indigestible fibre. Secondly, as we saw above, *Eucalyptus* leaves contain compounds such as phenols that are toxic, and tannins that reduce the digestibility of leaf nutrients and energy. The Common Ringtail partly solves both problems by setting its energy budget at a low level, thus reducing its intake of toxic chemicals, and by retaining scarce nitrogen and digestible energy by recycling its faeces. It is nevertheless still restricted to particular plant species that offer it a balance between palatability and sufficient energy and nutrients.

Surprisingly, availability of nest sites is a further factor limiting the distribution of the Common Ringtail Possum. Unlike other possums, these ringtails usually build their own nests among tangles of leaves and branches near where they feed. In scrubby or regenerating forest in the southern part of the range, they build nests in shrubs such as tea-tree and wattles where the understorey is dense. In open-forest, nests are built extensively in mistletoe (*Loranthus* spp.) tangles, so that this parasitic plant in effect determines where the possum itself can occur. In the northern part of its distribution in Queensland, nest-building is rare. Here, it is likely that the availability of conventional tree hollows dictates the local presence of this possum.

If the leaf-eaters in general are limited by their feeding and nesting requirements, why have they not evolved the same behavioral and morphological flexibility as the widespread and successful Common Brushtail Possum? Implicit in this question is the assumption that generalism is the best recipe for evolutionary success. But, this may not necessarily be true. Specialist folivores are more likely to efficiently exploit particular (and often limited) habitats and persist in them at the expense of their more generalist counterparts. This is the familiar analogy of the jack of all trades but master of none.

Lemuroid Ringtail Possum
Hemibelideus lemuroides Female
ATHERTON TABLELAND, QUEENSLAND

One example will suffice. The Short-eared and Mountain Brushtail possums (*Trichosurus caninus* and *T. cunninghami* respectively) are specialist occupants of tall open-forest and closed forest on the east coast of mainland Australia, and derive much of their food from understorey shrubs. Usually they will aggressively exclude the Common Brushtail Possum from its preferred habitat and, where there is coexistence, the Common Brushtail is the least abundant. In the absence of its competitors, both the northern subspecies of Common Brushtail (*T. vulpecula johnstoni)* and the Tasmanian subspecies (*T. v. fuliginosus)* extend their ranges into closed forest and forage, like the Short-eared and Mountain Brushtail possums, largely on ground level and understorey shrubs.

The exclusion of the Common Brushtail by its more specialist relatives from areas of preferred habitat raises a further question: how many folivores can persist in a given patch of forest? If we overlay distribution maps of all the folivores, it looks as if as many as eight species can overlap. As with the insectivores, however, field surveys show that local areas do not support this number; a rich community of folivores would be just four to five species. Not surprisingly, species densities of folivores are lowest in central regions where only the ubiquitous Common Brushtail Possum hangs on, and highest in the coastal and subcoastal forests of eastern Australia. Hot spots of abundance seem to be associated with patches of forest with high levels of nutrients in the leaves. In some remarkable work by Wayne Braithwaite and colleagues at CSIRO on the south coast of New South Wales, almost two-thirds of all folivores and insect-eating omnivores were discovered aggregating in just 9% of the forest where foliar nutrients were most concentrated.

More folivores occur together in structurally and floristically diverse forests where they can gain access to different food, nest or other resources. In the complex forests of northern Queensland, there is some evidence that folivores segregate by altitude and even by the height at which they forage in trees. Further south, unless foliar nutrients are unusually rich, folivores keep out of each other's way – partly by eating leaves from different species of trees. Species such as the Koala and Greater Glider specialize on *Eucalyptus* and smaller folivores eat a wide range of leaves from other trees and shrubs. Only one or two species occur in heath and woodland habitats where structural, if not floristic, diversity is low. As many as three species overlap in woodland areas of the Kimberley in Western Australia but here the structural diversity of the environment is enhanced by rock outcrops which provide homes for the Rock Ringtail and Scaly-tailed possums.

Grazers   Several species of rock-wallabies declined following the advent of European settlement and one species, the Toolache Wallaby (*Macropus greyi*), became extinct in the first half of the 20th century. With these exceptions, however, the grazing marsupials in general have not fared too badly since the arrival of Europeans and some species have increased their ranges as a consequence of changed land usage. Such expansions can provide useful insight into the factors that govern distributions. Consider the Eastern Grey Kangaroo (*Macropus giganteus*). This large kangaroo is usually restricted to

Rock Ringtail Possum
Petropseudes dahli Male
Arnhem Land, Northern Territory

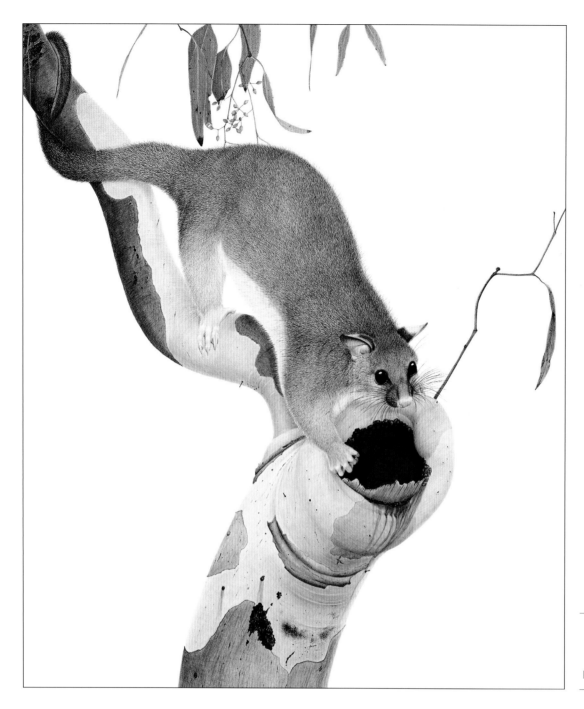

dense scrub and forest in eastern Australia but in the southern part of its range in Victoria and New South Wales it follows the 250 millimetre (10 inch) rainfall contour rather closely. Prior to 1970, the western limit of this species in New South Wales appeared to be the Darling River. By 1972, however, it had penetrated up to 290 kilometres (180 miles) west of the Darling into uncharted territory. Rainfall in the north-western part of the state in 1971 was 287 millimetres (11 inches), some three times more than in the previous 2 years of drought. Observations by Martin Denny, of Mount King Ecological Surveys, suggested that kangaroos were tracking the fields of fresh green grass that flourished

after the rain. Denny also predicted, correctly, that the distribution would shrink when the green feed had dried or been eaten out. This pattern of rain-induced expansion and subsequent contraction has since been documented several more times. From this we may surmise that the Eastern Grey Kangaroo is limited in the southern, inland part of its range by rainfall and the green feed that it stimulates. (It is of interest that the Western Grey Kangaroo (*Macropus fuliginosus*) is similarly restricted by rainfall in the south-western part of its range in Western Australia, extending its range inland in wet years and contracting it again during drought.)

Lumholtz's Tree-kangaroo
Dendrolagus lumholtzi Female
ATHERTON TABLELAND, QUEENSLAND

Parma Wallaby *Macropus parma* Female with pouch joey
GOSFORD, NEW SOUTH WALES

In the northern part of its range in Queensland, the Eastern Grey Kangaroo is not obviously restricted by rainfall but more probably by the availability of suitable habitat. On Cape York Peninsula it occurs at very low density in open forest, possibly benefiting from access to improved pastures from pastoral expansion in the region. It does not extend far into the tussock grass or hummock grass habitats of the inland. At Cunnamulla in south-western Queensland, early work by Graeme Caughley at the University of Sydney showed that the density and dispersion of this species is correlated with the density of vegetation. He concluded that this habitat preference would have been advantageous in reducing predation from native carnivores such as the Thylacine and Tasmanian Devil when they occurred on the mainland. More recent observations by Peter Banks at the University of New South Wales suggest that the introduced Red Fox now reinforces the selection of dense vegetation by kangaroos. Although the Eastern Grey Kangaroo generally prefers to forage near dense habitat, it is seldom found in closed coastal forests, perhaps because its movement there would be impeded. Several remarkable studies in Africa have shown that the shoulder heights of different species of forest ruminants are determined quite precisely by the heights of obstructing woody stems. A similar relationship may operate in Australia, where no large herbivores enter closed forest to any extent.

Allied Rock-wallaby
Petrogale assimilis Female
MAGNETIC ISLAND, QUEENSLAND

Of the large grazers, perhaps most attention has focused on the Red Kangaroo (*Macropus rufus*) and the Common Wallaroo (*M. robustus*). The distribution of the Red Kangaroo is tied to drier areas within the 500 millimetre (20 inch) rainfall contour but within this vast area, local populations favor areas where shade, grass and herbs are available. Pioneering research in the 1960s by Alan Newsome, then with the Conservation Council of the Northern Territory and later at CSIRO, evaluated the relative importance of food and shelter by charting local shifts in distribution during and after drought. During drought, two-thirds of all Red Kangaroos rested by day in woodland within 500 metres (1640 feet) of open plains and watercourses, where green herbage, but no shelter, was available. Three months after good rains had fallen, less than one-third of the kangaroos remained near their former drought refuges. Most had scattered instead to Mulga (*Acacia aneura*) woodlands over 1.5 kilometres (0.9 miles) away where shelter and temporary fresh green grass and herbage were available.

Newsome did not think that free drinking water was an important determinant of the local distribution of Red Kangaroos. He saw animals drink only during severe drought and noted that the water content of their preferred Woolly Butt Grass (*Eragrostis setifolia*) was high enough to maintain them. In Newsome's study the water content of the grass was 8–25% by weight (I have seen Red Kangaroos in western Queensland eating grasses with less than 10% water and still persist for many weeks). Water turnover in the Red Kangaroo is low, with losses minimized by several impressive adaptations. Animals pant and sweat to cool themselves only after sustained exercise, and they lick their fore-arms and lose heat by evaporative cooling. If caught in full sun, animals orient their bodies to minimize exposure, and their dense, pale fur reflects solar radiation quite effectively. Newsome concluded that fresh green food is the cornerstone of the distributional ecology of the Red Kangaroo and this has since been affirmed by other workers.

In contrast to the Red Kangaroo, the Common Wallaroo does not appear to be limited by green food but rather, in many areas, by the distribution of rock outcrops and

perhaps the availability of water. This species ranges over a vast area but its local distribution is often patchy. Its requirements for water are relatively low owing to its ability to produce concentrated urine and recycle urea (a key waste product in urine) through the kidney. Individuals seldom drink and derive most of their water from food; they are capable of withstanding considerable desiccation. The coat of this species is less dense than that of the Red Kangaroo and gives it less protection from solar radiation in summer; hence caves and shaded areas in rock outcrops are important for shelter and water conservation.

Early work in the arid Pilbara region of Western Australia by Tim Ealey of CSIRO confirmed the importance of water for the Euro (as the Common Wallaroo is known in the west). Here, the dams and artificial watering points provided for stock have allowed Euros to extend their

Short-eared Rock-wallaby *Petrogale brachyotis* Female
KIMBERLEY, WESTERN AUSTRALIA

distribution locally into quite desolate areas. Shelter is less important under these conditions because water lost by evaporative cooling (panting, sweating and licking the fore-arms) is replenished easily by drinking, and the animals are freed from their dependence on shaded retreats. Stocking practices have benefitted the Euro in other ways too. Since their introduction to the Pilbara in 1866, sheep have gradually depleted the more nutritious native pastures such as Woolly Butt and kerosene (*Aristida* spp.) grasses, and indirectly allowed the extensive spread of 'soft' Spinifex (*Triodia pungens*), which is a staple food for the Euro during drought. This spinifex contains too little nitrogen (0.06–0.5% by dry weight) to sustain sheep but is just adequate for the Euro. Despite several droughts that have afflicted the Pilbara since 1866, this adaptable grazer has been able to expand its local distribution from the protected hill country on to the open plains, while sheep have quite independently declined.

Among the smaller grazers, distribution patterns appear to be associated most closely with habitat. Rock-wallabies, for example, occur patchily throughout mainland Australia and are confined almost entirely to rugged, rocky habitats. Rock outcrops provide shelter from predators and extremes of temperature, and also yield a diverse and reliable supply of green food throughout the year. Some rock types are preferred over others. In South Australia, the Yellow-footed Rock-wallaby (*Petrogale xanthopus*) is found mostly on sandstone outcrops, perhaps because these erode to provide deeper soils for plant growth. In New South Wales, by contrast, the Brush-tailed Rock-wallaby (*Petrogale penicillata*) occurs on basalt, limestone and granite outcrops in addition to sandstone.

For some species of rock-wallabies, suitable outcrops may be 10–100 kilometres (6–62 miles) or more apart and separated by flat, open country. Historically, this would have posed considerable challenges for dispersal; now in many areas it is simply not possible because of the heavy toll taken of these small animals by the Red Fox. The preference for patchy habitats and the difficulty of moving successfully between them has contributed to bewildering complexity in the diversity of rock-wallabies. Early research by Geoff Sharman at Macquarie University and current work by Mark Eldridge at the Australian Museum has identified suites of species, subspecies, chromosome races and other variants throughout Australia, providing a superb example of a marsupial group that exhibits all stages in the evolutionary process of divergence and species formation. Nowhere is this better seen than in eastern Queensland. Nine species of rock-wallabies have interlocking distributions running from the southern border with New South Wales to central Cape York Peninsula. Two of these species, the Brush-tailed Rock-wallaby and Proserpine Rock-wallaby (*Petrogale persephone*), stand out from the crowd by virtue of their larger size and markings, but the remaining seven species are virtually indistinguishable from external appearance. These are the Cape York Rock-wallaby (*P. coenensis)*, Godman's Rock-wallaby (*P. godmani*), Mareeba Rock-wallaby (*P. mareeba*), Sharman's Rock-wallaby (*P. sharmani*), Allied Rock-wallaby (*P. assimilis*), Unadorned Rock-wallaby (*P. inornata*), and Herbert's Rock-wallaby (*P. herberti*). By contrast, great variation in size and color among populations of the Short-eared Rock-wallaby (*P. brachyotis*) suggest that this taxon has already diverged into different, as yet unrecognized races.

The three species of pademelons occur principally in closed forests and fringing tall open forests in eastern Australia. The Red-necked Pademelon (*Thylogale thetis*) uses closed forest by day but forages in adjacent pastures by night, apparently reducing its risk of predation by doing so. The Red-legged Pademelon (*Thylogale stigmatica*) also occupies closed forest in a broad sweep from Cape York to east-central New South Wales, and exploits open grassy areas in or on the edge of the forest in a similar manner to its relative. In heavily forested areas, both species can be seen emerging from dense

---

Tasmanian Pademelon *Thylogale billardierii* Female with pouch joey
CRADLE MOUNTAIN, TASMANIA

cover at dusk to forage on road verges. On one short (500 metre; 1640 foot) stretch of road in the Chichester State Forest, north of Dungog, I have counted as many as eight pademelons along the road edge, all intent on eating grasses, ferns or shrubby leaf material. The Tasmanian Pademelon (*Thylogale billardierii*) has the most southerly distribution of the pademelons, occurring in most forest types in Tasmania and on the larger islands in Bass Strait. It used to be abundant in forest and scrub on the coastal plains of Victoria, New South Wales and south-eastern South Australia but disappeared from the mainland around the turn of the 20th century. It is common enough in parts of Tasmania to be controlled as an agricultural pest.

Other small grazers such as the Banded Hare-wallaby (*Lagostrophus fasciatus*), Parma Wallaby (*Macropus parma*) and Quokka (*Setonix brachyurus*) prefer thickets or forest habitats with dense cover at ground level. In addition to reducing predation, such dense vegetation shelters animals from temperature extremes and thus reduces the metabolic costs of cooling and keeping warm.

Consider the Quokka. This short-tailed and rather anomalous wallaby can keep itself cool by panting even on days that reach 44°C (111°F) but to do this it needs access to freshwater. On Rottnest Island, just off the coast of Perth in Western Australia, where free water is limiting in summer, Quokkas gather around the few accessible seepages and fight for access to both the water and available shade. On the mainland, Quokkas are restricted to swamps. These provide access to abundant freshwater but, perhaps more importantly, provide refuges that are impenetrable to marauding Red Foxes.

The remaining marsupials in the grazing subguild are compact, powerfully built quadrupeds that excavate deep and complex burrows. They are the wombats. The most familiar species is the Common Wombat (*Vombatus ursinus*), which occupies forest and mixed scrub–pasture habitats in south-eastern Australia. Extensive studies by John McIlroy of CSIRO suggest that this species occurs locally where burrowing conditions are suitable and where coarse, fibrous grasses are available. Unlike kangaroos, wombats have rootless, continuously growing molars that allow them to cope with such tough fodder. Common Wombats are tolerant of cold and, at higher altitudes, can be sometimes seen trudging about in deep snow searching for food. At the University of Sydney's field station at Arthursleigh, on the southern tablelands

Species density

Grazers

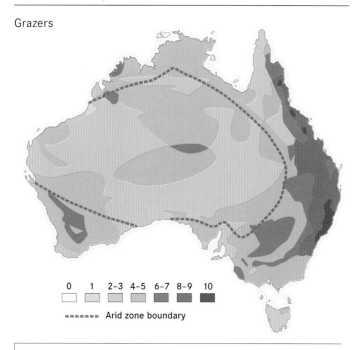

0   1   2-3   4-5   6-7   8-9   10

▪▪▪▪▪▪▪ Arid zone boundary

FIGURE 5 Species density map of grazing marsupials compiled by overlaying distribution maps for all species in the subguild. Lighter colors indicate areas where grazers are scarce or absent, stronger colors show areas where the distributions of up to 10 species overlap. The area within the dashed red line is the arid zone.

of New South Wales, animals routinely emerge to forage when temperatures are below zero and winds are howling; they are the only marsupials I have observed active there under such adverse conditions. However, they are intolerant of heat and use their deep burrows to escape it. One of the wombats in McIlroy's study suffered after limited exposure to a temperature of just 27°C (81°F) while in a trap, dying later from apparent heat (and possibly other) stress. Not surprisingly, in the warmer northern parts of its range, the Common Wombat occurs in cool, humid forests above 600 metres (1970 feet); it descends to coastal regions at more southerly latitudes on the mainland and in Tasmania. Food, climate and local burrowing conditions probably combine to shape this species' distribution, although the relative importance of each factor is difficult to specify.

In contrast to their forest-dwelling relative, the two species of hairy-nosed wombats occur patchily in semi-arid areas where free water is scarce and the summer temperatures soar above 40°C (104°F). Burrows are essential in these conditions. Take the Southern Hairy-nosed Wombat (*Lasiorhinus latifrons*). Long known from southern Western Australia and South Australia in areas receiving just 200–500 millimetres (8–20 inches) of

rain each year, and discovered recently in a similarly dry site in western New South Wales, this large marsupial uses its burrows for protection from predators and from the elements, for social interactions and just to 'switch off'. Burrows go down more than 1 metre (3.3 feet) and often branch into extensive underground complexes. They are humid and maintain a moderate but constant temperature of 25–26°C (77–78.8°F) in summer and 14°C (57°F) in winter. Animals rest in their burrows by day, often switching off by allowing their body temperature to drop several degrees to 31°C (88°F). With a low basal metabolic rate (roughly two-thirds that of the marsupial average), low water turnover and high body insulation, these wombats are very efficient at conserving both energy and water in their dry environment. The Northern Hairy-nosed Wombat (*Lasiorhinus krefftii*) is also critically dependent on deep burrows for its survival but its extreme rarity has precluded any detailed study of its burrowing habits.

Both species of hairy-nosed wombat produce single young and recruit them slowly into the population; this is offset by their longevity (up to 20 years in captivity). Early work by Rod Wells at the University of Adelaide suggests that a long life span may be advantageous in allowing animals to endure periods of drought. It is also conceivable that their slow, extended life history and reduced energy metabolism place hairy-nosed wombats at a disadvantage compared with kangaroos in the same habitats by retarding their ability to exploit good conditions when they arrive. Could the distributional boundaries of both species be set, in part, by competition? Several researchers have suggested that recent contractions in range of the hairy-nosed wombats are attributable to increased competition for food from rabbits and livestock, and this possibility seems quite plausible.

Given the diversity of grazers and the ubiquity of their food, can we say anything generally about which areas these marsupials favor and the factors that limit their distributions? A simple overlay map, constructed in the same manner as for the insect-eating marsupials, provides some immediate answers (Figure 5 on this page). Most species are found in forest and woodland habitats along the eastern seaboard; the fewest species occur in arid and semi-arid areas, on islands and on the western side of Cape York Peninsula. These patterns in overall species abundance appear to be explained largely in terms of habitat complexity and the availability of food.

Southern Hairy-nosed Wombat
Lasiorhinus latifrons Male
BLANCHETOWN, SOUTH AUSTRALIA

Honey Possum *Tarsipes rostratus*
Male (top right) and female
SOUTH-WESTERN WESTERN AUSTRALIA

In the forested country of north-eastern New South Wales, Peter Jarman and colleagues at the University of New England have described the presence of nine species of kangaroos and wallabies. They link this rich assemblage to the high diversity and interspersion of habitats in the region. Throughout much of central Australia, by contrast, habitats are structurally simple for grazers (but not for the smaller insect-eaters as discussed above) and only two species – the Red Kangaroo and Common Wallaroo – are widespread. Local increases in species densities do occur in the Macdonnell, Musgrave, Petermann and other ranges where the structural diversity of the environment is increased by rock outcrops. There are likewise more species of grazers on large islands, which have an increased range of habitats as well as a larger area, than on small ones. Although the larger grazers can eat more fibrous plant material with less energy and nutrients than their smaller counterparts, there is little quantitative evidence that grazers segregate strongly by diet. All species prefer to eat fresh green fodder when it is available and dietary differences between species often emerge only during periods of food stress such as drought.

The final 'herbivore' to consider is the anomalous Honey Possum. Although confined broadly to the south-west of Western Australia, this species reaches its highest numbers in coastal heaths with a high diversity of flowering plants. Its local distribution coincides in particular with proteaceous species, such as *Banksia* and *Dryandra*, that produce large quantities of nectar or with myrtaceous species, such as *Eucalyptus*, if its favored plant species are not flowering. Studies by Ron Wooller and others at Murdoch University have found pollen grains from different plant species adhering to the fur of Honey Possums at different times in the year, suggesting that these sweet-toothed marsupials shift their activity to exploit seasonal flowering patterns in their food plants. Wooller speculated that, because of their continuous visits to flowers and ability to carry pollen, Honey Possums may be important pollinators and could even be regarded, functionally, as marsupial bumble bees. Even without this splendid analogy, there is little doubt that the highly specialized diet of the Honey Possum ties it closely to the floral hot spot in the continent's south-western corner, and ensures that this marsupial has no equivalent among non-flying mammals anywhere in the world.

Life histories

Variations on a theme In view of the broad range of habitats occupied by marsupials and the wonderfully disparate ways that they exploit environmental resources, we might expect them to show equivalent diversity in their life histories. And they do. But before we explore this concept further, we need to ask what is meant by the term 'life history'. A person's life history is a catalogue of the important events that take place as they grow from childhood to adulthood through to old age. Major stages include puberty, employment, marriage, children and perhaps grandchildren later in life. The life histories of marsupials and other animals can be thought of in similar terms as the rules that govern when individuals mature and reproduce. Life histories have several components, the most important being when individuals first reproduce, the number of times they reproduce, the number of young they produce on each occasion, and death. These components interact and often constrain each other, and thus cannot be considered in isolation. Reproduction, for example, is a costly process, so heavy investment in a single large litter early in life may deplete the energy reserves of the parents and reduce their

chance of producing more offspring in future. The two sexes can also be expected to differ in their investment in young; in mammals, males often expend large amounts of energy in finding and inseminating females, whereas females usually focus on suckling and nurturing the young.

Marsupials exhibit spectacular variation in their life histories. Thus, while large kangaroos have been recorded living longer than 25 years in the wild, males of several species of small dasyurids never survive more than 12 months and die *en masse* after their first and only breeding season. Litter size is one in many herbivorous species but up to 12 in some of the small dasyurids, and breeding can be seasonal or continuous even within guilds, with 1-5 litters produced in a year. Male marsupials are usually larger than females, with the sexes of some dasyurids and macropodids showing a two-fold difference in body weight. Yet, female Honey Possums and Leadbeater's Possum conspicuously buck this trend by weighing 25% more than their mates. To interpret why there are so many variations on the life history theme among marsupials, it is convenient to consider the strategies exhibited within and between guilds, and the effects of both body size and seasonality in the food supply.

Carnivores Marsupials that specialize in eating invertebrates produce larger litters than any other Australian mammals. With the exception of the ever-anomalous marsupial moles, which have two nipples in their rear-opening pouch, all species of insect-eaters attempt to rear litters of three or more young, and produce one to three litters a year. If circumstances conspire to reduce a litter to just one or two individuals, lactation may stop as the suckling stimulus is too weak to keep milk flowing. Alternatively, the mother may kill and remove the young, presumably conserving her energy for a subsequent occasion when a full litter can be produced. Very few members of this subguild survive for more than 3 years. This 'live fast, die young' strategy is exemplified by many species of small dasyurids, in which litter size is four to twelve. In contrast, the large (700 gram; 25 ounce) Numbat rears just two to four young a year and may do so until 4-5 years of age. Larger flesh-eaters such as the Tasmanian Devil also have relatively sedate life histories. This carnivore carries two to four young annually in its shallow, rear-opening pouch, and may survive up to

6 years in the wild and 8 years in captivity. Fragmentary evidence suggests that the Thylacine also reared two or three young each year. Its longevity in the wild was never recorded but on the basis of its size we might expect it to have lived somewhat longer than any of its smaller relatives.

For all carnivorous marsupials, the seasonal availability of food is perhaps the most critical determinant of when to breed and how much effort to expend. Because of the high energetic costs of suckling large litters until weaning (65–120 days in the smaller species but as long as 180 days for the Numbat and 240 days for the Tasmanian Devil), it should benefit females to ensure that they are lactating at times when food is most readily available. In environments where invertebrates and other food resources peak during a restricted period each year, females usually achieve sexual maturity at 11 months and have time to produce just one litter. In these situations, most or all females breed and litter sizes are maximal. The males of such short-season species also expend a lot of effort in finding and copulating with females during the annual breeding period, to the extent that many die of their exertions soon after. The extreme manifestation of this strategy has been called 'big-bang reproduction' and is characterized by the abrupt and total death of all males before their young are born. All species of

Antechinus and *Phascogale* exhibit this remarkable strategy, as do the Kaluta (*Dasykaluta rosamondae*), some populations of the Dibbler (*Parantechinus apicalis*) and some species of *Dasyurus*. The Eastern Short-tailed Opossum (*Monodelphis dimidiata*) and Gray Slender Mouse Opossum (*Marmosops incanus*) of South America both appear to exhibit big-bang reproduction but it is otherwise unique among mammals. Indeed, until it was described in pioneering studies by Pat Woolley at the Australian National University in the 1960s, swift post-mating death was thought to be the lurid fate of a few unfortunate invertebrates such as male black widow spiders and praying mantises. The phenomenon has received a lot of attention and I return to it below.

In environments where food resources are abundant and available for longer periods each year, females are no longer confined to a short breeding season and can extend the time over which they are reproductively active. In species of *Planigale* and *Sminthopsis*, for example, females can produce two litters during the breeding season, perhaps even three if conditions are exceptional. They mature at just 5–6 months of age and can potentially breed in the season of their birth. To compensate for the cost of producing multiple litters, females reduce their expenditure by suckling only four to eight young each time. Reproductive effort is also

Kaluta   *Dasykaluta rosamondae* Female
<small>PILBARA, WESTERN AUSTRALIA</small>

Fawn Antechinus
Antechinus bellus Female
Near Nourlangie, Northern Territory

less for males and many that survive to maturity are able to breed in two or more consecutive seasons. If food resources are available reliably and consistently throughout the year, we might expect temporal constraints on reproduction to be even more relaxed and thus see year-round breeding. Surprisingly, there have been very few year-round studies in non-seasonal habitats to confirm this expectation but it does seem to hold for several species of dasyurids in New Guinea and also for the Common Planigale in monsoonal habitats in the Northern Territory. Although monsoonal rains are highly seasonal, the floodplain habitats occupied by planigales in the Top End are highly productive and must harbor reliable populations of invertebrates in all seasons.

Given that the time of breeding is so important, what cues do animals use to trigger reproductive activity? This question is particularly relevant for seasonal breeders because they must initiate mating many weeks before the energetic demands of lactation become acute. Breeding in some carnivorous marsupials seems influenced by changes in temperature or rainfall but such cues are probably not reliable and provide only a general guide to improving conditions later in the season. Instead, most species appear to use photoperiod, or day length, as their major trigger. Changes in day length are very consistent and predictable in any locality irrespective of the prevailing weather or how much food might be available at

the time. In several species of dunnarts, for example, reproductive activity begins in spring when day length is increasing and slows or ceases after the summer solstice. This day-length effect can be confirmed by exposing captive animals to increased artificial light, which prompts testicular enlargement in males and oestrus in females even at times outside the usual breeding period Using increased day length (or decreased night length) to initiate breeding activity makes intuitive sense because longer days herald improving conditions in spring and summer. But this cue is not sufficient to account for the precise year to year timing of breeding in carnivorous marsupials such as antechinuses and phascogales, nor can it explain the surprising observation that most species of *Antechinuses* in eastern Australia breed progressively later in spring at lower, more tropical latitudes. If increasing day length alone is important, why wouldn't animals breed at any time between mid-winter and mid-summer when days are lengthening? And why is breeding delayed at low latitudes, where the spring flush of food resources needed by lactating females could usually be expected to arrive early?

Photoperiod is still crucial in driving the observed pattern of seasonal breeding in antechinuses, and perhaps phascogales too, but it is used in a way that is not immediately obvious. The first part of this puzzle was cracked in 1982 by Steve Van Dyck at the Queensland

Museum, who provided a simple but elegant explanation for the unusual south-north flush in invertebrate food resources that occurs along Australia's east coast. Van Dyck proposed that invertebrate populations increase early at higher latitudes in southern New South Wales and Victoria because they respond to warmer temperatures after the cold winter months. At lower latitudes in Queensland, by contrast, where winter temperatures are mild, invertebrate flushes are driven by increased rainfall which usually arrives in late spring or early summer.

The photoperiodic trigger for breeding in east coast antechinuses turns out to be not day length as such but rather the rate at which day length changes during the spring. At low latitudes, day length differs little between the seasons, so changes in day length are small and take place at a leisurely pace. At high latitudes, by contrast, the change from short winter days to long summer ones means that day lengths in spring change quickly, so a given rate of change in day length is attained earlier at more southerly latitudes. If the average time of ovulation of females is recorded from populations of east coast antechinuses, different species are found to ovulate when a specific but different threshold in the rate of change is reached. For the Yellow-footed Antechinus ovulation is triggered when day lengths increase at 77–97 seconds per day but thresholds for other species are higher: 92–99 seconds per day for the Subtropical Antechinus, 99–111 seconds per day for the Brown Antechinus and 127–137 seconds per day for the Agile Antechinus. Responding to these different rates of change allows each species of antechinus to rear litters in spring or summer, and to do this earlier at southerly latitudes to take advantage of the earlier pulse of food there. Even the most northerly distributed species, the Fawn Antechinus (*Antechinus bellus*) and Cinnamon Antechinus (*A. leo*), appear to respond to rate of change of day length but do so at the slower rates of 32–47 seconds per day. Intriguingly, the spring flush of food appears to arrive later at higher latitudes on the west coast of Australia and populations of the Yellow-footed Antechinus there breed correspondingly later in the more southerly parts of their range. These populations must be responding to some cue other than rate of change in day length.

Sandstone False Antechinus
Pseudantechinus bilarni Male
<small>WESTERN ARNHEM LAND, NORTHERN TERRITORY</small>

It seems astonishing that these small marsupials can perceive and respond to tiny increments in the vernal day length. We do not know precisely how they do it but there is some suspicion that the pineal gland, which sits at the base of the brain, may be involved in some way. A further suggestion has been made by Hugh Tyndale-Biscoe in his recent book *Life of marsupials*. Tyndale-Biscoe noted that it takes male antechinuses about 2 months to reach full sexual development, and females almost as long, so that animals are perhaps reading and responding to declining rates of change of day length before the winter solstice. Some experimental work on captive Agile Antechinuses indicates that females can be brought into oestrus early if exposed to shortening day lengths, thus supporting Tyndale-Biscoe's idea. However, it may not apply universally, as populations at low latitudes breed 3 months or more after the winter solstice and must therefore initiate reproductive development when day lengths are increasing slowly in early spring. Although rate of change of day length clearly triggers reproduction, its mode of action may differ between species and locations.

Solving the second part of the photoperiod puzzle and putting the finger on rate of change required both detective work and good luck. Early in my doctoral research I wanted to document reproduction in two species of antechinus to see if they staggered their times of breeding to reduce competition for scarce food resources. This meant spending many cold days and nights in the field in winter to check when males died and females gave birth, checking the literature to see when other researchers had recorded animals breeding and hassling colleagues for their unpublished observations. At about the same time, I read a paper proposing that African rock hyraxes respond to the rate of change of photoperiod and wondered if something similar might explain the unusual pattern of reproductive timing in antechinuses. Then, at a meeting of the Australian Mammal Society in 1982, I had the good fortune to meet Bronwyn McAllan from the University of Adelaide, who was already working on the rate of change idea. Combining Bronwyn's models with my data set quickly allowed us to solve the long-standing mystery of what governs the remarkable synchrony and pattern of reproductive timing in antechinuses. More recently, with Mathew Crowther at the University of Sydney, we have been able to show that the rate of change model can be applied to eight of the ten described species of *Antechinus*.

The strange and beguiling ways of antechinuses do not end there. As we saw above, antechinuses and some of their relatives are big-bang reproducers, and this aspect of their life history – especially the death of males after mating – has attracted a great deal of attention. How do males die? Although ovulations occur quite precisely when a certain threshold is reached in the rate of change of day length, both sexes appear primed for reproduction in the lead up to the breeding season by an internal 'clock'. This annual rhythm prompts increases in body condition before breeding and, in males, stimulates an increase in circulating stress hormones. Cortisol is the most potent of these hormones; it has the twin effects of suppressing appetite and promoting conversion of protein into sugars so that animals can engage in sustained activity. This helps males to interact aggressively with each other, search for mates by day and night, and mate with many females (up to 16 have been recorded) during the mating period. Copulations may last 2–12 hours. Despite increasing the activity of males and maximizing their access to mates, elevated cortisol levels have negative side effects. The hormone suppresses the immune system and inflammatory responses and thus predisposes animals to increased risk of disease and infection by parasites. Death follows swiftly and inevitably.

The change in the appearance of males over the two-week period of mating is breathtaking: they lose up to 40% of their body weight, get cut and gashed, lose fur and become increasingly listless to handle as the end draws nigh. Autopsy of males after mating reveals massive hemorrhaging from ulcers in the stomach and intestines, organ failure, and elevated loads of internal and external parasites that have taken advantage of the immune system collapse. From this period in the year until the young become independent in summer, field sampling reveals only females. Hence, populations of all species are low in spring and do not reach peaks until late summer and autumn when young emerge from maternal nests.

Big-bang reproduction was long thought to be confined to antechinuses, phascogales and the Kaluta, but I had the honor of discovering it in another marsupial, the Dibbler, in 1986. The Dibbler is one of the most beautiful of all the small marsupials but its great rarity means few people are lucky enough to have seen one. By 1967 it was thought to be extinct. Then, wildlife photographer Michael Morcombe stumbled by chance upon a remnant population near Albany in Western Australia, and subsequent searches led to other small populations being discovered. Eighteen years later, Dibblers turned up unexpectedly on two tiny islands off the coast of Jurien, a small fishing town north of Perth. I quickly sought permission to study Dibblers there, took a lesson in how to handle a small boat and headed to Jurien with great glee. The first trapping session, on Boullanger Island in March 1986, was fantastic: 30 Dibblers in just 3 days – about as many as anybody had seen in the previous century! But a huge shock awaited me on the next trip in early April. Dibblers were still there but there were far fewer of them, and every single animal we caught was a female.

What had happened? My first thoughts were very dark: Dibblers must have such delicate sensibilities that my presence on the island was too much for them and they had died of stress or run, lemming-like, into the sea to escape my traps. But why should only males be so sensitive? I set traps on new parts of the island where animals had not previously been captured and still failed to find any males. Close inspection of the females showed that they were growing rings of long white hairs around the edge of the pouch, losing hair around their nipples and showing a general reddening of the pouch area – all sure signs of pregnancy. I began to suspect that males may have died after mating but still harbored deeply depressing thoughts that I had somehow contributed to it. Then on 2 April – ominously close to April Fools Day – I spotted a male lying moribund on the sand that looked literally shagged out. It had external cuts and abrasions, had lost several patches of fur and its penis was drooping on the ground. I quickly phoned my colleague Adrian Bradley, an expert on male die-off in dasyurids at the University of Western Australia. He raced to Jurien, and performed an autopsy on the now-fresh carcass that confirmed the animal had died with the same post-mating symptoms as a male antechinus. Oh, joy: Dibblers were big-bang-reproducers! That night there was an unseemly celebration over the corpse of the departed male. On the next trips I confirmed that females carried young and captured juveniles of both sexes later in the year as they became independent. After numerous escapades, boating accidents and near misses on the treacherous Indian Ocean, I gave up the Dibbler work at the end of 1988. I didn't want to be the next male to die.

Knowing how males die after mating is one thing but it does not explain the more difficult question of why death occurs. This question is still hotly debated and several answers have been proposed. The earliest hypothesis suggested that males altruistically remove themselves from the population to reduce competition for food with females and their young. However, this idea is not tenable as it is envisages a situation that is unstable from an evolutionary point of view: altruists would quickly lose out to 'cheats' that survived and fathered young in subsequent breeding seasons. Another hypothesis proposes that, with the seasonality of food allowing just a single litter to be reared each year, males should invest all their effort in siring young at their first opportunity because their chance of surviving to a second breeding season is very low. This idea suggests that males compromise their immediate survival prospects to increase their chance of paternity, and seems plausible because it views male behavior as selfish rather than altruistic. However, females in many big-bang species often survive to a second breeding season and, given this, it is unclear why only males should expire so abruptly.

Two further hypotheses build on the assumption that strong advantages accrue to males that maximize mating effort, even if the cost of this is premature death. The first recognizes that females in all big-bang species store sperm after mating and may not release it to fertilize their eggs for as long as two weeks after copulation. Females mate with many males during the mating period, so that litter-mates usually have two or more different fathers. Multiple mating reduces the risk that a female ends up with a dud or incompatible male, and helps to ensure that she will have a full complement of young. From the male perspective, the mating period is too long to stay with a single mate and guard her from the approaches of other males. The only way to ensure paternity is thus to inseminate as many females as possible during the mating period, even if the effort of doing this causes death. The second hypothesis notes that females themselves do not always survive to wean their young. Again, from the male perspective, mating with many females could thus be seen as 'insurance' that some of their offspring will be born to females that survive. Very recently, Kristen Wolfe and colleagues at the University of Western Australia have confirmed that male Dibblers can actually survive die-off if rich food resources are available to them. This suggests that post-mating male death in big-bang species is more flexible than has been thought. It also highlights the Dibbler as

Dibbler *Parantechinus apicalis* Male
Shown with flower of *Banksia baxteri*
CHEYNE BEACH, WESTERN AUSTRALIA

an excellent study species to evaluate the merits of the competing hypotheses that purport to explain this extraordinary phenomenon.

Omnivores Because they do not rely on a single kind of food that may be seasonally limited, many omnivores are able to breed over extended periods or even throughout the year. Most have two or more periods of oestrus during the year, carry moderate-sized litters of one to five young, and live for 2 years or longer.

Insect-eating omnivores Most marsupials within this subguild give birth over several months of the year, with the production of litters timed so that females are lactating and young weaned when food resources are most abundant. Small species such as pygmy-possums can bear as many as two to three litters over the year, with four or five young carried each time. Productivity seems to be greater at low latitudes. For example, Eastern Pygmy-possums (*Cercartetus nanus*) in New South Wales have been recorded with pouch young from September to April, whereas both this species and the Little Pygmy-possum (*Cercartetus lepidus*) in Tasmania curtail their breeding after just 4 months in January. In fact, there is some evidence that Eastern Pygmy-possums might breed year round if conditions are suitable. During a wildlife class that I took at the Royal National Park, near Sydney, in June 2005, I confidently told the class that any pygmy-possums we might find during the week would be either juveniles or adults that had completed reproduction for the year. Of course, the first two pygmy-possums that we captured were females with small pouch young … banksias were flowering profusely at the time and presumably provided a sufficiently reliable supply of nectar to allow extended breeding that year. Young quit the pouch at no more than 30 days of age, are weaned at 50–60 days and achieve sexual maturity at 5 months, giving these little possums one of the highest reproductive potentials among Australian marsupials.

Larger insect-eating omnivores take life more slowly. Sugar Gliders, for example, breed from June to November or December, producing one or two young that stay in the pouch for at least 70 days. These are deposited in a group nest for another 40–50 days until weaned. Unlike pygmy-possums, which tend to disperse soon after weaning, newly independent Sugar Gliders remain within extended social groups until 7–10 months of age, often accompanying their mother on foraging forays. Females can be sexually mature and breed at 8 months but males do not achieve maturity until a year of age. The interval between successive litters is 140 days, almost double that for pygmy-possums. Despite their slower production of young, Sugar Gliders that survive the trials of their first year will often breed until 4–5 years of age; exceptional animals have been recorded living for 9 years. Life history events are very similar in the Squirrel Glider (*Petaurus norfolcensis*), so much so that fertile offspring result if these two glider species are allowed to interbreed.

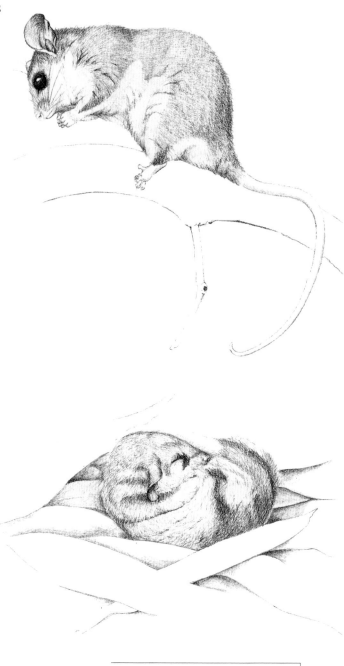

Eastern Pygmy-possum
Cercartetus nanus Female
Eastern New South Wales

Breeding in the other wrist-winged gliders is similar to that in the Sugar Glider but the different life history stages appear to be progressively delayed in the larger species. Yellow-bellied Gliders typify this pattern of retardation. Females produce single young (twins on rare occasions) once a year in winter or spring, and suckle them in the pouch for 90–100 days and in a tree den for a further 40–60 days. Sexual maturity is reached at 18 months in males and 2 years in females. Yellow-bellied Gliders may balance their slow rate of reproduction with increased survival and longevity but this expectation awaits a suitably long-term field study.

There is, of course, one exception to these generalizations and it comes in the form of the Mountain Pygmy-possum. This small (40–60 gram) burramyid is restricted to boulder fields in the Australian Alps, and is a slow and highly seasonal breeder. Females produce just one litter of four young during the short alpine summer. Although the young are weaned at 70–75 days and fatten quickly before the onset of winter, they are not sexually mature until 12 months of age and do not achieve adult weight until aged 2 years. The strong seasonality of breeding and low annual fecundity are clearly con-straints imposed by the short alpine growing season but, in compensation, wild animals have exceptional longevity and presumably produce a litter each season. Males live to a venerable 4 years of age, females to a truly astonish-ing 12 years. It can be granted that these animals spend half their lives in hibernation but the Mountain Pygmy-possum nonetheless holds the longevity record for terrestrial mammals of comparable size.

Fruit-eating omnivores   Field studies on the Musky Rat-kangaroo indicate that breeding begins in October and lasts for up to 6 months. As this period encompasses the wet season in north-eastern Queensland, it may be surmised that lactating females gain nutritional and energetic benefit from access to fruit that is washed from the forest canopy and underground fungi that appear toward the end of the wet season. Litters of one, two or occasionally three young are born each year, remain in the pouch for some 21 weeks and are left thereafter in the mother's nest where they continue to be suckled for several weeks more. Sexual maturity is attained at 2 years. Information on the life histories of the two Australian species of cuscus is too scant to allow generalizations to be made. However, breeding appears to be extended – not surprising if it is assumed that these canopy-dwelling marsupials have year-round access to fruit.

Long-footed Potoroo *Potorous longipes*
Female and juvenile
EASTERN VICTORIA

Fungus-eating omnivores All species in this small subguild are similar in some aspects of their life histories, such as in giving birth to single young, entering oestrus and mating again very soon after birth, in having embryonic diapause, and in having the potential to breed in most or all months of the year. Beyond these similarities, two distinct life history strategies can be identified that relate to differences in the habitats that are occupied. Bettongs, on the one hand, are the speed breeders. Females gestate for 21–24 days and suckle young in the pouch for up to 115 days. The young mature at 7–10 months and have a life expectancy up to 4 years in the wild. Potoroos, on the other hand, are much more sedate. Gestation lasts 38–42 days, pouch life 140–150 days, and young do not reach sexual maturity until aged 1–2 years. These differences allow potoroos to squeeze in just two to two and a half litters a year to the bettongs' three. As a trade-off for their relaxed rate of reproduction, potoroos can live up to 7 years in the wild.

Bettongs occupy a broad range of habitats, including grassland and woodland, that tend to be prone to disturbances such as wildfire. Their speedy life history allows them to quickly and efficiently exploit these productive but unstable environments. In contrast, the living potoroos occupy stable forest and heath habitats with dense groundcover, where annual rainfall is high (750–1200 millimetres; 29–47 inches) and predictable. The slower turnover in their life history allows potoroos to saturate the habitat while spreading the costs of reproduction over more extended periods.

Generalist omnivores These marsupials are impressively flexible in both their diets and reproduction, and they are also the fastest breeders in the Australasian region. At equatorial latitudes in New Guinea, the Papuan Bandicoot (*Microperoryctes papuensis*) and Common Echymipera (*Echymipera kalubu*) have been recorded breeding year round, and scattered evidence suggests that several other tropical species carry young in most months. At low latitudes in Australia, both the Golden Bandicoot (*Isoodon auratus*) and Northern Brown Bandicoot have extended breeding seasons. At higher latitudes, most bandicoots are capable of producing young in all months in captivity but in the wild exhibit a hiatus in reproduction from late summer to winter. Day length probably initiates breeding in the Northern Brown Bandicoot but increases in temperature and rainfall are probably also important triggers because

they stimulate plant growth and thus herald the arrival of increased food resources. If bandicoots are plied with supplementary food in the field, they can often be persuaded to continue reproduction deep into autumn and early winter.

Reproductive output is high in all members of this subguild. In the Northern Brown and the Long-nosed bandicoots, for example, gestation lasts 12.5 days, young are weaned at 60–70 days and females can breed when aged just 3–3.5 months. Litter sizes average only two to four, although occasionally six or seven young may be found in the pouch. If conditions are favorable, four or even five litters can be produced over the course of a year. Bandicoots and bilbies are exemplars of the 'live fast, die young' strategy. Despite the high rate of production, most young die before or soon after weaning, so that recruitment into the population is usually no more than 10–15%. For individuals lucky enough to survive past weaning, annual survival seldom exceeds 25% and few adults survive to a third or fourth year.

The bandicoot life history poses several questions. Most obviously, we might ask why so few young bandicoots survive to adulthood and whether we can predict the few that will make it. We should ponder why so many young are produced if survival is so poor. In many marsupials the period of obligate attachment to the mother's nipple is a secure one and there is a good chance of the young surviving until the nipple can be relinquished. There are exceptions, of course, such as if the mother dies or becomes so stressed that she dumps the young to improve her own immediate chance of survival. But in bandicoots, family planning starts at this early stage. Litter size in these omnivores diminishes almost inevitably over time, with a third to a half of the young born in any litter often failing to reach the end of pouch life. What happens? In their interactions with humans, bandicoots give the impression of being 'highly strung' and can respond poorly to being trapped. Unless traps are checked at night and animals handled carefully under cover of darkness, females may evict some or all of their young from the pouch and trample them in their efforts to escape. It has been suggested that females do the same in the wild if stressed by a close encounter with a predator or competitor. This does not seem very compelling. Unless a female was heavily encumbered by her young and could flee a predator more quickly by dumping them, it is hard to see any real advantage in this behavior. Also, the litters of long-term captive females routinely diminish over time, even though these individuals are not stressed in any obvious way.

If there is little advantage to the mother in pruning her brood, how can the litter shrink? While working on Southern Brown Bandicoots near Perth, I sometimes found young dead in the pouch but still attached to a nipple. These young never bore tooth marks or injuries that might have suggested maternal interference but they often had two things in common. Firstly, they had spun around on the nipple, twisting it so that the flow of milk had been reduced or entirely blocked. Secondly, they were more than twice as likely to be at the front of the pouch than the back. These observations surprised me: the front end of the pouch is more enclosed and protected than the back, so that if mothers injure their young in any way through clumsy grooming or attempts at eviction, the rear-most young should be most affected. However, there is another explanation. The front part of the pouch is certainly protected but as a consequence provides such limited room to move that a pouch young snarled-up on a nipple might find it difficult to twist back into position. If it could not reposition itself quickly, the milk supply through the nipple would remain choked and it would starve.

Intrigued by this strange possibility and curious to know how pouch young could get twisted on the nipple, I looked at my results more closely and was stunned to see that young at the front of the pouch survived better if there were no young on the posterior nipples. This was a one-way effect: survival of young on the rear-most nipples was the same whether anterior young were

Southern Brown Bandicoot, Nuyts Islands form *Isoodon obesulus nauticus*
Female with young
FRANKLIN ISLAND, SOUTH AUSTRALIA

present or not. Could these observations mean that young on the posterior nipples reduce the survival of their siblings? Potentially, yes. Pouch young wriggle about actively on the nipples within days of attachment and probably jostle each other increasingly as they grow. The rear-most young could easily twist their siblings out of position on their nipples and reduce the space available for them to disentangle. If this scenario is correct it would represent the only example of siblicide in marsupials, perhaps even among mammals more generally. The survival of young on the anterior nipples is higher in large than in small females, supporting the idea that physical space in the pouch may be limiting. Further study is needed to elucidate the true impact of siblicide in the pouch but for now it remains yet another enigmatic aspect of bandicoot biology.

Young bandicoots are probably not in the clear even at the end of pouch life but their survival in the wild during the next life history stage – suckling while in the maternal nest – has never been quantified. Nestlings survive well in captivity but are exposed to dangers such as inclement weather, snakes, Cats and Red Foxes outside. Radio-tracking shows that bandicoots spend most of the night foraging away from the nest, thus heightening the risks to their young still further. Unlike the Bilby, which constructs deep and well-protected burrows, bandicoots generally excavate shallow scrapes under leaf litter or vegetation for their young. It would not be surprising if many were lost to the elements and to natural enemies at this stage. Losses post-weaning are similarly unknown. The transition to independence is a risky period for most marsupials and there is no reason to suspect that young bandicoots would be any different.

Given the many hazards that young encounter as they grow, it has been suggested that lots of young have to be produced so that a lucky few will make it. This kind of 'hope-for-the-best strategy' works well for many invertebrates, fish and amphibians that can produce eggs cheaply and abundantly but it doesn't square so well in species where each young costs a lot of time and maternal resources to produce. Bandicoots and many other marsupials with more than two nipples do in fact produce more young than can attach but this ensures that all the available (and not recently used) teats are occupied and the energetic demands of the suckling young then increase steeply as they grow. More likely, the bandicoot life history can be interpreted as a response to the disturbed, patchy

and fire-prone habitats that they often occupy. In such heterogeneous environments the almost continuous production of young allows rapid colonization of new habitats as they regenerate after disturbance and become available. Bandicoots can thus be viewed as pioneer species. Indeed, no other Australian marsupials exploit 'regeneration niches' so successfully and this allows bandicoots exclusive access to new food and shelter resources in many disturbed situations. Think of this when next you see the conical holes that some enterprising bandicoot has dug in your lawn or flower bed!

Herbivores Because green plant food seems to be so readily and abundantly available, we might expect herbivorous marsupials to show much commonality in their life histories. There are, in fact, considerable similarities in traits such as litter size (usually one but occasionally two or more), numbers of nipples (two to four) and even longevity (some individuals in most species studied have been recorded living at least 5 years). There is also a bewildering array of subtle and more striking variations in traits between species, such as seasonality of breeding, length of gestation, presence of postpartum mating and diapause, rates of growth and social organization. To make sense of this variation, I consider the life histories of selected leaf-eaters and grazers below.

Leaf-eaters One of the most familiar and best-studied of the browsing marsupials is the Koala. *Eucalyptus* leaves form the bulk of the diet of this distinctive species but they are low in protein and other nutrients, and high in indigestible compounds, tannins and other toxic chemicals. Although they are highly specialized for eating this poor quality food – and have few peers in this regard among the leaf-munchers – Koalas still have to work hard physiologically to extract sufficient energy and nutrients to grow and reproduce. To reduce the costs of food processing Koalas sleep and rest for some 20 hours a day and have slowed their rate of metabolism so that it ticks over at about 70% of the level expected for a marsupial of its size. However, although this lethargic lifestyle ensures that Koalas can eat their daily greens, it also acts as an important constraint on their rate of reproduction.

Koalas mate in spring and early summer, and give birth to a single young (occasionally twins) after a gestation of 35 days. The cubs (so named owing to the affectionate but horribly erroneous old common name, 'Koala Bear') are weaned at about a year, emerging to forage independently the following spring or summer when *Eucalyptus* leaves are fresh and most nutritious. They become sexually mature at 2–3 years. Individual females do not enter oestrus until the previous year's cub has been fully weaned and thus cannot produce more than one young a year. Moreover, if the previous year's young is suckled until after the summer solstice, there is some evidence that decreasing day length inhibits further oestrus and mating activity. These constraints result in an average reproductive rate, or number of young produced annually by females, of just 0.65–0.70. In marked contrast to the 'live fast, die young' set, this low output is balanced by the survival of most young to weaning (about 80%) and by a productive lifespan for females of at least 10–12 years.

Other leaf-eaters such as the Short-eared Possum and Greater Glider have low reproductive rates similar to the Koala (0.68–0.73 young per female per year). These species also breed seasonally, achieve sexual maturity at 2–3 years and wean single young once a year. Longevity in the Greater Glider is not known but in the Short-eared Possum lifespans of up to 17 years have been recorded. In contrast, two other species of leaf-eater, the Common Brushtail Possum and Common Ringtail Possum, achieve much higher reproductive outputs. These species can breed twice each year, reach sexual maturity within a year or just over, and wean their young after only 6–7 months. The annual reproductive rates are correspondingly high. The Common Brushtail Possum produces 0.9–1.4 young per female each year, and females live 10 years or more. Common Ringtail Possums rear 1.8–2.4 young on average each year (they produce at least two young per litter). Longevity is compromised as a consequence of the elevated reproductive output; very few Common Ringtail Possums survive to see their sixth birthday.

The faster-breeding possums in general are better able to exploit disturbed habitats, including suburban backyards, where additional food is often available, such as soft green shrubs, ornamental flowers and fruit. Both the Common Brushtail and Common Ringtail possums appear able to elevate their reproductive output by exploiting these rich dietary supplements. There is also some evidence of variation in reproductive output in less disturbed forest habitats, where reproductive success is enhanced in 'hot spot' areas with concentrated foliar nutrients. The slow-breeding Greater Glider, Short-eared Possum and Koala are much more restricted to stable forests, and colonize new habitats slowly. Their populations often vary little over long periods. These species are particularly susceptible to both human-induced disturbances such as logging, as well as to drought, fire or forest die-back. They can persist temporarily in small patches of intact forest but the long-term outlook for such specialists is undoubtedly poor in these situations.

Short-eared Possum *Trichosurus caninus* Female
LAMINGTON PLATEAU, QUEENSLAND

Greater Glider *Petauroides volans* Male
Near Tumut, New South Wales

Western Brush Wallaby *Macropus irma* Female
SOUTH-WESTERN WESTERN AUSTRALIA

Grazers Despite the large number of species in this subguild and the range of habitats they occupy, all kangaroos, wallabies and wombats produce single joeys after relatively long gestations of 27–37 days, and most retain them in the pouch for 6 months or more and wean them after 7–8 months. In general, reproductive rates range from 0.5 to 1.8 young per female per year, slightly lower than those of browsing marsupials. Some variation in reproductive rate arises from differences between species in the rate of growth and length of lactation but a large and significant component is due to the degree of seasonality of reproduction. We can distinguish three main strategies.

In the first strategy, exemplified by 'obligate seasonal breeders', most or all mature females give birth during a single restricted period each year. This strategy is exhibited by the Tammar Wallaby (*Macropus eugenii*), Western Grey Kangaroo and the Tasmanian subspecies of the Red-necked Wallaby (*Macropus rufogriseus rufogriseus*). In the last, after birth, mostly between January and March, the young remain in the pouch for about 270 days before being weaned at a year or less. Females enter oestrus and mate again within hours of giving birth but then hold the resulting embryo in a quiescent state for up to a year. It is reactivated and born 6–8 weeks after the summer solstice on 22 December, some time after its older sibling has stopped suckling. In the Tammar Wallaby, the quiescent embryo reactivates in the first half of the year only if the older joey is lost. Growth is suppressed by day length in the second half of the year and reactivated only after the summer solstice on 22 December.

Because of the short window of opportunity for reproduction, the maximum reproductive rate of obligate seasonal breeders is 1.0. Species exhibiting this strategy predominate in temperate regions of southern Australia. Here, fresh grasses and forbs are abundant in spring and summer, thus providing females with nutritious food when their young are emerging from the pouch and the demands of lactation are high. Intriguingly, the strict once-a-year breeding of these species cannot be over-ridden even after years in captivity under artificial light. However, it does shift by 6 months in animals kept in the northern hemispere; births there occur in the northern autumn from August to October. This 'hard-wired' breeding pattern is presumably adaptive in environments where resource peaks arrive very predictably and on time each year.

The second strategy is exemplified by 'facultative seasonal breeders' such as the Western Brush Wallaby (*Macropus irma*), Banded Hare-wallaby and Quokka. Like their obligately seasonal relatives, these smaller grazers breed mostly in summer and autumn. However, under certain conditions they can also give birth in the second half of the year and thus greatly increase their reproductive rate. On Rottnest Island, for example, female Quokkas enter oestrus and mate immediately after giving birth; the resulting embryo is held in diapause but can be mobilized and born within 25 days if the first joey is lost. Not only this, female Quokkas provided with extra food, or otherwise in good condition, will sometimes enter oestrus and produce young during winter or spring. As joeys remain in the pouch for only 190 days and are weaned less than 2 months later, two young can be produced within a year and elevate the annual reproductive rate to 1.8 young per female. The predominantly seasonal breeding of these grazers suggests that they occupy largely seasonal environments but their ability to produce out-of-season young indicates that they are attuned to good conditions at any time.

Most grazers are 'continuous breeders', and can produce young at any time of the year. This third life history strategy is exhibited especially by many of the large kangaroos, such as the Agile Wallaby (*Macropus agilis*) and Antilopine Wallaroo (*Macropus antilopinus*) of the tropical lowlands of northern Australia, as well as by the Northern Nailtail Wallaby (*Onychogalea unguifera*), the diminutive Nabarlek and probably all other rock-wallabies. Although these small grazers occupy a wide range of climatic zones, the rock outcrops that they use must provide a protected and constant supply of food resources throughout the year. Mainland populations of the Red-necked Wallaby (*Macropus rufogriseus banksianus*) and Quokka also breed continuously. Both occur in forest in high rainfall areas that provide higher and more consistent levels of food than are available to their respective island populations. The ability to exploit resources year-round allows these grazers to achieve reproductive rates of 1.8 young per female per year.

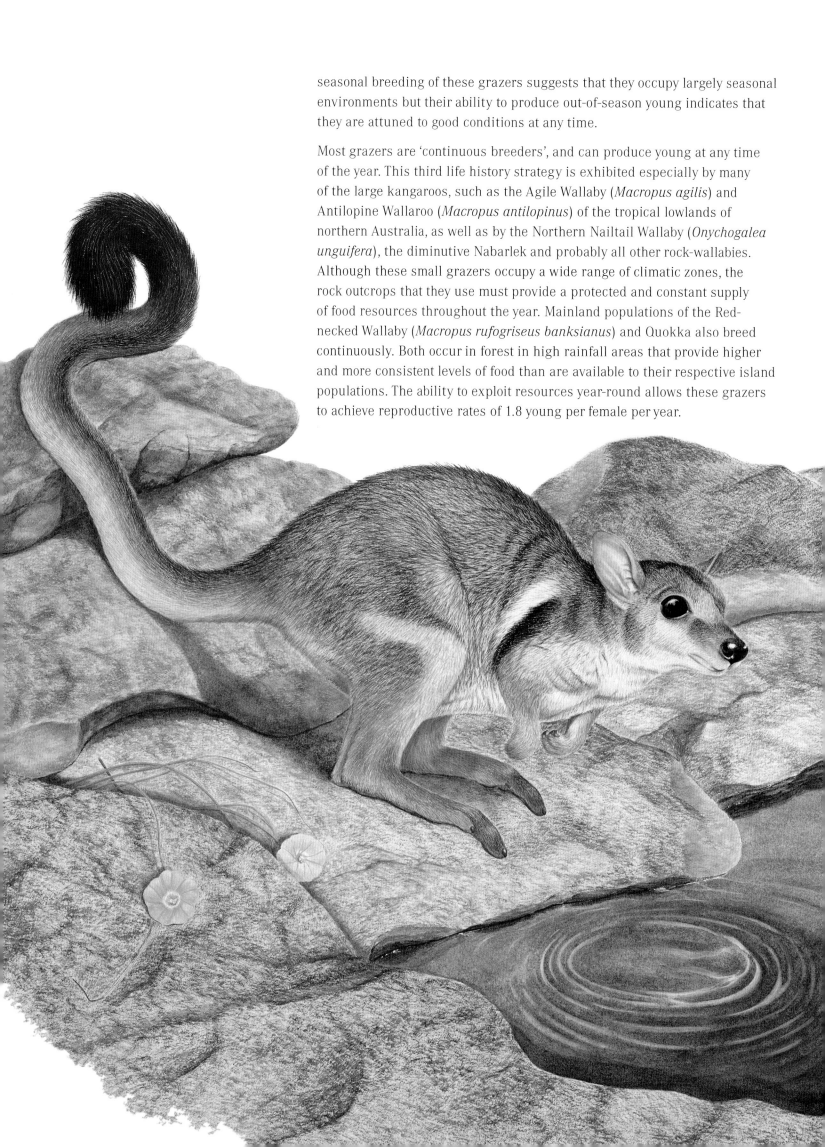

Continuous breeders are also found in the central deserts of Australia but in these arid habitats they can reproduce continuously only when green food resources are plentiful after rain. Red Kangaroos and Common Wallaroos are good examples. As do other continuous breeders, both species enter oestrus and mate very soon after giving birth, hold the resulting embryo in a quiescent state until ready and then reactivate it for birth 30–32 days later. Young Red Kangaroos vacate the pouch after 235 days and young Common Wallaroos a month later; females can give birth the following day and will then suckle the two different-aged joeys for about 4 months until the older one is weaned. This allows Red Kangaroos to achieve an annual reproductive rate of 1.6 young per female, and Common Wallaroos a rate of 1.5.

During drought, breeding slows or ceases, although females with existing young usually continue to suckle. As drought deepens, females either lose interest in mating and stop breeding altogether, or continue to produce young that have little chance of survival. Half the young of these latter females die by the age of 2 months and none survives longer than 8 months. When they die, the embryo in diapause is activated, and when it is born the females mate again and put another on hold. This strategy, in which females undergo frequent pregnancies, is possible because it doesn't cost much, energetically, to produce and replace small pouch young. Perhaps so, but why not just shut down altogether? When rains fall and the good times roll, the pouch young of the breeding females survive the 2-month barrier and over 90% are quickly and successfully weaned. Females that had stopped breeding take a couple of weeks to ovulate after good rains have fallen and thus lag far behind the persistent breeders in turning green food into offspring. Sometimes drought is so severe that even the hardiest females stop breeding and males cease production of sperm. After rains fall, sperm production restarts slowly and some females may remain unfertilized even if they return quickly to oestrus. Under these conditions, populations recover very slowly from drought.

Anyone who has lived through a long, dry spell in central Australia can attest to the utter desolation that it causes. Small mammals can seek shelter below ground and turn to invertebrates for food and moisture but kangaroos

Nabarlek *Petrogale concinna* Male
ARNHEM LAND, NORTHERN TERRITORY

have no choice but to stick it out. Some move long distances to find improved conditions but many stay put. The great pioneer of Red Kangaroo research, Alan Newsome, most recently of the CSIRO Division of Sustainable Ecosystems, has often regaled me with stories of how these astonishingly tenacious beasts persist in the harshest parts of the continent's interior and he is still in their thrall more than 40 years after first fathoming their secrets. Little wonder that the Red Kangaroo is a mythically important creature for Aboriginal people. I too have joined the fan club after seeing Red Kangaroos struggle through severe droughts in the Channel Country and Simpson Desert of western Queensland, only to explode in numbers after good rains fell.

The big picture In the early 1980s I spent two very fortunate years as a postdoctoral visitor in David Macdonald's research group – then called the 'Fox-lot' – at the University of Oxford. The many eccentric and wonderful characters of the Fox-lot poked around in the private lives of different carnivore species to understand their behavior and ecology. My role in this enterprise was to investigate the food resources available to the Red Fox within Oxford itself, and complement a detailed study on fox movements and behavior by Patrick Doncaster, now at the University of Southampton. Pat and I set traps for small mammals and invertebrates in every green patch that we could find in Oxford. My additional task was to collect as many fox scats as possible to confirm what the animals were eating. The results astounded me. Small mammals occurred in higher densities than any marsupials I had ever encountered or read about, and some, such as the Common Shrew (*Sorex araneus*), were pumping out an astonishing three to four litters of young over the notoriously short English summer. No seasonally breeding marsupial comes close to that kind of productivity. The shocks did not stop there. The jars that we set for invertebrates in summer were full to the brim a week later with spiders, beetles, bugs, larvae and other would-be prey for insectivores. Similar pitfall traps that I used to set in forest near Canberra would be a quarter full, at the very most, after a whole month. These various prey items were all eaten by foxes and sustained a large population of this opportunistic predator within the Oxford city limits. I have since been able to trap small mammals and invertebrates using similar methods in other parts of the world, and the results look like those from Oxford.

Northern Nailtail Wallaby
Onychogalea unguifera Male
ARNHEM LAND, NORTHERN TERRITORY

From these experiences and from reading the work of others, two conclusions are clear and inescapable: the Australian environment is generally very unproductive; and it effectively retards the reproductive rates of organisms that exploit it, with profound consequences for the development of their life histories. Let us explore both conclusions further.

Compared with many other parts of the world, soils in Australia are deficient in key elements such as phosphorus and nitrogen that are needed for growth and productivity. This deficiency arises from the fact that much of the surface of the continent is stable and ancient. The lack of volcanic activity has minimized the formation of new soils, and the antiquity of the environment has provided ample time for the leaching and loss of nutrients. Add to this the general aridity of the Australian continent and productivity is understandably miserable. Even in good years there is a muted flush of green growth in spring compared to that in more fertile regions overseas, the growth rates and biomasses of grasses and shrubs lag behind those in comparable savanna regions elsewhere, and agricultural output can be sustained only by the use of massive amounts of fertilizer.

Of all the elements that are scarce in the Australian environment, nitrogen is perhaps the most critical for animals. It is a component of protein and thus essential for all stages of development, growth, body maintenance and reproduction. Like other animals, marsupials cannot extract nitrogen from the air but must acquire it on a regular basis from their food. If nitrogen is scarce and hard to obtain, we might expect to see animals making the most of what they can find. Marsupials do this in several ways. Leaf-eaters and grazers pick and choose parts of plants that contain most nitrogen, and aggregate in patches of the environment where this element and others are locally more abundant. Many omnivores opportunistically eat insects, which are rich in nitrogen; for carnivores, invertebrates or vertebrates are the mainstay of the diet. Adelaide scientist Tom White has gone so far as to argue that the relative shortage of nitrogenous food is the single most important factor limiting animal abundance in Australia and elsewhere, and I think this position has a lot of merit.

Many marsupials have mechanisms to conserve nitrogen once they have ingested it. For example, kangaroos maintain populations of bacteria in the fore-stomach that synthesize protein and digest this as it moves through the gut. Not only this, they recycle the main breakdown product of protein, urea, by reabsorbing it in the kidney. Retention of urea is most effective if the diet is already protein deficient, and this in turn minimizes the loss of precious nitrogen in the urine. In contrast, Common Ringtail Possums maximize their retention of nitrogen by reingesting their faeces. This allows them to digest bacterial protein that has been synthesized in the caecum and would otherwise be lost. Another mechanism that allows nitrogen to be retained is a bit harder to explain. Faced by stress, many mammals produce a hormone surge that stimulates the release of glucose into the blood stream, and this allows them to respond by fighting or fleeing. If the stress is prolonged, they mobilize protein from their tissues which is then broken down and excreted in urine. But many marsupials do not show this latter response. In general, these are herbivores such as kangaroos and wombats that face nitrogen limitation in their diets already; the carnivorous marsupials show the typical mammalian response. Resistance to the usual effects of stress hormones clearly conserves nitrogen but we do not know what other mechanisms kick in to allow the stress to be reduced.

Despite these behavioral and physiological tricks, scarcity of nitrogen in the environment must set an ultimate limit to the rate at which protein can be made. In this case, we might expect to see animals slowing their rates of growth and reproduction to levels commensurate with their ability to find and process nitrogen from the environment. Does this happen? In general, marsupials have metabolic rates 30% lower than those of ecologically similar placental mammals and their body temperatures, averaging 35.5°C (95.9°F), are almost 3°C (5.4°F) cooler. Turning down the thermostat slows the rate of growth and

reduces the rate at which nitrogen needs to be acquired. If the daily nitrogen requirements of marsupials and eco-logically similar placental mammals are compared, those of marsupials are usually considerably lower. Indeed, on a weight-for-weight basis, the Koala has one of the lowest nitrogen requirements known for a mammal, and it is able to persist on leaves that contain as little as 1% nitrogen by dry weight.

How much slower do marsupials grow and reproduce than their placental counterparts? Both mammalian groups grow quickly and at similar rates as embryos, but it takes marsupials about 1.5 times as long to wean their young as it does placental mammals. Steve Thompson at the Smithsonian Institution found that small marsupials (less than 400 grams or 14 ounces) mature later than ecologically similar placental mammals and also have slower rates of population growth. In accordance with our observations of Common Shrews, above, population growth in these small species is constrained largely

Antilopine Wallaroo *Macropus antilopinus* Male
COBOURG PENINSULA, NORTHERN TERRITORY

Spectacled Hare-wallaby *Lagorchestes conspicillatus* Female
This species remains widespread in northern Australia and has survived
changes in land use that have threatened the Rufous Hare-wallaby
QUEENSLAND

because they produce fewer litters per year. Large mar-supials (over 10 kilograms; 22 pounds) also grow slowly at the population level even though their development to maturity is similar to that of their placental counterparts. Medium-sized marsupial and placental mammals take similar times to reach sexual maturity and wean their young but marsupials in this category show surprisingly quick population growth. In Thompson's analyses, shrews had the highest reproductive potential, with rates three to four fold higher than those of ecologically similar marsupials.

One of the problems of comparing marsupials with pla-cental mammals is that they diverged from each other a long time ago, so any differences they show in life history patterns could just reflect their different evolutionary histories. We can reduce this 'ancestral baggage' problem by comparing Australian marsupials with their counterparts from South America. If marsupials in the latter region have evolved in a more productive environment, we might predict that they would show higher rates of growth or reproduction than those from Australia. And, in general, they do. Small opossums such as the Shrewish Short-tailed Opossum (*Monodelphis sorex*) have up to 24 or 25 nipples and presumably raise large litters; others such as the Fat-tailed Opossum (*Thylamys elegans*) sometimes rear two litters of up to 15 young per year. Larger opossums (*Didelphis* species) can rear over 20 young annually. These efforts surpass even the most fecund marsupials in the Australian region and reinforce the notion that animals in this group face real environmental constraints.

Further insights can be made if we restrict ourselves to comparisons within Australia itself. Firstly, if environmental productivity acts as the ultimate constraint on life history evolution, we might predict that species occupying high and low productivity areas within the continent will differ in their growth and reproduction. The Brush-tailed Phascogale provides a good test case, as populations occur in the low productivity forests in south-western Australia and also in higher productivity forests in the Great Dividing Range and slopes. Research by Susan Rhind at Murdoch University in Perth has confirmed that animals in the western populations grow more slowly and are some 30% less massive than their counterparts in Victoria. Litter sizes of phascogales are predictably smaller in the west than in the east, with western females possessing six to eight nipples and usually rearing litters of six, and eastern females possessing eight nipples and rearing full litters.

Secondly, if the Australian environment is unproductive, it should constrain the life histories of other animals. In fact, we might predict that constraints should be increasingly obvious for groups of animals that have experienced the longest confinement within the continent. This prediction too receives much support. Consider the native rodents. The first wave of invaders arrived some 4–5 million years ago and their descendants now occupy all parts of the continent. With a single exception, the Tree Mouse (*Pogonomys* sp.), all the old endemic rodents have just four nipples and usually raise litters of two to four young. Gestation periods are 30–44 days, young are weaned at 21–42 days and become sexually mature when aged 3–10 months. The second wave of invading rodents arrived perhaps a million years ago, and radiated to form a small group

of seven living species. All are in the genus *Rattus*, the true rats. They rear litters of four to eight young after gestations of 3 weeks and wean them just 3 weeks after birth. The youngsters in some species can breed when aged just over a month. The reproductive rates of these new endemics are much faster than those of the old. Of course, species of *Rattus* are renowned for their speedy breeding and lack of family planning, so comparisons between the two groups of endemic rodents again come with the confounding problem of ancestral baggage. Nonetheless, the growth and reproductive rates of the old endemic Australian rodents are almost glacial in comparison with rodents elsewhere in the world, and strongly suggest that they have faced an environmental brake on their activities.

Similarly retarded rates of growth and reproduction characterize many members of Australia's old endemic frog, reptile and bird faunas but not their more-recently arrived relatives. There are likely to be further patterns to uncover among the native fish and land invertebrates but this speculation will have to await further study. At this stage I will invite debate by positing a very general rule pertaining to the continent's terrestrial vertebrates:

The longer a group has been present on the Australian continent, the more likely it is that its surviving members will show retardation in their growth, reproduction and overall life history.

There will be exceptions, of course, and these will be instructive to review, but it seems to me that this rule holds quite generally. As we have seen, the relationship is most likely to arise from the low productivity of Australian soils, and in particular from the scarcity of available nitrogen. The ability of individuals to harvest this essential element may be the ultimate rate-limiting step in the evolution of life histories, and this is of profound consequence for interpreting many of the patterns and differences that are observed between Australian faunas and those elsewhere.

Do marsupials have advantages in unproductive environments? Several researchers have proposed that a prime advantage of the marsupial way of life is that mothers can dump their young if environmental conditions

Yellow-footed Rock-wallaby *Petrogale xanthopus* Male
FLINDERS RANGES, SOUTH AUSTRALIA

deteriorate or if they are threatened by predators or other enemies. This does occur, especially in kangaroos, and allows mothers to terminate their investment in young before the costs of lactation become too great. Once a placental mother has made a decision to breed, by contrast, she must usually carry the young to term even if conditions fail and food becomes scarce. However, many marsupials breed in seasonally predictable environments and may squeeze out just one or two litters in their lifetimes. For these species dumping the young would provide little benefit and, in fact, this happens very rarely.

Is it possible, instead, that the general advantage of being a marsupial is that it is adaptive in an impoverished, low-nutrient environment? The time from conception to weaning is 1.5 times longer in marsupials than in ecologically similar placental mammals, perhaps because the transfer of nutrients is slower via lactation than across the placenta. In an environment where nutrients take longer to harvest, slowing the growth of the young should be quite advantageous in allowing females to maintain both their condition and maternal reserves. The marsupial mode facilitates this. In contrast, the rapid *in utero* growth of the young in placental mammals may be advantageous when conditions are good but lead to depletion of maternal reserves when nutrients are scarce. In these circumstances placental mammals may be at a disadvantage compared to their marsupial counterparts.

There is some tantalizing fossil evidence that supports this line of thinking. In 1992, while digging at Tinga-marra in south-eastern Queensland, Henk Godthelp, Mike Archer and others at the University of New South Wales, uncovered the lower molar of a small ground-dwelling mammal. From the shape of the tooth and the structure of its enamel they identified the original owner as a placental, *Tingamarra porterorum*. But, this was no recent fossil; it was aged about 55 million years and probably belonged to the now-extinct order Condylarthra. A few years later teams led by Tom Rich at Museum Victoria described two more placental-like mammals, *Ausktribosphenos nyktos* and *Bishops whitmorei*, from Flat Rocks in southern Victoria. Although the precise identity of these mammals has been disputed, they probably bore a passing resemblance to modern-day

hedgehogs. Most astonishingly, they date back over 100 million years. More recent placental fossils have also been described, including teeth and pieces of tusks from elephant-like animals but their status as true members of the Australian fauna remains in doubt. We can be sure of one thing: ground-dwelling placental mammals (or mammals very much like them) thrived on the Australian landmass before marsupials arrived. They did not persist afterwards. We do not know exactly when or why these placental mammals succumbed but it is reasonable to speculate that they were less able to cope with the harsh and unproductive environment of Australia than their marsupial counterparts.

Ancient history is all very well, but if marsupials do have an adaptive advantage in the Australian environment, we have to ask why rabbits, house mice, foxes and so many other imported mammals do so well now. The answer to this is simple: since European settlement people have disturbed and modified the environment to suit themselves, their crops and their livestock. This restyling has laid forests to waste, created vast areas of 'improved' pastures, added permanent water to dry parts of the landscape and fertilized other parts that were non-productive. These changes have dramatically favored the newcomers. The settlers were also fanatically determined to change the composition of the mammal fauna. On the one hand they actively brought familiar species with them to breed and release into the environment. Rabbits, for example, were released on dozens of occasions over 70 years until they finally gained a foothold on Thomas Austin's property near Geelong in Victoria. On the other hand, settlers made ruthless attempts to exterminate marsupials and other native nuisances. We explore the motivations of this behavior in the next chapter but note here that this 'out with the old, in with the new' mentality would have tipped the balance in favor of the introduced placental mammals against their marsupial rivals over vast areas. In other words, the success of recently imported placental mammals in Australia can be seen as an artefact, owing much to human cupidity.

I don't wish to take these musings too far but I am tempted to speculate that the marsupial ascendancy in Australia is more than a happy accident of biogeography.

Green Ringtail Possum *Pseudochirops archeri* Male
ATHERTON TABLELAND, QUEENSLAND

Interactions between marsupials and their environment

The environment is clearly a major determinant of many aspects of marsupial biology, affecting where species occur, their local abundance, their use of food and other resources and even their life history strategies. As we have seen, there is also compelling evidence to suggest that the marsupial way of life is advantageous in the low-productivity Australian environment, and that placental mammals are often not competitive unless assisted by habitat change, pastoralism and other human activities. However, the environment–marsupial relationship is not just one way. Several studies over the last few years have shown that marsupials can affect not just their own immediate environment but also that of many other organisms. In some cases the effects are quite trivial and difficult to measure; in others they are profound and have dramatic consequences for plant productivity, soils, nutrient cycles, and even for the frequency and intensity of fires.

Lesser Hairy-footed Dunnart
Sminthopsis youngsoni Male (left) and female
EDGAR RANGE NATURE RESERVE, WESTERN AUSTRALIA

An obvious way in which marsupials affect their environment is by using natural products for food, shelter or living space. When a kangaroo forages, it reduces the immediate availability of grass and forbs for other grazers but possibly increases productivity in the longer term by stimulating growth of fresh new shoots by its food plants. The 'overcompensation hypothesis' proposes that moderate levels of grazing cause plants to produce more leaf or root tissue than they would if they remained uneaten, in turn providing more food for other grazers and perhaps inducing grasses to crowd out potential competitors such as shrubs. There is some evidence that grey kangaroos stimulate compensatory growth in their food plants. However, this idea remains contentious, especially in pastoral areas where kangaroos are regarded as pests. It is likely that increases in plant growth depend not just on herbivory but also on rainfall and the rate of return of nutrients to the soil.

If grazing pressure is very high, production of new plant growth will not be able to offset the losses and pastures may become severely depleted. The effects of such overgrazing can be seen along creek lines and around natural and artificial waters, especially during drought and in situations where kangaroos coexist with rabbits and grazing stock. The impacts of overgrazing are also evident in many areas where the total grazing pressure from kangaroos, stock and rabbits has increased since predators have been removed. In the Western Division of New South Wales, for example, predation by wild dogs and Dingoes has been drastically curtailed by a concerted campaign of poisoning, shooting and trapping, and by construction and maintenance of the celebrated 'dog fence' that runs along much of the border of New South Wales with South Australia and Queensland. Counts of kangaroos and Emus on both sides of the dog fence by Graeme Caughley, Alan Newsome and other researchers have shown that the abundance of these large native species is up to 40 times higher in the Western Division, where wild dogs are absent, than on the other side of the fence. Stock densities too are very high in the Western Division. Although parts of this region can spring to life after good rain, the high grazing pressure exerted there for over 100 years has helped destroy many native plant communities and cause widespread loss of topsoil. To me, the Western Division is a depressing place. Once home to almost 40 species of marsupials, the area now has just 23, some vanishingly rare. Other than the large kangaroos, many of the survivors are confined to outcrops of rock, riparian strips or small areas that are protected from overgrazing.

The environmental effects of native and introduced grazing species are usually combined, making it difficult to specify the influence of kangaroos alone. Effects of feeding by other marsupials are easier to identify. Let us consider the Koala, a species whose leaf-eating habits are not paralleled by any introduced mammals. Koalas occur in low densities in many areas, their numbers kept in check by continuing fragmentation and loss of habitat, predation by dogs and, in more settled areas, collisions with motor vehicles. In places where numbers are not contained, however, such as on French and Phillip islands in Victoria, overbrowsing has led to severe defoliation and even death of large areas of *Eucalyptus* forest. At Framlingham Forest, near Warrnambool in western Victoria, leafless trees also bear silent witness to heavy Koala

Eastern Quoll *Dasyurus viverrinus* Male
NEAR LAUNCESTON, TASMANIA

browsing pressure. Standing on the main road through this patch of forest, I have counted 14 emaciated Koalas within a circle of just 100 metres radius; this equates to an unsustainable density of some 4.5 animals per hectare (2.5 acres). Similar overbrowsing occurs in other parts of mainland Victoria and on Kangaroo Island in South Australia. Overbrowsing by Australian marsupials is most notorious in New Zealand, where large areas of forest have been depleted by the introduced Common Brushtail Possum. Local densities of over 25 possums per hectare (2.5 acres) have been recorded, giving the trees little chance of survival in the longer term.

If kangaroos are restricted to creek lines during drought, or Koalas and possums are confined to fragments of forest, their dietary options are likely to be very limited and overgrazing, or overbrowsing, then becomes inevitable. We might expect hungry animals to switch from eating preferred but scarce foods to alternative ones if these are available and easier to find. This may account for occasional reports of Koalas eating the bark of trees and kangaroos eating leaf litter. What happens if food is not limiting and animals have the chance to be choosy? The environmental effects of selective food consumption have been little studied in herbivorous marsupials but probably help to maintain plant species diversity. In dasyurids, however, there is some tantalizing evidence that selective feeding can drive changes in the composition of invertebrate communities. Observations of Brown Antechinuses near Sydney and of Lesser Hairy-footed Dunnarts in the Simpson Desert show that soft-bodied spiders and cockroaches are favored dietary items, and hard-bodied beetles, small crustaceans and ants are avoided. In the absence of the dasyurids, most invertebrate prey taxa increase, with spider populations showing particularly dramatic expansions of up to 400%.

In situations where a prey species is super-abundant, dasyurids pursue it avidly to the virtual exclusion of other foods. This has been documented for Dusky Antechinuses eating Bogong Moths in the Australian high country and for Eastern Quolls eating Army Worms and Corbie Grubs in pastures in southern Tasmania. I have also seen Agile Antechinuses gorging on Christmas Beetles on *Eucalyptus* trees near Canberra, and recorded them gaining almost 30% more body mass than their counterparts in non-infested areas. It is interesting to speculate whether these rapacious little predators play a role in reducing the severity of outbreaks of these and similar insect pests, and thus have a generally beneficial effect on their local environment.

Consumption of various prey species reduces their abundance and thus has a direct effect on their populations and the structure of the communities to which they belong. The indirect effects of foraging are also pervasive and impact on the environment in a host of surprising ways. In central Australia, for example, the predatory Brush-tailed Mulgara enhances the local diversity of other dasyurid marsupials by over two species, on average, apparently by suppressing the numbers of smaller competitor species that would otherwise predominate and exclude other dasyurids. My experiments in the Simpson Desert showed that removal of the mulgara allowed one dasyurid species to predominate at the expense of all others, thus reducing local diversity to just one. The species that prevailed, the Lesser Hairy-footed Dunnart, was up to 80% more abundant in mulgara-exclusion plots than in control plots; it probably out-competed its smaller relatives. These 'keystone predation' effects have been demonstrated only rarely in terrestrial environments.

On Barrow Island, off the north-western coast of Western Australia, the Golden Bandicoot provides another good example of how foraging can impact indirectly on the environment. This species spends much of its foraging time in search of termites, which comprise over half of its diet. This bandicoot is so numerous on Barrow that it will be encountered on a short stroll in most areas and often returns trap success rates of over 50%. Given these observations, it is reasonable to suspect that Golden Bandicoots suppress termite abundance and so contribute indirectly to the large mounds of dead, uneaten spinifex grass that characterize much of the Barrow Island environment. Golden Bandicoots are now extinct elsewhere in arid Australia and in those areas abundant termites consume dead spinifex before it can accumulate into mounds. Even casual inspection of the mounds on Barrow shows that they provide homes for other small mammals, birds and certain species of reptiles. The mounds are highly combustible when dry and act as pyres of fuel that could start a conflagration in the event of lightning strike. (This is not likely on Barrow Island at present, as the establishment of a working gas and oil field demands active fire management.)

The former existence of mounds in other parts of arid Australia may well have influenced the likelihood, intensity and extent of wildfires.

Finally, the foraging behavior of the Yellow-bellied Glider also has indirect and positive effects on other species. The V-shaped notches that this species cuts on certain *Eucalyptus* trees produce copious flows of sap; the oozing wounds are used in turn by other sweet-toothed marsupials and numerous insects.

We have so far considered the effects of marsupials on the environment as being primarily consumptive. However, foraging activities and even the act of feeding itself can have surprising and profoundly unexpected effects. These arise not from killing prey or from reducing their populations but rather from redistributing biologically important resources in the environment. These effects have been called 'ecological services', and they include moving seeds, spores and pollen to new sites where they can become active.

The movement of seeds by marsupials has been known for a long time, with the best examples coming from the fruit-eating didelphids of South America. In a detailed study by Pierre Charles-Dominique in French Guyana, several species of tree-dwelling marsupials were shown to eat fleshy fruits and excrete the intact seeds some distance from the parent plants. Fruits with seeds of less than 20 millimetres (0.8 inches) diameter were preferred, and at least five fruit-bearing species of plants seemed to rely on marsupials as their only dispersal agents. Didelphids did not move seeds larger than 20 millimetres (0.8 inches), and instead simply ate the fruit pulp that surrounded them. Further work on the Monito del Monte (*Dromiciops gliroides*) in southern Argentina has confirmed that this small marsupial is the sole dispersal agent for the parasitic mistletoe, *Tristerix corymbosus*. After eating the fruits, monitos often excrete the intact seeds above ground on a new tree that will potentially host the mistletoe after it germinates. What is remarkable about this system is that the seeds will germinate only if they pass through the gut of a monito. In view of the antiquity of the monito's lineage, and that of the mistletoe, it is tempting to speculate that this specific marsupial–plant association is itself a very ancient one.

Golden Bandicoot *Isoodon auratus* Male
Canning Stock Route, Western Australia

As we have seen, there are few fruit-eating omnivores among Australia's marsupials and certainly no specialist frugivores; hence, we might expect the role of Australian marsupials in transporting seed to be quite minor. However, recent studies suggest that we need to rethink this conclusion. Working on the Musky Rat-kangaroo in northern Queensland, Andrew Dennis at CSIRO showed that animals deliberately collect and bury large, fleshy fruits for later use. These caches are so well hidden in fresh soil that animals seldom remember all their locations, thus allowing many seeds to germinate later when conditions are suitable. These indomitable hoarders move up to 900 fruits per hectare (2.5 acres) of forest each month, consuming some but conveniently burying over three-quarters just below the soil surface. In a second study, Mark Garkaklis and Maria Murphy at Murdoch University showed that Brush-tailed Bettongs play a major role in dispersing the seeds of Western Australian Sandalwood (*Santalum spicatum*). Sandalwood seeds are too large to be blown by wind and need to be moved away from under the parent plant before they can germinate and grow. Bettongs pick up fallen seeds and bury them singly in caches up to 80 metres (262 feet) away, promoting local germination and also helping Sandalwood to regenerate over large areas. There is currently debate in New Zealand about whether Common Brushtail Possums act as dispersal agents for large tree seeds, while in New Guinea some fruit-eating species of cuscus are suspected to disperse and perhaps even cache tree seeds. These possibilities are plausible and fascinating but await more detailed study.

Two further considerations suggest that marsupials may be under-recognized as seed dispersers in Australia. In the first instance, dietary studies show that many species eat fruits and seeds, even though their contribution to the overall diet is small or seasonal. The seeds are often excreted intact and, if they are able to germinate after passage through the gut, may benefit from being transported. Secondly, when working in the field, it is common to see or capture marsupials with seeds sticking out of their fur. Many plants produce seeds with hooks, spikes or sticky appendages that readily attach to fur, and again these may benefit if the seeds fall out on fertile soil away from the parent plant. I have made casual observations of Quokkas, Red-necked Wallabies (*Macropus rufogriseus*), Common Brushtail Possums and even dunnarts inadvertently carrying seeds but I

know of no study that has quantified how important such transport might be for individual plants or plant communities. This is an exciting area for future study.

Fungus-eating omnivores disperse fungal spores in the same way as seed-eaters. They eat the sporocarp, or fungal fruiting body, and excrete the intact spores elsewhere. Sporocarps are not known to be cached. The fungi eaten most frequently by marsupials are underground species that form important associations with the roots of plants. These associations, termed ectomycorrhizal, provide great benefit to plants because the fungi act as specialized extensions of the roots and assist the plants in concentrating critical nutrients from the soil. Experimental work by Andrew Claridge and colleagues at the NSW Department of Environment and Climate Change has confirmed that spores from potoroo faeces germinate readily in soil and then form ectomycorrhizal associations with *Eucalyptus* species. Plants with the root association grow much more vigorously than those without it. As fungus-eating marsupials are commonly present in areas that have been disturbed, such as after fire, they probably play a key role in vegetation recovery by disseminating spores and speeding plant growth.

In addition to moving seeds and fungal spores around, marsupials are now being appreciated increasingly as vectors of pollen. The most obvious pollen transporters are the insect-eating omnivores which visit flowers to obtain nectar, pollen and perhaps insects, and also the Honey Possum. Insect-eaters such as antechinuses are also being seen increasingly as pollen vectors, despite debate about whether their visits to flowers are for the floral resources as such or simply to gain access to flower-visiting insects. Whatever the motivation, the mode of pollen transport is the same for all marsupials: animals pick up pollen on their fur when rummaging among flowers and then simply carry it with them as they move.

To be effective as pollinators, marsupials have to carry sufficient pollen to transfer some to the receptive stigmas of different flowers. Casual observations suggest that large quantities of pollen can lodge in the fur. Both Eastern Pygmy-possums and Western Pygmy-possums (*Cercartetus concinnus*) that I have captured have had their faces covered in pollen from low-lying shrubs. I have also spotlighted Sugar Gliders foraging in flowering *Acacia* shrubs looking as if they have been dusted with yellow talc; strips of sticky tape placed on the fur of such flower visitors will often reveal many thousands of pollen grains. Experimental exclusion of several species of burramyid and dasyurid marsupials from flowers has resulted in lower fruit-set, thus confirming their role as pollinators. Although marsupials are less mobile than other pollinators such as birds and bats, there seems little doubt that they contribute importantly to plant abundance and diversity on a local scale.

The ecological services provided by marsupials are not restricted to the transport of plant and fungal resources. Another critically important but greatly under-recognized service provided by many species is their modification of the physical environment by activities such as digging and nest construction. This

Western Pygmy-possum
Cercartetus concinnus
Male (lower right) and female
Keith, South Australia

class of activity is called 'ecosystem engineering', and some marsupials, especially the medium-sized species, are very good at it.

Let us begin by considering the engineering effects of bandicoots and bettongs. The conical holes left by foraging Bilbies, and Southern Brown, Northern Brown and Long-nosed bandicoots act as traps for seeds and organic debris, enhancing the patchiness and local richness of vascular plants. The holes probably also allow greater infiltration of water into the soil and facilitate the continuation of nutrient cycles, although these aspects have so far been little-studied. The digging activities of the Brush-tailed Bettong have even more startling results, as described in pioneering studies by Mark Garkaklis at Murdoch University. Although averaging only 1.3 kilograms (2.9 pounds), these little Aussie diggers each excavate some 38–114 holes every night while searching for food, and displace 4.8 tonnes (5.3 short tons) of soil annually. The scratchings and digs have dramatic effects on the ability of water and nutrients to penetrate the soil, increasing infiltration to about 10 centimetres (4 inches) but reducing it below that depth. If bettongs are absent from a site for only 3–4 years, soils become so hard and water repellent that plant growth and establishment are severely reduced. Brush-tailed Bettongs used to occur over large areas of inland New South Wales and South Australia where soils are now hard-baked and unproductive. Descriptions in the narratives of early European explorers such as Joseph Hawdon, Sir Thomas Mitchell and Captain Charles Sturt suggest that soils in these areas were once rich and friable. It is tempting to speculate that the loss of bettongs contributed to the massive degeneration of soil quality that we see now.

If the Brush-tailed Bettong can be called a digger, the slightly larger Burrowing Bettong can be viewed as a true mining engineer. Although this species disappeared from the Australian mainland in the late 19th and early 20th centuries, its old warrens can still be seen in parts of the semi-arid rangelands and stand as an enduring legacy. Work by Jim Noble at CSIRO has uncovered many old warren complexes. In general, these structures are roughly circular in shape, 20–100 metres (65–330 feet)

in diameter, and surrounded by walls up to 50 centimetres (20 inches) high that enclose a central lens of calcrete. When occupied by bettongs, burrows within the warren structure would have acted as traps for soil and organic material, and perhaps have provided nutrient-rich 'hot-spots' that accelerated plant growth. Even now, with the structures long abandoned, the warrens reduce erosion of topsoil and greatly increase variability in soil richness and the diversity of plant communities at the landscape scale. Aboriginal people have spoken of warrens being used as shelter by Western Quolls in the continent's central deserts and I have seen extant bettongs on Barrow Island sharing their excavations with such unlikely bed-fellows as Common Brushtail Possums and Golden Bandicoots. At Scotia Reserve in western New South Wales, where Burrowing Bettongs, Brush-tailed Bettongs and Bilbies have been reintroduced, the soil in many areas is turned over regularly and honeycombed with holes. It looks much healthier than the bare and scalded landscapes on the surrounding properties that lack the marsupials. The loss of such species over vast areas of inland Australia has been a tragedy not just for the species themselves but for the catastrophic simplification of ecological processes that it has caused – and the massive loss of landscape function, productivity and diversity.

Other marsupials probably also engineer their environments. Wombats are likely to have especially marked effects owing to the size of their burrows but smaller diggers such as Brush-tailed and Crest-tailed mulgaras, Kowaris and marsupial moles probably also play some role in altering soil structure, nutrient recycling and infiltration of water. Kangaroos do not dig but the shallow 'hip-holes' that they create at favored resting spots may act as accumulation points for organic matter. Unfortunately, the importance of such effects remains speculative at this point and provides an exciting direction for future research.

Crest-tailed Mulgara *Dasycercus cristicauda* Female
Shown eating Panther Skink *Ctenotus pantherinus*
NORTHERN TERRITORY

Marsupial Moles

Underground is a good place to be if you live in the arid sandy deserts of central Australia. Whereas the surface is dry and temperatures there fluctuate wildly – baking on summer days and freezing on winter nights – underground the climate is much less extreme. Most small to medium sized animals in the desert spend at least some time sheltering underground but Marsupial Moles have taken this behavior to the extreme and appear able to live their entire lives underground. They are arguably the world's most fossorial mammals.

Unlike most fossorial mammals, Marsupial Moles don't construct hollow tunnels in which to live and run about underground. They tunnel and backfill as they go, carving a hole in the lightly cemented sand ahead with their extraordinary spade-like fore-feet, and pushing the loosened sand back behind themselves with their slightly webbed hind-feet. At the same time they squeeze their tubular body forward a few centimetres at a time. They are completely blind, and their calloused nose and forehead is presumably used as a ram. They have strikingly well developed shoulders, and their creamy-white fur is fine and silky and must pose little resistance to their tunnelling. At only 40–60 grams (1.4–2.1 ounces), Marsupial Moles have such modest oxygen requirements that they subsist by breathing the air that flows between sand grains. While their unusual form of locomotion is slow and laborious, they also seem tuned to a frugal life and save energy and resources by allowing their body temperature to reflect that of the surrounding sand, as if they were reptiles.

The two recognized species of Marsupial Moles are similar in most respects. The species are often featured in Aboriginal mythology and the names assigned to them that originated in Indigenous language – Kakarratul (*Notoryctes caurinus*) and Itjaritjari (*N. typhlops*) bear a more poetic tone than the respective Northern and Southern Marsupial Mole labels that are more commonly used.

Marsupial Moles are very unlike other marsupials and some evidence suggests they branched off 64 million years ago not long after dinosaurs became extinct. In a striking example of convergent evolution, Marsupial Moles have developed similar appearance and adaptations to some placental moles, particularly the Namib Golden Mole which also inhabits desert dune fields. The Namib Golden Mole is adept at sand-swimming through loose sand but routinely runs about on the surface to forage. In contrast, Marsupial Moles seem to flounder in loose sand, which is relatively rare in Australian deserts, and struggle to drag themselves along the surface.

Marsupial Moles are at home underground where they feed on small animals such as insects, and seem especially partial to eggs, larvae and pupae. In captivity, they have also been observed to devour centipedes, spiders and geckoes and may eat similarly sized animals in the wild if given the opportunity; their ability to tunnel rapidly may make them a formidable threat to any small animal that is slow moving or dormant.

Southern Marsupial Mole *Notorcyctes typhlops* Male
JITIRLPANDA OUTSTATION, NORTHERN TERRITORY

Although Marsupial Moles have been known to science for over a century, we are just beginning to learn about these enigmatic little creatures. Until recently, they were only known from the very occasional find on the surface. Many were collected by Aborigines in the early 1900s when museums paid handsomely for these difficult-to-procure animals. Since then, specimens and sightings have averaged 5–10 per decade, suggesting that the species are either very rare or very elusive. Their remains are often found in the scats of Red Foxes, and to lesser extent Cats, at frequencies that are alarming in comparison to the rarity of sightings.

Recent survey and monitoring techniques are based on the abundance of sand-filled tunnels. These signs appear widespread and common and suggest that Marsupial Moles may be much more abundant than previously thought. There is, however, still uncertainty about how long these signs persist in the sand; they may reflect abundances from several decades ago. Nonetheless, averages of 20–60 km (12–37 miles) of recognizable tunnel per hectare (2.5 acres) are common in central dunefields. It is possible that Marsupial Moles might be so abundant that they profoundly affect their environment both by their consumption of prey and by turning the soil as they tunnel.

Both species are officially regarded as endangered and are likely to remain so until further studies clarify the significance of their abundant underground signs. It seems strangely fitting that these wonderfully bizarre animals should have us still confused about their abundance and conservation status.

Joe Benshemesh
SCHOOL OF BIOLOGICAL SCIENCES, LA TROBE UNIVERSITY, MELBOURNE

an uneasy relationship

For the last 20 years I have given lectures to second year university students on the biology of birds and mammals, and usually conclude the lecture series with a discussion on conservation. To get everybody thinking, I often ask the students to list marsupials that have become extinct, and others that are at risk and face a gloomy future. The Thylacine (*Thylacinus cynocephalus*) comes up every time and is clearly the best known of the missing marsupials. Occasionally someone mentions an extinct bandicoot or wallaby but I am always surprised when students vote the Koala (*Phascolarctos cinereus*), Bilby (*Macrotis lagotis*) or tree-kangaroos (*Dendrolagus* species) on to the list. About 17 species and subspecies of marsupials have disappeared since European settlement in Australia but these three species are not among them. Students' lists of threatened marsupials are usually also quite short. One class managed to name 10 of the 34 species that are currently considered to be at risk but typically the list peters out when just three or four species have been identified. By contrast, students can readily name threatened mammals from other parts of the world, with the plight of the Black Rhinoceros (*Diceros bicornis*), big cats, hunting dogs, tapirs, lemurs, orang-utans and other primates being known to most.

Sometimes I field enquiries from people outside the university who have seen an odd or different animal and wonder what it might be. These mystery beasts usually turn out to be Common Brushtail Possums (*Trichosurus vulpecula*), bandicoots or, increasingly, introduced Black Rats (*Rattus rattus*), but occasionally I am able to confirm the identity of smaller and more cryptic marsupials such as pygmy-possums, young gliders and antechinuses. Many people are amazed to find that marsupials come in such varied shapes and sizes and to learn that Australia is home to so many species. Some have been so moved that they have enrolled in classes to find out more; others have volunteered their time on field trips to see these marsupials at first hand.

My feelings about these encounters are mixed. On the one hand, it is a great privilege to help people discover the enthralling native fauna of Australia and to understand more about the lives and needs of its furred and feathered inhabitants. Teaching does not get better than that! On the other hand, I often wonder why the continent's unique biota seems to be so poorly known and under-appreciated in the first place. What are the attitudes of Australians to marsupials and have these changed over time? If marsupials are not generally appreciated, what are the reasons for this and what are the consequences for these mammals in future? My personal view is that Australians have a 'bipolar' relationship with the native fauna and with the environment more broadly. This chapter explores the relationship between humans and marsupials in more detail, and asks how the fragile balance between these very different mammals can be sustained.

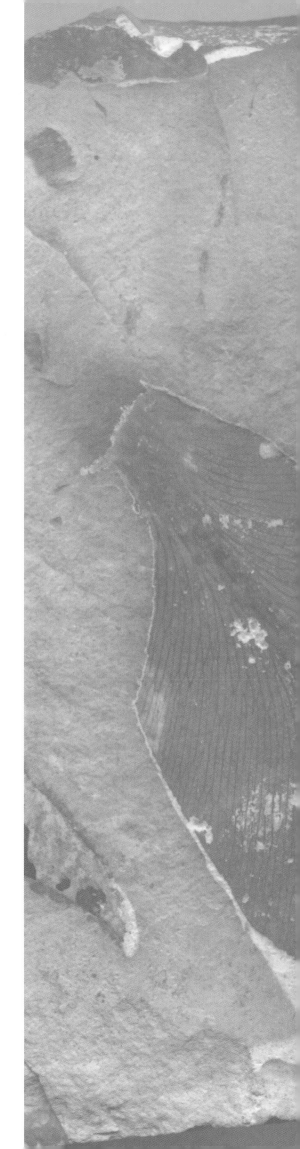

Thylacine
Thylacinus cynocephalus Male
TASMANIA

The first Australians

Archaeological evidence indicates that the first humans arrived in Australia
at least 45,000 years ago. These early immigrants probably clung to coastal
regions and inland rivers where food resources were assured, taking some
thousands of years to exploit the continent's arid interior. The establishment
of land bridges during a period of severe climatic cooling between 18,000
and 20,000 years ago reduced sea levels by up to 160 metres (525 feet) and
created extensive land connections across much of northern and southern
Australia. These bridges allowed people to become established on islands
off the continent's southern coast, including Tasmania and Kangaroo Island,
and marked the completion of human colonization of the present land area
of Australia. It remains unclear how many waves of immigrants arrived
in Australia but some evidence points to three main tides, the last just
5000 years ago.

The earliest ancestors of today's Aboriginal people would have encountered an environment that was strikingly different from the present one. The newcomers would have lived alongside very large marsupials, flightless birds, lizards and snakes. The largest of the marsupials, *Diprotodon optatum*, was solid enough to rival the modern rhinoceros in stature. Some distant relatives, such as *Palorchestes*, were probably about the size of modern cattle; other marsupials, such as the blunt-faced kangaroos *Procoptodon* and *Simosthenurus*, stood up to 3 metres tall. With the demise of the giants soon after the first wave of human immigrants, later settlers would have experienced a relatively depauperate environment dominated by much smaller vertebrates. Many of these species are present today.

The larger beasts are remembered in dreamtime stories; the downsized survivors remain part of the every-day life of many Indigenous people and form an important component of their local culture and resource base.

Most species are used for food. The pelts of some are used for clothing or for ceremonial purposes, the skins of others are used in the construction of percussion instruments, and sinews and other tissues are used for tools and a myriad other purposes. Traditional knowledge of the marsupial fauna has been forged over many hundreds of generations, allowing deep insights into where different species occur, what they eat, when they breed and how they live. This knowledge is organized within an overarching framework, the dreaming, and is used both to exploit marsupials and other animals as resources, and to honor and maintain them as integral parts of the world. As a much newer Australian I do not have the understanding nor the right to explain the dreaming but it does seem to represent an intimate spiritual bond between the land and all living things. I have learnt much from discussions with Kado Muir, a member of the Ngalia people of the western desert, and from his perspective below.

Some of the early European naturalists appreciated Indigenous knowledge and used it to increase their understanding of the strange new land they had entered. Others appropriated it shamelessly. In his 1898 book *Spinifex and sand*, for example, explorer–naturalist David Carnegie writes that he unceremoniously kidnapped Aboriginal people and made them serve as guides to waterholes. Later naturalists have generally accorded much greater respect to Indigenous knowledge. In *The red centre*, first published in 1935, Hedley Finlayson describes the 'extraordinary intimacy' and 'command [of] every relevant detail of its life-history and habits' that hunters display when tracking mammalian quarry. Andrew Burbidge, of the Western Australian Department of Environment and Conservation, has described how he and colleagues obtained unprecedented insight into the distribution, abundance and habits of desert mammals from discussions with Aboriginal elders; many other researchers have followed this lead. Partnerships between Aboriginal people and western scientists are now helping to 'look after country' in parts of central, western and northern Australia, and thus to conserve and maintain the fauna and flora of these regions.

Quokka *Setonix brachyurus* Male
ROTTNEST ISLAND, WESTERN AUSTRALIA

An Indigenous Australian perspective on marsupials: the Red Kangaroo – Marlu

The group 'marsupial' is not a definition arising from Indigenous science in Australia. Indigenous Australian knowledge arises from the cultures, the environment and the experiences of Indigenous peoples over an immense timeframe, extending across millennia – modern archaeology suggests over 50,000 years before the present. It is unlikely that Indigenous Australians categorize native animals according to the methods adopted and used by modern western science. Categorizing these animals as marsupials is a construction of a knowledge system that arises from an environment which did not have marsupials as natives to that environment. From the earliest days of European contact in Australia, its animals were considered a novelty, an anomaly or even a fake as in the case of the Duck-billed Platypus.

Therefore, to invite a discussion of marsupials as a category of fauna is, from an Indigenous Australian's perspective, an artificial construction. That is not to say that Indigenous Australians do not care about the category of fauna known as marsupials; rather, Indigenous Australians know and have an understanding of this category of fauna from an entirely different perspective in which it is not possible to separate marsupials into a particular category. Indigenous Australians knew and understood the native mammals of Australia, with a level of knowledge that melded spirituality, kinship and physiology into a holistic system of knowledge. Indigenous Australians continue to hold knowledge and understanding of native species based on this ancient and wise knowledge system. The kangaroo was considered a strange new discovery to the eyes of European settlers. To Indigenous Australians, the strange new animals were the four-legged beasts introduced to Australia.

For the purposes of this discussion on Indigenous Australian perspectives on marsupials, the focus is on one example, the Red Kangaroo (*Macropus rufus*). All across Australia different Indigenous cultures have stories about Red Kangaroo. Here, the perspective is that of the Ngalia people of the deserts of Western Australia.

In the desert regions of Western Australia, Aboriginal people believe that all things were created in the dreaming. There are dreaming stories explaining the creation of the world; there are dreaming stories explaining the creation of people; and there are dreaming stories explaining the creation of animals. In all of the dreaming stories there is an understanding that a powerful bond exists between the spiritual, natural and human worlds. The evidence of the creative events enacted in the dreaming is today found transformed into the landscape, to which the spiritual bond extends. This spiritual bond unites all living and natural things, and links them all back to the spiritual source.

In the dreaming, ancestral beings – who were the spiritual ancestors of man and animals alike – lived according to their daily patterns. The world in which they lived was one devoid of physical, moral and social laws. In the beginning the dreaming was a period of great confusion and chaos. The ancestral beings shifted between animal and human forms; they did not live according to the laws we have today. After some time, some of the ancestral beings decided to bring law and order to the universe. Some ancestral beings travelled great distances across Australia bringing the laws, traditions and customs of modern Aboriginal people to the land. Other ancestral beings created precedents for the laws through their actions. Still others decided to take on the physical form of certain living things, often through the culmination of a series of dramatic and climactic events.

The laws, customs, traditions and beliefs were enacted in the dreaming. People today understand, live and abide by the laws of the ancestral beings. In many cases the laws, customs, rituals and

knowledge are of a very sacred nature and cannot be shared and spoken of openly, without first meeting the requirements of initiation, maturity and responsibility.

Marsupials are major dreaming characters. The Red Kangaroo, called Marlu in the western desert regions, is one such dreaming character. Across the deserts of central and Western Australia, Aboriginal men know and understand certain laws, rituals, ceremonies, designs and knowledge as being associated with the Marlu dreaming. It is not possible to speak openly about this particular dreaming as it would cause offence to tribal laws. It is important to know that the Marlu is a very important animal to desert Aboriginal people; it is in fact a sacred animal and is highly respected.

Marlu is not only a sacred animal it is also an important part of the Aboriginal diet, and in the diet of Aboriginal people it captures three key elements. Firstly, it is widely believed by Aboriginal people that eating Marlu is good for your health. In many communities throughout the Western

Australian desert regions, Marlu is an essential part of household diets. It is said that Marlu is rich in omega-3. It is not a fat meat like sheep or cattle and it is central to the emotional and social well-being of many Aboriginal people.

Secondly, Marlu has strong cultural and spiritual attributes, as it is still widely believed that one must follow the tribal laws that respect the sacrifice of Marlu, who dies so that humans might live. There also are certain Aboriginal people who cannot eat Marlu. First are those who share the spirit of the Marlu through their totem. A person is born with a totem and if their totem is the Marlu then that person is obligated throughout their life to look after and take care of the Marlu. That person must not kill Marlu, that person must not eat Marlu and in ceremonial life that person must perform the rituals necessary to maintain the spiritual energy of the Marlu. This totemic responsibility extends to all Aboriginal people who have a variety of different totemic associations. It may be Emu, Dingo, birds or even features in the landscape representing dreaming ancestral beings. This kind of relationship is called a personal dreaming. The other people who cannot eat Marlu are those who have recently had a death in their family. It is Aboriginal law in the desert that if there is a death, then all family members must stop eating Marlu and start eating Tjunikapartu or the Euro (*Macropus robustus*). This food restriction will continue for a year long mourning period. At the end of this mourning period the family can then resume eating Marlu.

All hunters who hunt Marlu know the habits of the animal intimately. There are common Aboriginal dances that show or depict how Marlu behave when feeding. These dances serve to instruct young men on the habits and behavior of Marlu. This knowledge is an important part of becoming a strong and effective hunter. For instance a hunter must learn to recognize the different tracks of a Red Kangaroo, a Western Grey Kangaroo (*Macropus fuliginosus*) and a Euro. A hunter must learn to observe the way in which a kangaroo sits, with its back to the wind to smell danger. A hunter must watch the ears of a kangaroo which twitch toward every sound, listening for danger. And, of course, a hunter must approach the kangaroo without being seen.

If a hunter catches a kangaroo, then there are other laws in the desert relating to the preparation, cooking and carving of the Marlu. Traditional desert law says a kangaroo must not be skinned. It must be prepared for cooking with the skin on. A fire pit is dug out and a fire is lit. Once the fire burns down the kangaroo is placed on the flames to singe the fur before having the tail cut off and the legs cut off. The tail is cooked by the hunter for his own consumption. The legs are removed in a way that allows the hunter to extract tendons or sinew which are then used, after preparation, to fix spears and other tools. In the burnt down fire, the coals and hot sand are scraped aside forming a hole in the pit. The kangaroo is then placed in the fire for cooking. It is Aboriginal law that the nose and upper legs must never be buried. Once the kangaroo is cooked, it is cut up according to a special 'cut'. Each cut of meat is then divided among the tribe according to their relationship to the hunter. In this way everyone in the tribe is fed and there is no argument about who gets what. As each man in the tribe, whose totem is not the kangaroo, will get a kangaroo at some stage then all the hunters will share in different cuts of meat depending on their relationship to the other hunters.

In the southern part of Western Australia there are two big Aboriginal nations. The south-west is the traditional lands of the Noongar Nation. The Noongar people have many Western Grey Kangaroos in their lands, for which they have big dreaming stories. They also eat the Western Grey Kangaroo. In the south-eastern or desert regions of Western Australia the lands are part of the Wongutha Nation, to which the Ngalia people belong. The Wongutha peoples have the Marlu and Euro. Wongutha people eat Marlu or Euro if the tradition requires it. It is one of the great teasing jokes between these two nations of Aboriginal people that Wongutha people will never eat a Western Grey Kangaroo, also known as a Kulbirr (Goolbid). Wongutha people call the Kulbirr a stinker!

In the modern age more people throughout the world are becoming aware of the great qualities of kangaroo meat. In the past the commercial harvest of kangaroo was based on servicing the pet meat market; today it is based on servicing a human consumption market. It is now widely found on supermarket shelves and is also becoming popular overseas. Aboriginal people are a little uncomfortable with this trend for two key reasons. Firstly, Aboriginal people feel strongly that the kangaroo is a sacred animal and must be treated with respect. New consumers of kangaroo meat must be made aware that there are strong spiritual associations between Marlu and Aboriginal people and it would be a better world if all humans respected the animals who die so we might live. Secondly, Aboriginal people are concerned that commercial harvests of kangaroo may adversely affect the primary source of fresh meat in many Aboriginal households. This may be through mismanagement or it may be through regulations that deny traditional rights to hunt Marlu. In each case Aboriginal people are concerned that an essential supply of their bush meat is maintained.

In the Western Australian desert Marlu is the main marsupial of spiritual, social, cultural and economic significance to Aboriginal people. There are also stories and significance associated with the many other marsupials but none as in-depth and significant to as many people as that of the Red Kangaroo or Marlu.

Kado Muir
Ngalia Heritage Research Council (Aboriginal Corporation)

Red Kangaroo *Macropus rufus*
Male with female and juveniles
Borefield Track, South Australia

The European explorers and naturalists

In the early 16th century, expansionist notions saw the great European powers of the day beginning to move to distant parts of the world in search of more land and resources. Portuguese and Spanish mariners, in particular, were active in the southern hemisphere at this time, so it is little surprise that Iberians were among the first Europeans to encounter marsupials. There is no definite record of who made the very 'first contact' but the Spanish explorer Vicente Yáñez Pinzón must be high on the shortlist of contenders for the honor. Sailing from Palos on the Andalusian coast in November 1499, Pinzón put to shore in eastern Brazil in January 1500 and the discovery of pouched mammals was made soon after. Pinzón himself was sufficiently impressed by the presence of such biological oddities that he ordered the capture of several marsupials to accompany the explorers on their voyage home.

Elsewhere, Portuguese explorers were encountering new environments in lands as far flung as the east coast of Africa, India, China and the Far East, including the Celebes (Sulawesi) and islands of the Moluccas. Although tales of strange new animals in the Far East abounded during the earliest years of the 16th century, the first unequivocal record of a marsupial from the region dates to about 1544. This original observation seems to have been made by António Galvão, Portuguese Governor of the Moluccas, who lived on the island of Ternate between 1536 and 1540. Galvão was a keen chronicler of life in the Moluccas, and his detailed observations on the appearance, reproduction and tree-dwelling habits of a creature called a 'kusus' or 'kuso' by the local people leave no doubt that he was describing a cuscus. As the Blue-eyed Cuscus (*Phalanger matabiru*) is the only member of its family on Ternate, this small, strikingly blue-eyed creature can be safely assumed to be the first Australasian marsupial documented in print.

European exploration of Australia itself did not get underway until the 17th century. As in other regions, the impetus for reconnaissance was mostly to satisfy the lust for new land, material resources and trading partners; however, mariners of the era were none the less incredulous about the strange new animals and plants that they saw. Perhaps to make sense of the new world that they were opening up, these explorers often described marsupials in terms of mammals that they were familiar with at home. A good example of this is a description of the Tammar Wallaby (*Macropus eugenii*) on Houtman's Abrolhos, Western Australia, by Dutch seaman Francisco Pelsaert in 1629:

We found in these islands large numbers of a species of cats, which are very strange creatures; they are about the size of a hare, their head resembling the head of a civet-cat; the forepaws are very short, about the length of a finger, on which the animal has five small nails or fingers, resembling those of a monkey's forepaw. Its two hind legs, on the contrary, are upwards of half an ell in length, and it walks on these only, on the flat of the heavy part of the leg, so that it does not run fast. Its tail is very long, like that of a long-tailed monkey; if it eats, it sits on its hind legs, and clutches its food with its forepaws, just like a squirrel or monkey.

In this short passage, Pelsaert invoked the behavior or appearance of no less than five placental mammals to describe the Tammar Wallaby.

Later voyagers to Australia's west coast saw and reported other chimaeras. In 1658 Dutch mariner Samuel Volckertzoon described a 'wild cat resembling a civet-cat but with browner hair' in reference to the Quokka (*Setonix brachyurus*); in 1696 his countryman Willem de Vlamingh thought it more 'a kind of rat as big as a common cat'. De Vlamingh was so taken with this odd creature that he named its island home 'Rottenest' (rat nest) in its honor. Another mixed-up mammal was described from Dirk Hartog Island three years later by the buccaneer William Dampier. This 'sort of raccoon' was almost certainly the Banded Hare-wallaby (*Lagostrophus fasciatus*). Captain Cook's voyage to Australia between 1768 and 1771 led to the discovery of a quadruped 'resembling a polecat' (Northern Quoll, *Dasyurus hallucatus*) as well as two strange hopping mammals (Common Wallaroo, *Macropus robustus*, and Eastern Grey Kangaroo, *Macropus giganteus*).

After settlement at Sydney Cove in 1788, interest in marsupials continued. The Long-nosed Potoroo (*Potorous tridactylus*) was figured in the work of Governor Phillip in 1789, the same year that a pair was sent for exhibition in London. Five years later the English capital had breeding kangaroos on display. Within 30 years of settlement, 23 species of marsupials had been formally described and Henri de Blainville had shown definitively that these strange creatures could be distinguished

Banded Hare-wallaby
Lagostrophus fasciatus Male
BERNIER ISLAND, WESTERN AUSTRALIA

Long-nosed Potoroo *Potorous tridactylus* Male
NORTH-WESTERN TASMANIA

than in the time of Gould and his contemporaries. Interpretations of how animals live are made using tools that Gould could not have dreamed of. Discoveries now come from painstaking observations of animals behaving naturally in the wild, from tightly controlled experiments on captive individuals, from statistical evaluation of complex datasets or from inspecting DNA sequences on a gel at a lab bench. The thrill of exploration remains. Centuries after intrepid Europeans were first awed by the other-worldly creatures of the Antipodes, the mammalogists who now study the marsupials are no less fascinated by their subjects and are overjoyed when they gain new insight into how they behave and exploit their world.

The settlers and pastoral pioneers

Of course, Europeans did not seek new lands just to discover different forms of life but rather to expand their empires. Once established at Sydney Cove and later at other settlements, the main aim of explorers was to map the new country and assess its material assets and potential for productive use. Marsupials were just a small and incidental part of this picture. Some explorers, such as the ill-fated Burke and Wills, appeared not to notice marsupials or failed to comment on them if they did. Others were quite disparaging of the continent's flagship fauna, calling them 'ugly' or even 'useless'. Still others were lax. Many specimens of mammals collected by Sir Thomas Mitchell during his explorations in New South Wales and Victoria cannot be found; some were probably lost from poor curation or oversight but at least one unfortunate specimen was caught and then fed to a servant's dog!

As settlers occupied new land or followed the tracks of the explorers with their domestic animals and livestock, conflicts with the land's original inhabitants became more frequent and severe. Skirmishes between the settlers and Aboriginal people have been documented elsewhere but the campaigns waged by the newcomers against the marsupials and their environment are perhaps less well known. Brush-tailed Phascogales (*Phascogale tapoatafa*) were viewed as 'blood-thirsty vampires' owing to their ability to kill and (apparently) exsanguinate poultry; quolls also were viewed as 'very ferocious' and capable of attacking companion animals and committing atrocities in chicken coops. Bettongs and bandicoots welcomed the settlers by eating the produce

from other mammals by the unique structure of their reproductive system. The pioneering work of John Gould and his collectors, most notably John Gilbert, added immeasurably to our knowledge. Between 1841 and 1860 Gould described 33 species and subspecies of marsupials, including almost a third of the currently recognized kangaroos and wallabies. He also took time out to produce three volumes entitled *The mammals of Australia*, which include beautiful and much sought-after plates of the various species. Contemporary writings and enchanting lithographs by other authorities such as George Waterhouse and Gerard Krefft helped to ensure that marsupials stayed in the eye of the public.

Discoveries of new Australian marsupials have continued to the present and knowledge of many aspects of marsupial biology is accruing at an impressive rate. However, the nature of exploration has changed. Except perhaps for forays into remote parts of New Guinea and Indonesia, present-day field research is less dangerous from their vegetable gardens and suffered opprobrium as a result; the larger Thylacine and Tasmanian Devil (*Sarcophilus harrisii*) were seen from the earliest days as wanton killers of sheep. By the mid-19th century medium-sized marsupials generally were viewed with suspicion in pastoral districts, and the digging habits of wombats, bilbies and bettongs did little to endear them to their new neighbors. As flocks of sheep and cattle rose in size and importance, populations of kangaroos also burgeoned and came to be seen as mortal enemies and competitors. Petitions were sent to state governments from residents in agricultural districts seeking solutions to the marsupial menace. One, from New England in 1878, called for special legislation to 'grapple with the evil' ravages of marsupials that were over-running the land. The concept of 'terra nullius' (land of no-one), long applied to deny Aboriginal people their right to the land, could at this stage be extended to other, less familiar life forms. Soon after settlement, battle was thus joined with all things native.

Small cottage industries sprang up to make traps or devise other ways of controlling pesky native species in the early 19th century but the first official shots were fired in 1852 when legislation was passed in New South Wales to destroy native dogs. This encouraged the setting of strychnine baits. In 1879 a *Marsupials Destruction Bill* was tabled but it lapsed and paved the way for the *Pastures and Stock Protection Act* the following year. This Act identified native dogs as pests but it also included rabbits and 'Any kangaroo wallaroo wallaby or paddamelon'. District lists of pests were more extensive, adding bandicoots, bettongs, Common Wombats (*Vombatus ursinus*), Bilbies and even Brush-tailed Rock-wallabies (*Petrogale penicillata*). The Act allowed for the payment of bounties on scalps and condoned the unfettered destruction of listed pests. Millions of marsupials were killed over the next few years as a result. Unpublished research by my colleague Gus Bernardi and me indicates that between 1881 and 1900 alone, some 21.4 million kangaroos and wallabies were killed in New South Wales, as well as 3.2 million smaller marsupials. Other Australian states also mandated the elimination of marsupials to support pastoral interests. The Thylacine was put to death for this reason in Tasmania, and populations of many species of bandicoots, bettongs and larger marsupials were depleted in concerted poisoning, trapping and shooting campaigns elsewhere. During the

53-year tenure (1877–1930) of the *Marsupial Destruction Act* in Queensland, for example, some 27 million medium-sized marsupials were killed. We consider the effects of these pogroms in the chapter, Conservation.

The marsupials were not without their defenders. As early as 1836, Charles Darwin noted declines of kangaroos and Emus on the southern tablelands of New South Wales and blamed this on the impacts of the settlers' greyhounds. Naturalists George Bennett, John Gould and Gerard Krefft later predicted the extinction of several species of marsupials because of the expansion of settlement and the rapacious acquisition of land for livestock. By the turn of the 20th century there was increasing concern for the future of marsupials and 10 species, plus the Short-beaked Echidna (*Tachyglossus aculeatus*) and Platypus (*Ornithorhynchus anatinus*), were given protection in 1903 in New South Wales under the *Native Animals Protection Act*. Not everyone agreed with this initiative. During debate before the passage of the Act, the Hon Henry Dangar, a member of the Legislative Council, distinguished himself with a memorable comment about kangaroos: 'The sooner they were all exterminated the better it would be. They were hideous, useless brutes … We had passed measures offering the pastoralists a premium for the destruction of these pests, and here we were passing a bill to protect them.' Alexander Campbell, who represented the electorate of Kiama, took most exception to the Common Brushtail Possum, calling it '… the greatest enemy the settler has to face in new districts …' on account of its ravages of cereal crops.

Despite the continuing blood lust in some quarters, sentiments about native animals mellowed further in the early years of the 20th century. Naturalists called for the protection of both land and native species, and groups such as the Wild Life Preservation Society of Australia sprang up to lobby for conservation interests. In New South Wales the *Birds and Animals Protection Act* was introduced in 1918, the *Pastures Protection Act* of 1934 excluded marsupials from the definition of noxious animals for the first time and the *Fauna Protection Act* was passed in 1948. The latter Act was the first to formalize the concept of 'natural reserves' and also

Brush-tailed Phascogale *Phascogale tapoatafa* Male
In background, leaves and fruit of Apple-top Box
Eucalyptus angophoroides
Victoria

emphasized the importance of educating the young about wildlife. More recently, the *National Parks and Wildlife Act 1967*, the *Endangered Fauna (Interim Protection) Act 1991* and the *Threatened Species Conservation Act 1995* have formalized the protection of marsupials and native biota to a degree that would have had the Hon Mr Dangar frothing at the mouth. Parallel legislation exists in all Australian states and territories, and was developed at more or less the same time as in New South Wales.

I had the privilege of being a member of the inaugural scientific committee established under the *Endangered Fauna (Interim Protection) Act*, and of chairing the expanded scientific committee under the *Threatened Species Conservation Act* for the first seven years of its operation. One of the key tasks of the two committees was to identify and list threatened native species. As these lists were published in newspapers and comments invited from the public, the process also provided a superb opportunity to gauge the attitudes of the community towards conservation. Many people who wrote to the committees strongly supported the protection of native species, others thought this was an excellent notion provided that it did not impinge on their activities and some were openly hostile. Farmers, cereal growers, orchardists and other primary producers were sometimes worried, quite understandably, that listing species they viewed as pests would limit their ability to control them. Pastoralists also expressed concerns. But my greatest surprise was the attitude of the housing and development industry. The scientific committee that I chaired was often inundated with submissions about why certain species or communities of species should not be considered at risk, even when they were obviously at death's door. Many letters of objection were drafted by industry lawyers for added weight. The industry looks upon land set aside for conservation as 'sterilized' because it is not available to be built upon. For myself, I cannot imagine a more sterile habitat for marsupials or anything else than many hectares of concrete and bitumen.

Strangers in a strange land

Our attitudes towards marsupials have changed greatly since the wild days of early settlement but as a society we still seem to retain ambivalent feelings about them. Recent legislation provides a wonderful expression of our desire to conserve native animals and their habitats.

Scientific and professional societies provide guidance about conservation management, schools and universities recognize the importance of environmental education, wildlife carers rescue and nurse sick or injured animals, and advocacy groups draw attention to improvements that we can make in our personal and corporate behavior. Cruelty to animals is not tolerated, and anti-social acts deservedly condemned. As I write (1 August 2007), for example, Sydney's *Daily Telegraph* newspaper carries a piece decrying the finding of butchered kangaroos on a beach in northern Queensland. Yet, we are also curiously apathetic. Over the last decade state governments sanctioned the clearing of millions of hectares of native vegetation to grow more crops and run more livestock. As we will see in the next chapter, this wanton spree killed many millions of marsupials and other vertebrates but the slaughter raised hardly an eyebrow. There was similarly little coverage of the valiant last stand of the Eastern Barred Bandicoot (*Perameles gunnii*) at Hamilton in Victoria in the early 1990s and no outrage when a madman deliberately released the immensely destructive Red Fox (*Vulpes vulpes*) into Tasmania in 1999 or 2000.

We might expect attitudes towards marsupials to vary regionally, perhaps along the traditional bush versus city divide that was characterized so vividly in Russel Ward's 1958 book, *The Australian legend*. But evidence for this is scant. Questionnaire surveys by RG Davies and colleagues at the then NSW National Parks and Wildlife Service in 2001 and 2002 confirmed that some city dwellers harbor romantic notions about wildlife, with 55% liking the thought that Koalas might live around their homes and 26% prepared to extend a welcome to possums. However, 17% of survey respondents were unimpressed that their backyards might be used by Koalas and 39% thought this way about possums. Earthworms, small birds and butterflies scored higher than the marsupials. I am sure that even more divergent feelings would be elicited by bandicoots and bettongs. In the bush in western New South Wales, Queensland and the wheat belt of Western Australia, I have been very fortunate to meet people with extraordinary knowledge of the native fauna and flora, and great empathy for the natural environment. Others, sometimes neighbors, have been uninterested and even shown visceral dislike for their situation. I suspect that attitudes in Tasmania are similarly polarized. During a visit with colleagues to the island state in November 2005 to review the status of the Red Fox, I met people who were appalled by the importation of this pest and others who thought it would be 'one in the eye for greenies'. Sceptics and hoaxers abounded.

In recent and extensive surveys, Dan Lunney, Mathew Crowther, Jessica Bryant and Ian Shannon of the NSW Department of Environment and Climate Change further explored people's attitudes to marsupials as pests, ranking responses from low support for this idea to intense agreement. Koalas generally evoked positive responses but were considered to be pests in the central west and far south coast of New South Wales where they are all but extinct. Spotted-tailed Quolls (*Dasyurus maculatus*) also tended to be viewed as pests more strongly west of the Great Dividing Range, where they are very rare, than in the tablelands and east of the ranges. Wombats were regarded as real villains by many people, perhaps because they are perceived as potential competitors with livestock and because they seem to dig burrows along fences and at the bases of farm gates. Surprisingly, though, agreement as to the pest status of wombats was most intense in parts of central and north-western New South Wales where these distinctive marsupials have never been recorded. Perhaps it isn't wombats that are muddle-headed after all!

What is the origin of our ambivalence, our national bipolarity, towards marsupials and the natural environment? To help them survive the rigors of their strange new land, the early settlers brought with them companion animals, livestock, crops, tools and other materials that had served them in their English homeland. Old but trusted ways of using the land were imposed on terra nullius, no matter how difficult or inappropriate these proved to be. Pastoralists in particular were rewarded for their pioneering efforts and in the early years gained much wealth from their large flocks of sheep and cattle. Mirroring the attitudes of landed gentry in England, many sought rights to own native animals and to import more for amenity and for sport. Acclimatization societies proliferated in the 1860s and were dedicated to importing 'useful' animals and plants from anywhere else in the world. As Eric Rolls put it in 1969, in his excellent book *They all ran wild*, '… foreign animals … were to live with us to be shot, hunted, coursed, worn as fur, eaten, admired, or listened to'. The first legislation to protect animals in Australia, the New South Wales *Game Protection Act 1866*, was targeted squarely at conserving introduced mammals

and native game birds to support sporting interests. Unless they could be hunted, there was little room for marsupials in this world view.

The Australian environment and its inhabitants got short shrift in other spheres too. Artists and painters, for example, have commonly depicted local trees with emerald rather than bluish-green foliage, and bent their shapes to resemble the oaks and elms of England. They also rejoiced in showing the development and opening up of the land. In their 1978 book, *100 masterpieces of Australian landscape painting*, William Splatt and Susan Bruce selected works that '… penetrate deeply into the reactions between man and his environment'. Indeed. Of the 63 paintings from the period up to 1900, 22 illustrate modified landscapes, another 20 show livestock or poultry, and three show tree cutters with their axes. Kangaroos are depicted just once, in *Landscape near Ballan, Victoria* (1877) by Henry Rielly. Other marsupials do not feature, although several species of native birds can be recognized. Scenes of pastoral idyll continue to be depicted in paintings after 1900 but livestock and logging feature in just five of them. Marsupials are conspicuously absent.

Australian literature has also tended to laud pastoral heroes and to reinforce the notion that the bush is a tough place, habitable and useful only when tamed. In *The man from Snowy River*, Banjo Paterson writes lyrically about wild horses in Australia's high country, and of the courage and great horsemanship displayed by the protagonist in retrieving a lost colt from their midst. In the iconic and poignant *Waltzing Matilda*, the main players are a swagman and the sheep he has stolen. Many of Henry Lawson's short stories are also populated by tough and resourceful characters that survive the adversity of life in the bush. It was left to later writers, such as Oodgeroo Noonuccal and Judith Wright, to explore the environmental damage that the settlers and squatters had wrought. Beyond the heroes of fiction, we also celebrate the accomplishments of real-life bushmen and pioneers such as Albert Facey, and the Costellos, Macarthurs and Duracks. Biographies of the 'Cattle King' Sir Sydney Kidman, the expert bushman and remarkably audacious cattle thief Harry Redford who was known as 'Captain Starlight', and Ned Kelly and other bushrangers, have enduring appeal. This is a wonderful opus but its emphasis on battling or exploiting the natural environment does little to foster appreciation for native wildlife.

Few poems concern marsupials but one that does is *The hospital kangaroos* by Peter Kocan. The first three stanzas express the enduring the link between kangaroos and the bush, and their use of new environments such as meadows and roadsides. The poem continues ominously:

In their eyes the apathy
Of a played-out breed, broken
By the monotony of survival.

In their gestures a caution
Sprung less from fear
Than primordial habit. They dwell at the fringe

Of our lives, tolerated
For cuteness or curiosity, prey
To camera and larrikin's gun.

They are the real spirits of the place,
Waiting mutely, year to year,
Like sufferers who perceive an end.

Kocan's view of kangaroos may be dark because he wrote about them while incarcerated for attempted murder in a psychiatric hospital. Nonetheless, it is easy to imagine that many Australians share similar feelings.

What can we conclude from looking at popular culture? Advertisements for rugged four-wheel drive vehicles usually show them conquering the roughest of outback tracks, driving through creeks or pulling the last tree stump out of a threadbare paddock. The drivers and passengers seldom get out to inspect the view or the damage they have caused. When they do, in one memorable advertisement, they are terrorized by a blue-tongued lizard and flee back to the city for safety. In film, the natural environment is depicted as a place of intrigue and mystery, as in *Picnic at Hanging Rock* and *Walkabout*, a place of beauty but hidden danger, as in *Japanese story* and *Wolf Creek*, and as a mythological place full of lovable rascals, as in *Crocodile Dundee*. After surveying the breath-taking but desolate landscape of inland Australia, a kindly camel driver in *Gallipoli* tells the film's heroes that the enemy is '… welcome to it'. And in the *Mad Max* trilogy the spectacular environs of Broken Hill and southern Victoria are used to imagine the end of the world – the broken, dystopian society in Australia when oil runs out. Wildlife is usually depicted sympathetically in film, although the graphic footage of a kangaroo cull in *Wake in fright* is bloody and brutal. These few examples suggest that the visual media impressively

Rufous Bettong *Aepyprymnus rufescens* Male
North-eastern New South Wales

convey the stark beauty of Australia's environment. But they also highlight its treacherous and unforgiving nature, its dangerous inhabitants and confirm the constancy of the battle that must be waged to subdue it.

Taken together, these brief and selective vignettes suggest to me that new Australians have not severed their umbilical connection with the mother country and are not yet comfortable in their own skins. Perhaps living in a strange new land and pretending it is the same as the old has given rise to our bipolarity. Whatever the case, the contrast in attitude between the first Australians and the recent arrivals could not be more stark. In *Going native*, Mike Archer and Bob Beale call on us to rediscover old ways and kindle new ways of thinking about how to use our environment with respect. Their ideas are simple but ingenious. It should, for example, be made easier to keep native species as pets and harder to keep our now-familiar companion animals – the same cats, dogs, rabbits and house mice that run amok in the bush. We should value natural resources and use them sustainably, and in general adapt ourselves to the land rather than forcing it to accommodate us. I also advocate recognizing the pioneers who first respected the new land and appreciated its strange marsupial inhabitants. Among these are the early naturalists John Gould and Gerard Krefft, their intrepid and visionary successors Hedley Finlayson, David Fleay, Ellis Troughton, David Stead, Francis Ratcliffe and Frederic Wood Jones. More recent standard bearers include Harry Butler, Mike Archer, Tim Flannery and Hugh Tyndale-Biscoe. If you don't know these names, check them online; all have extensive entries. We often venerate the rich, the conspicuous, the exploiters and despoilers who return little of value to Australia. Perhaps we can heal our bipolarity by acknowledging those who have given us a true sense of place.

Northern Bettong *Bettongia tropica* Male
NORTH-EASTERN QUEENSLAND

Bridled Nailtail Wallaby

This nocturnal wallaby was once widespread and abundant in the inland woodlands of eastern Australia between the Murray River in Victoria and Charters Towers in Queensland. Expeditioners in the 1850s were delighted that they could catch as many as they wanted in one afternoon, just by using the Aboriginal method of stringing a net across a patch of bush and flushing them into it. The invasion of feral predators

Bridled Nailtail Wallaby *Onychogalea fraenata* Female with pouch joey
<small>EASTERN QUEENSLAND</small>

and stock, and increasing clearing of forest for farmland soon took their toll and by around 1930, the Bridled Nailtail Wallaby had apparently disappeared.

Then in 1973 a fencing contractor, preparing a bush cattle property in Central Queensland for imminent clearing, glimpsed a small, grey wallaby with stripes behind the shoulders and recognized the species from a watercolor print in a *Woman's Day* magazine article about extinct wildlife. Queensland Parks and Wildlife Service staff confirmed the sighting, and prevented what would have been the final blow to the species by halting clearing. The service bought that and the adjacent property for a reserve.

How did a small number of Bridled Nailtail Wallabies manage to survive on an area of only 11,000 hectares (27,000 acres) about 10 km (6 miles) wide? The favorable patchwork of open grassy woodland and dense brigalow scrub on these properties was one factor and the local absence of Red Foxes was another. In the 19th century, the species' disappearance followed the invasion of the Fox northwards. When I drove around the reserve spotlighting from a vehicle, a scared young Bridled Nailtail Wallaby about the size of a kitten flung itself down flat on the track just ahead, lying still, pressed to the ground behind a grass tussock. Adults behave the same way when terrified and it's not much defence against a Fox or Cat. The absence of Foxes may be explained by the particular abundance of Dingoes in the area (the nearest town is called Dingo) and Australia-wide evidence indicates that Dingoes can protect vulnerable native mammals by eating or displacing Foxes. However, Dingoes do not suppress the Cat population here, and feral Cats are still a major predator of Bridled Nailtail joeys once they begin to leave their mother's pouch.

Drought is the main killer. Each dry year causes worrying declines in this small, endangered population. In the early 1990s, for example, the wild population dropped from around 1000 to an estimated 200 animals.

The wallaby is also favored by an extraordinary reproductive ability. They breed very readily in captivity and can raise three joeys a year. Females start reproducing when they are younger than six months, barely out of the pouch themselves. For the rest of their lives, they breed continuously by mating when they have a three month old pouch young, and giving birth again the day after it abandons the pouch at four and a half months. The older joey still suckles by putting its head into the pouch containing its newborn sibling attached to a different teat. Instead of following its mother around like a young kangaroo, this tiny and defenceless joey spends all day and part of each night hiding by itself under a shrub or leafy branch near the ground, or in a small log – the need for dense patches of groundcover and timber is clear. The mother rests hundreds of metres away, drawing attention from the hiding place. She visits furtively at dusk and at intervals through the night to suckle it. The youngster which was enjoying milk on demand is suddenly going without during the day in the harsh climate of central Queensland, where days around 40°C are frequent in summer and freezing nights are common in winter.

A single population at one small site is at high risk from disasters. Between 1996 and 1998, the risk to the population was spread by releasing just over 100 captive-bred and wild Bridled Nailtail Wallabies in Idalia National Park. The park is in western Queensland on the western edge of the wallaby's former range and Foxes are strictly controlled. The gamble on the arid site paid off when the population grew to rival the wild central Queensland one in size in only three years. Since then, a smaller translocated population has also been established at Avocet Nature Reserve near Emerald in central Queensland. Drought has since reduced numbers at both sites but the Bridled Nailtail Wallaby has the ability to bounce back again when rain returns.

Diana Fisher
SCHOOL OF BOTANY AND ZOOLOGY,
AUSTRALIAN NATIONAL UNIVERSITY, CANBERRA

CONSERVATION

a fragile balance

Today, Australians are justifiably proud of their achievements in many fields of endeavor, from scientific advances and technological breakthroughs to developments in the arts and prowess in the sports arena. Many feel that the nation 'punches above its weight' in these areas, attaining successes and records that elude larger nations. But one record is less well known and is certainly not one to brag about. Over the last 200 years or so, Australia has lost at least 28 species and subspecies of mammals, the highest rate of extinction for any country or continent over the same period, and 17 of them are marsupials (see Table 3 on page 166). The remainder comprises nine species of native rodents and two bats. There are signs that many living species are on borrowed time and may vanish if we lower our guard. Could we face a time when Australia is no longer the land of marsupials and, if so, what are the implications of such an extinction event? To evaluate this potentially apocalyptic scenario, we need to understand why conservation is important, how conservation status is assessed, how threatened species are identified and what is causing these species to be threatened. It is critical that we also look at what can be done to retrieve species that are on the brink of oblivion.

Why conserve marsupials?

As a biologist, I take the importance of the natural world for granted and see it as imperative that we protect biological diversity from the more destructive effects of human activity. But this is not a universal view and it is still, surprisingly, quite new. Indeed, it could be argued that conservation was not part of the mainstream agenda in Australia until the 1970s when Jack Mundey led the 'green bans' against rampant development in Sydney, and Bob Brown galvanized the campaign to stop the damming of the wild Franklin River in Tasmania. The conservation ethos competes with other ideals even now. So why is conservation important and why should we be concerned if we push a few more marsupials to extinction?

One argument is that marsupials have a right to exist independently, and should not depend on the capricious whims of humans for their future survival. This perspective is sometimes dismissed as being too simple-minded and philosophical when humans are so clearly the dominant species on the planet but it is nonetheless compelling from both ethical and moral points of view. To me, it is arrogant in the extreme to deny life to another species simply because it does not serve us or cannot stay out of our way.

Another argument is that marsupials are aesthetically pleasing and should be conserved because of their intrinsic interest and the simple joy that they can bring to people. Children are awed by the sight of possums, gliders and kangaroos, whether the animals are up close in an animal park or glimpsed moving about in the wild. My own daughter, Alice, delighted as a child in extracting small mammals from traps that we had set in the field and in

Yellow-bellied Glider
Petaurus australis Female
ATHERTON TABLELAND, QUEENSLAND

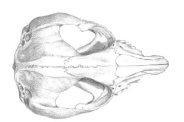

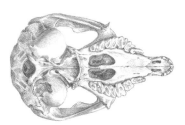

Central Hare-wallaby
Lagorchestes asomatus skull only
CENTRAL AUSTRALIA

TABLE 3 Recently extinct species and subspecies of Australian marsupials

Species	
Common Name	**Scientific Name**
Thylacine	*Thylacinus cynocephalus*
Desert Bandicoot	*Perameles eremiana*
Pig-footed Bandicoot	*Chaeropus ecaudatus*
Lesser Bilby	*Macrotis leucura*
Broad-faced Potoroo	*Potorous platyops*
Nullarbor Dwarf Bettong	*Bettongia pusilla*
Desert Rat-kangaroo	*Caloprymnus campestris*
Central Hare-wallaby	*Lagorchestes asomatus*
Eastern Hare-wallaby	*Lagorchestes leporides*
Crescent Nailtail Wallaby	*Onychogalea lunata*
Toolache Wallaby	*Macropus greyi*
Subspecies	
Common Name	**Scientific Name**
Western Barred Bandicoot (mainland subspecies)	*Perameles bougainville fasciata*
Tasmanian Bettong (mainland subspecies)	*Bettongia gaimardi gaimardi*
Burrowing Bettong (mainland subspecies)	*Bettongia lesueur graii*
Brush-tailed Bettong (SE Australian mainland)	*Bettongia penicillata penicillata*
Banded Hare-wallaby (mainland subspecies)	*Lagostrophus fasciatus albipilis*
Rufous Hare-wallaby (SW Australian mainland)	*Lagorchestes hirsutus hirsutus*

seeing how the pointy-nosed marsupials behaved compared to their placental counterparts. I remain transfixed at the sight of an antechinus spiralling rapidly up a tree after being disturbed, at kangaroos stirring languidly in the long grass as the sun sets and at possums squabbling in the trees outside the house. These are sights to impress even hard-nosed sceptics.

It is a short step from admiration to inspiration. Marsupials feature prominently in Aboriginal art and culture, as we have seen. And, despite their generally low profile in the culture of new Australians, marsupials still appear in enchanting stories for children such as *The magic pudding* by Norman Lindsay and *Possum magic* by Mem Fox, and in poems such as *Antechinus* by AD Hope and *Kangaroo* by DH Lawrence. They serve as faunal emblems for the Northern Territory and all mainland states except New South Wales (whose emblem is the Platypus, *Ornithorhynchus anatinus*). The extinct Thylacine (*Thylacinus cynocephalus*) features on the Tasmanian state coat of arms, while the equivalent symbol for Australia features a Red Kangaroo (*Macropus rufus*) and Emu (*Dromaius novaehollandiae*) supporting the national shield. Marsupials also serve as logos for high profile companies such as Qantas and Tasmania's Cascade Brewery, as well as for many small local businesses. If you wish to establish your Australian credentials, what more recognizable symbol could you choose than a marsupial?

Other arguments for conserving marsupials appeal to the most utilitarian among us because they highlight how we use the animals to improve our lives or make money. This approach, while selfish, places value on marsupials and turns them into resources that we are more likely to maintain. We have seen already that marsupials perform a remarkably broad and important range of 'ecosystem services' such as dispersing seeds and maintaining soil quality. If they are taken out of the system, we risk not just the loss of productivity

Pig-footed Bandicoot
Chaeropus ecaudatus Female
MUSGRAVE RANGES, NORTH-WESTERN SOUTH AUSTRALIA

and landscape function that now characterizes so much of Australia's semi-arid inland but a host of other far-reaching and irreversible effects that could preclude any attempts at restoration in the future. And we have seen that they provide excellent opportunities for us to learn about ecology, behavior, physiology, genetics, evolution and the development of the Australian environment. Because marsupials represent an alternative path in mammalian evolution, they also present unique insights for biomedical researchers. Currently, we are using marsupials to understand the development of the immune, reproductive and nervous systems, the expression and determination of sex in the young, mechanisms for the repair of DNA and for the processing of cholesterol and other dietary fats, how to cope with stress, and much more. We are extracting compounds from their milk to overcome bacteria that have become resistant to our standard antibiotics and sequencing the genetic code of selected species to probe the origins of genetic diseases. The potential use of marsupials in the biomedical field is enormous but greatly under-appreciated at present.

Still sceptical? Then let us consider the bottom line. Marsupials make money for us and, for an entrepreneurial few, the profits are great. The tourist industry, for example, relies in part on the attraction of marsupials to entice visitors Down Under. Surveys of foreign tourists leaving Brisbane and Sydney airports by Tor Hundloe and Clive Hamilton of the Australia Institute found that nearly a quarter had been influenced to visit by Australia's unique wildlife and two-thirds said that nature-based activities were important for their Australian experience. Most nominated the Koala (*Phascolarctos cinereus*) and kangaroos as their most popular creatures. Taking the 11% of visitors who said they would not have come were it not for the continent's unique wildlife, Hundloe and Hamilton estimated that wildlife tourism contributes a very hefty $1.8 billion a year to the Australian economy. About $1.1 billion (and some 9000 jobs) can be attributed to the Koala alone. This study was completed in 1997; the current economic benefit of wildlife tourism is probably much higher. If we add the returns gained from the harvesting of kangaroos (over 2 million animals a year, yielding leather and over 6 million kilograms (13 million pounds) of export meat), sales of possum skins (up to 250,000 a year), souvenirs, novelty items and related products, marsupials clearly add thousands of millions of dollars to the national economy on an annual basis.

These are well-trodden arguments but a final one has been articulated less often. This says that we cannot always know the value of a species – or any other resource – in the future. Extinction now may mean that a future gene, drug, food, ecological process or other benefit will never be realized. We can call this a missed opportunity cost. It is difficult to value because it is hard to reliably predict where research and understanding will take us in future, and which species will prove to be of most value. From a precautionary point of view we should therefore strive to maintain the species that currently persist. After all, extinction is forever.

Ningbing False Antechinus
Pseudantechinus ningbing Female
KIMBERLEY, WESTERN AUSTRALIA

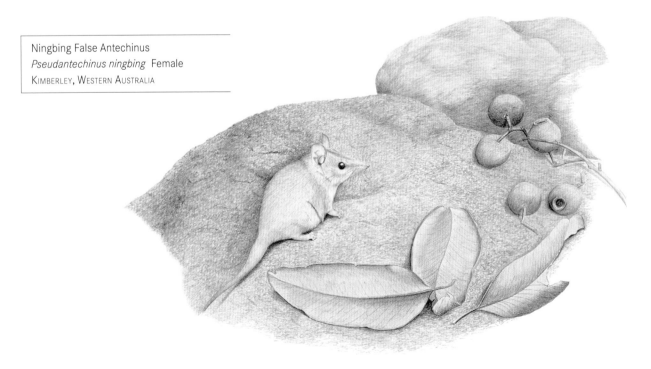

Assessing conservation status

The task of deciding which species are secure and which are at risk can be undertaken at several levels but the gold standard for assessment has been developed over many years by the World Conservation Union or IUCN. This organization specifies eight categories for pigeon-holing species, and also provides detailed and rigorous criteria that allow species to be objectively classified. Let us review each of the categories to see how marsupials are faring.

The first category, Extinct, is used when there is no reasonable doubt that the last individual of its kind has died. To remove reasonable doubt, investigators need to carry out exhaustive surveys in known or expected habitat throughout a species' original distribution and to do this at times when the species could be expected to be present. If they fail to find any living individuals, the species can be pronounced dead. Although the criteria for extinction are quite strict and unambiguous, species do occasionally resurrect themselves. In 2004, for example, the Ivory-billed Woodpecker (*Campephilus principalis*) turned up in the United States for the first time in 60 years. How this large and spectacular bird – known occasionally as the 'Lord God Bird' or 'Good God Bird' from the comments of astonished observers – managed for so long to evade the prying eyes of ornithologists is not known. The discovery prompted a media frenzy and great excitement among conservation biologists all over the world. However, such phoenix-like events are exceedingly rare and it is most reasonable to regard extinction as a terminal event that extinguishes a species and the evolutionary history that is carried in its genes, forever.

The Thylacine is a good example. Following a dedicated campaign to exterminate this distinctive and unique marsupial in Tasmania, the species withered and then vanished in the early part of the 20th century. The last known Thylacine died from exposure in pathetically inadequate conditions on the night of 7 September 1936, in a small cage at the Hobart Zoo. We do not need to relate the story leading to the loss of this once-proud species here. The whole disgraceful episode, including the culpable behavior of politicians, graziers, scientists and others at the time, has been laid bare already in the excellent book, *The last Tasmanian Tiger*, by Robert Paddle. You will need to be willing to suspend your disbelief as Paddle uncovers the extraordinary mismanagement that drove the largest marsupial carnivore to its death. Seventy years on from the fiasco at Hobart Zoo, we can be sure that the Thylacine qualifies as extinct under the IUCN criteria. There has been no confirmed sighting, specimen record or other solid evidence that the species persists, despite intensive searching for it by many researchers in habitats where Thylacines were once known to occur. Of course, the species is iconic and, like Elvis, is often reported to be alive and well. Eye witness accounts of Thylacines have come from all over Tasmania and from Western Australia and Victoria over the last few years but none has ever been substantiated.

Other extinct marsupials have commanded less attention than the Thylacine, partly because they are smaller and perceived as less charismatic but also because we know less about them. The Central Hare-wallaby (*Lagorchestes asomatus*) for example, is known to science only from a single skull that was collected from a carcase in 1932. The Crescent Nailtail Wallaby (*Onychogalea lunata*) used to be distributed widely across the inland and was well known to Aborigines and early European settlers. By the middle of the 20th century, before any systematic attempts to study it had been made, it had virtually disappeared. The Toolache Wallaby (*Macropus greyi*), often regarded as the most beautifully marked member of its family, crashed to extinction in western Victoria and south-eastern South Australia at about the same time as its smaller relatives. Although anecdotal sightings continued to be reported until the 1980s, hunting and habitat loss took a terrible toll on this species; the last known survivor died a lonely death in 1939 in an enclosure at Robe in South Australia.

The Desert Rat-kangaroo (*Caloprymnus campestris*) also vanished from sight in the first part of the 20th century, the last report being made in 1936. Virtually all that we know of this elegant little potoroid comes from an account in 1932 by Hedley Finlayson, perhaps the most intrepid naturalist to explore Australia's arid centre. Among other aspects of their biology, Finlayson was impressed that Desert Rat-kangaroos could hop fast enough to tire a galloping horse, and could maintain their speed, astonishingly, for 'up to 12 miles' (19 kilometres). Along with others, I have searched the arid gibber plains, alluvial flats and dry river courses of

western Queensland and northern South Australia where the species once occurred, and marvelled that this small surface-active marsupial could persist in such harsh conditions. Its loss, as with all the other vanished marsupials, has significantly impoverished the natural and cultural resources of Australia.

Sometimes when a species is sliding inexorably towards extinction in its natural habitat, it is possible to remove some of the last individuals and maintain them in captivity. Such species can be categorized as Extinct in the Wild. If the species breeds successfully in captivity it may be moved to a new area when numbers are sufficient in an attempt to re-establish wild populations. These will often be well away from the species' original range and even after a successful translocation to a new area, the

taxon is still regarded as being Extinct in the Wild. No species of marsupials are so listed but two subspecies occupy this category. The first is the South Australian Tammar Wallaby (*Macropus eugenii eugenii*). Once an abundant inhabitant of Eyre and Yorke peninsulas and at least four small islands off the South Australian coast, this little wallaby fell victim to land clearing, deliberate persecution from landholders, wildfires and probably predation from Red Foxes (*Vulpes vulpes*) in the late 19th and early 20th centuries. It soon disappeared from the wild. Before this happened and some time after 1861, a small consignment of South Australian Tammars was transported to New Zealand to form part of a menagerie maintained by Sir George Grey, who had been reappointed Governor of New Zealand. The wallabies

Toolache Wallaby *Macropus greyi* Male
SOUTH-EASTERN SOUTH AUSTRALIA

thrived on Grey's island residence, Kawau Island, near Auckland, where they are now regarded as a local pest. In 2003–2004, 82 wallabies were captured and returned to South Australia. Many remain in pens at Monarto Zoological Park but 46 wallabies were released into Innes National Park on the southern tip of Yorke Peninsula between 2004 and 2006 in a repatriation attempt. The success of the program is not yet known but its assessment forms the basis of an exciting PhD project by Leah Kemp at the University of Adelaide. Two other subspecies of Tammar Wallaby also occupy restricted ranges but neither *Macropus eugenii decres* on Kangaroo Island nor *Macropus eugenii derbianus* in Western Australia is at risk.

The second marsupial that is Extinct in the Wild is the unnamed central mainland subspecies of the Rufous Hare-wallaby or Mala (*Lagorchestes hirsutus*). Once distributed over about a quarter of the continent, this attractive but tiny wallaby succumbed to changes to its habitat following European settlement, wildfires, and predation from feral Cats (*Felis catus*)

Rufous Hare-wallaby
Lagorchestes hirsutus Female
TANAMI DESERT, NORTHERN TERRITORY

and Red Foxes. The last known wild population in the Tanami Desert was wiped out by fire in late 1991. Acting with considerable foresight, scientists at the Northern Territory Parks and Wildlife Commission removed five animals from the Tanami in 1980 with the intention of breeding them in facilities at Alice Springs. Small numbers of animals were brought in to supplement the captive population over the next 6 years, allowing the colony to thrive. Animals from this colony have since been used to stock a small, predator-free area near Willowra in the Tanami (the 'Mala Paddock') and to start a further semi-captive colony in predator-proof enclosures at Watarrka National Park some 450 kilometres (280 miles) south-west of Alice Springs. In 1998 animals from the Mala Paddock were used in turn to found a new wild population on Trimouille Island off the coast of north-western Australia; and in 2005, wallabies from Watarrka were shipped to Uluru–Kata Tjuta National Park to found a new colony there. The Trimouille program is particularly interesting as this island was used for nuclear weapons testing by the British in the 1950s. Despite some residual radiation, the Rufous Hare-wallabies there have thrived.

The next three categories defined by the IUCN are Critically Endangered, Endangered and Vulnerable. For convenience, these are often grouped under the general term, Threatened. The worst category for a living species is Critically Endangered. The risk of extinction in the wild for such species is extremely high. As defined by the IUCN, this category includes species with: populations that are declining at a rate of 80% in just 10 years or three generations; geographical distributions covering less than 100 square kilometres (38.6 square miles); small populations containing fewer than 250 mature individuals; or a 50% probability of extinction in the wild within 10 years or three generations, as predicted in quantitative analyses. To reliably predict the likelihood of extinction it is necessary to know a great deal about a population's size, the age at which individuals reach sexual maturity, their reproductive output and survival rates at different ages, and the degree of variation that can be expected in these parameters between years. Using this information, population models can then be constructed to predict a species' viability for 10 years or more into the future. Other criteria are used also, such as whether the species is known to fluctuate or to aggregate in a single place, and whether the species faces known threats.

At present, two Australian marsupials fall within this most worrying of categories. The first is Gilbert's Potoroo (*Potorous gilbertii*). Discovered originally in the 1840s near Albany in Western Australia, this small and cryptic species remained unseen for most of the 20th century and was thought to be extinct until it reappeared in 1994. It is known from just a single site, on the headland of Mount Gardner in Two Peoples Bay Nature Reserve, close to the site of its original discovery. The population probably numbers about 30 animals and occupies a tiny area of less than 5 square kilometres (1235 acres). This site is isolated, at risk of being burnt, susceptible to invasion by predators such as Red Foxes and Cats, and suffering die-back of vegetation because of the root-rot fungus, *Phytophthora cinnamomi*, which has become established there. In such a small population, with a limited choice of mates, breeding between close relatives can be expected. As if this catalogue of woes was not enough, few young are recruited into the adult population so the opportunity for growth is currently limited. For these reasons, Gilbert's Potoroo is the most Critically Endangered species of marsupial in Australia or, indeed, anywhere else.

The second marsupial that is judged to be in critical danger is the Northern Hairy-nosed Wombat (*Lasiorhinus krefftii*). This large herbivore has been recorded from just 3–4 disparate localities in eastern Australia, from Deniliquin in southern New South Wales to Epping Forest in central Queensland. It may have been declining at the time of European settlement. It is now found only at Epping Forest, where it occupies an area of less than 10 square kilometres (2471 acres). The population was estimated at just 67 animals in 1985 and appears to be on an upward trajectory; in 2003 the population estimate was 113. In 2000-2001 about 10% of the population was killed by Dingoes (*Canis lupus dingo*) but the subsequent erection of a dog fence should alleviate this threat in future. Despite the upward trend in numbers, the most recent surveys suggest that there may be as few as 25 breeding females in the population, with perhaps 10 more that are not breeding. These small numbers and the geographical confinement of the population are causes for grave concern and readily justify this wombat as being classified as Critically Endangered.

The next most extreme IUCN category, Endangered, refers to species with a very high risk of extinction in the wild. These are species with: populations declining at a

Gilbert's Potoroo *Potorous gilbertii*
Female (top) and juvenile (bottom)
Near Albany, Western Australia

rate of 50% in 10 years or three generations; geographical distributions covering less than 5000 square kilometres (1930 square miles) populations with fewer than 2500 mature individuals; or a 20% probability of extinction in the wild within 20 years or five generations. Fifteen species are listed in this category in Australia and another eight in the New Guinea region.

One species that exemplifies the Endangered category is the Western Barred Bandicoot (*Perameles bougainville*). This diminutive bandicoot was probably found over much of southern Australia, exploiting semi-arid habitats from Shark Bay, across the Nullarbor, to central New South Wales. Living animals are now known only from Bernier and Dorre islands in Shark Bay, and from a population that was reintroduced in 1995 to Heirisson Prong on the adjacent mainland. The island populations are not declining. However, being confined to two small islands and to a mainland site that requires continual vigilance to control the scourge of Red Foxes and Cats, the Western Barred Bandicoot remains at great risk and meets the criteria for being endangered. In particular, it is highly susceptible to catastrophic events such as wildfire, introduction of disease or release of predators onto the islands, which could all drive the populations to low numbers very quickly. Examples of other endangered species are the Bridled Nailtail Wallaby (*Onychogalea fraenata*), Leadbeater's Possum (*Gymnobelideus leadbeateri*), and the Northern and Southern Marsupial

Moles (*Notoryctes caurinus* and *N. typhlops*, respectively). Two of the most beautiful small dasyurids are also endangered, the Dibbler (*Parantechinus apicalis*) and Red-tailed Phascogale (*Phascogale calura*).

The third category for threatened species is Vulnerable. Marsupials on this list face a high risk of extinction in the wild and have: populations declining at a rate of 30% in 10 years or three generations; geographical distributions covering less than 20,000 square kilometres (7720 square miles) populations with fewer than 10,000 mature individuals; or a 10% probability of extinction within 100 years. More marsupials are included in this formal threat category than in any other: 17 species in Australia, with another 14 in the New Guinea region.

A good example of a Vulnerable species is the Kowari (*Dasyuroides byrnei*). Confined to the harsh and remote gibber deserts of western Queensland and northern South Australia, this tenacious little predator would appear to have few enemies. However, changes in land use and, probably, the establishment of feral Cats in Australia's arid regions have seen its distribution decline. Another current threat comes from, ironically, the Great Artesian Basin Sustainability Initiative established by the state and federal governments in 1999–2000. This initiative aims to allow efficient use of water from the Great Artesian Basin for the pastoral industry. It provides funds for piping water into dry areas that are currently

Northern Hairy-nosed Wombat
Lasiorhinus krefftii Male
EPPING FOREST, QUEENSLAND

on the mainland and known from less than half a dozen specimens collected near Kalumburu in the northern Kimberley of Western Australia. With so few records, why are these species considered to be Vulnerable and not Endangered or Critically Endangered? Both species occupy islands offshore and, although their populations are probably small, they are not subject to obvious threats that could drive numbers to lower levels.

A further IUCN category, Near Threatened, allows for placement of species that do not meet the minimal requirements for listing as Vulnerable but which none-theless exhibit trends in population size, distribution or other indicators of status that warrant concern. This category contains an extraordinary 26 species and a further 15 subspecies from Australia. The most positive category, that of Least Concern, contains just over half (53%) of Australia's marsupials – species with large populations such as some of the forest-dwelling dasyu-rids, species with broad geographical distributions such as some of the possums and gliders, and also the large kangaroos that have benefitted from the broadscale conversion of forest to pasture. It also contains several timid and elusive species such as the Long-tailed Pygmy-possum (*Cercartetus caudatus*) and the Black Wallaroo (*Macropus bernardus*). Although little researched and seldom seen, these marsupials occupy protected habitats that secure their populations from any obvious external threats.

In addition to the delineated categories of risk, a few marsupials are considered too poorly known to assign their status with any reliability, and these are grouped within the final IUCN category of Data Deficient. Just three Australian marsupials have this label but in the New Guinea region, 25 species are considered to be Data Deficient. Much of the New Guinean fauna is genuinely poorly known, at least to western science, because of difficulties of access and limited opportunities for field survey. For example, the ringtail possums *Pseudochirulus caroli*, *P. canescens* and *P. schlegeli* are known from a handful of specimens collected in a small number of localities. There is not enough information on population size, distribution or threats to be confident about whether they are secure or at risk. Similarly, little can be said of

the preserve of the Kowari. As cattle move into the newly watered areas, accompanied inevitably by feral Cats and Red Foxes, the quality of habitat for the Kowari will decline and drive the population to still lower levels.

Another Vulnerable species likely to be affected by initia-tives to water the desert is the Bilby (*Macrotis lagotis*). This iconic and beautiful marsupial persists in suitable patches of habitat in the Northern Territory and Western Australia but also co-occurs with the Kowari in the Channel Country of Queensland. It fares poorly in the presence of cattle and is also susceptible to predation from introduced mammals.

While the Kowari and Bilby are reasonably well known, other Vulnerable species are not. The Carpentarian False Antechinus (*Pseudantechinus mimulus*), for example, is exceedingly rare on the Australian mainland and in the last 100 years has been recorded only in the vicinity of Mount Isa in north-western Queensland. Butler's Dunnart (*Sminthopsis butleri*) is also vanishingly scarce the Seram Bandicoot (*Rhynchomeles prattorum*), a distinctive but enigmatic species collected in 1920 in rugged rainforest on Mount Manusela, Seram. Several surveys in the second half of the 20th century did not record it but it is not clear whether suitable habitats were explored or whether the species has really disappeared.

In Australia, one Data Deficient species is the Kultarr (*Antechinomys laniger*). Although sometimes portrayed as a hopping animal on account of its long hind-legs, the Kultarr is in fact an agile and highly maneuverable quadruped that bounds from its hind-legs to its fore-legs when running. It has a very wide distribution in arid and semi-arid Australia but has disappeared from Victoria and has declined, apparently, in all other mainland states. It is probably at risk from habitat degradation from overgrazing, and from predation by Red Foxes and Cats and, in some areas, flooding of habitat. Like most small dasyurids that occupy arid habitats, Kultarrs do not dig their own burrows but either usurp the burrows of digging animals or shelter under debris on the ground surface. They may thus be unusually susceptible to drowning after floods or heavy rain. Populations are reported to have crashed after heavy rains in one

study in western Queensland in the late 1960s and to have disappeared after local flooding from Kinchega National Park and the Nyngan region in New South Wales in 1989–1990.

Nevertheless, the Kultarr is recorded sporadically in surveys in apparently suitable vegetation and often vanishes from, or appears in, local areas for no apparent reason. This inconsistency in finding Kultarrs makes it difficult to identify trends in population size and distribution, and to be precise about what factors affect them. Since 1986 I have trapped intensively using pitfall traps in what might be suitable habitats for Kultarrs in New South Wales, Queensland and Western Australia but have captured this elusive species only once – at Boorabbin National Park, east of Perth, in the goldfields region near Coolgardie. The sandplain heath at this site looked similar to heath at many other sites that I visited and provided no obvious clue as to why I should find Kultarrs there and nowhere else. On occasion, I have also spotlighted Kultarrs moving nimbly at night on the developmental road that connects Windorah to Bedourie in western Queensland but still cannot predict the conditions when they will reveal themselves. This is a truly enigmatic beast that has earned its Data Deficient label.

Australia's two other Data Deficient species are the Chestnut Dunnart (*Sminthopsis archeri*) and White-footed Dunnart (*Sminthopsis leucopus*). The latter species has a patchy distribution in Tasmania, Victoria and southern New South Wales, with a single and apparently small outlying population in north Queensland. It has been recorded in a wide range of habitats and, although apparently terrestrial, has been

Long-tailed Pygmy-possum *Cercartetus caudatus* Male
ATHERTON TABLELAND, QUEENSLAND

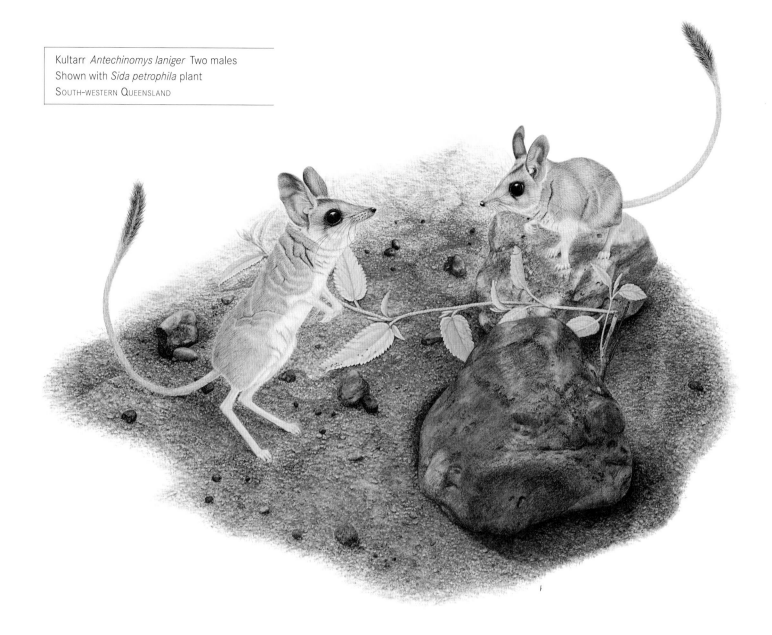

Kultarr *Antechinomys laniger* Two males
Shown with *Sida petrophila* plant
SOUTH-WESTERN QUEENSLAND

discovered in Tasmania to nest occasionally in trees. Like the Kultarr, this small and cryptic species turns up erratically in field surveys and threats to its survival are not known. It tolerates logging and burning of its habitat but appears to decline when forest regrowth becomes too dense. Studies of its biology are complicated by uncertainty about how best to catch it and by the possibility that White-footed Dunnarts are overlooked in the field because of their similarity in appearance to the Common Dunnart (*Sminthopsis murina*).

The Chestnut Dunnart is even less well known than its white-footed relative. Although the first specimens were collected in 1898, it was not recognized as a species distinct from other dunnarts until 1986. Even now, it remains known from just over a dozen specimens collected from New Guinea, one from west of Townsville and another seven from Cape York Peninsula. The Data

Deficient category was designed for species such as this! Following recent assessments, it is likely that both species of marsupial moles will join the select set of Data Deficient marsupials, whereas the White-footed Dunnart may be transferred to Vulnerable.

The IUCN assessments show that many marsupials have declined in abundance and distribution, with almost half being assessed as either Near Threatened or Threatened. This is quite a grim picture. Can we be confident that the assessments are reliable? The IUCN is an independent organization with a large international membership of professional scientists, fauna and flora experts, and other specialists. The criteria that it sets for placing species within the various status categories are set using input from the international scientific community, and are in fact the product of much research, debate and revision by that community. To assess status, the IUCN appoints experts who have personal experience with particular species or groups of species. There may be anywhere from just two or three people in a specialist group to several dozen, and it is the task of these groups to keep the status of 'their' species under review. In Australia, the Marsupial and Monotreme Specialist Group has comprised 50–65 members at different times, with other specialists being co-opted to the group as needed. Large, formal meetings to review the status of all Australian marsupials took place in 1995 and again in 2005, and were venues for robust discussion and exchange of data. On occasion there was joy if a species could be downgraded from a high risk category to a lower risk one, despair if another species had continued to slide. I attended both meetings and came away exhausted, but also enlightened and enriched, by what I had learnt. In addition to the formal meetings, application can be made to change the status of any marsupial at any time when new information has been obtained to justify it. Overall, this seems to be a very effective structure for evaluating the status of marsupials and other species, and for this reason we can confidently take the IUCN evaluations to be the best available.

Threats to marsupials

Why have marsupials fared so poorly in the last 220 years? We saw in the Cultural History chapter that marsupials were targeted for deliberate destruction in many areas but also noted that the zeal for outright extermination has been tempered in recent decades by increasing appreciation for our native fauna. Indeed, rather than succumbing to direct human persecution, there is much evidence that marsupials have been the unintended victims of a suite of changes imposed on the Australian environment by Europeans and later settlers. Many of these changes were wrought within the first 100 years of settlement and at about the same time, making their relative contributions difficult to disentangle. Nonetheless, we can list some of the most damaging changes and speculate on how they may have affected different groups of marsupials. For some changes, such as those caused by the pastoral industry

White-footed Dunnart
Sminthopsis leucopus Female
SOUTH-WESTERN VICTORIA

and the arrival of the Red Fox, we can be sure that the magnitude of impact was – and continues to be – large. For others, such as the widespread use of poisons and pesticides, or the prevalence of disease, there is much less certainty. To focus our review of the changes that took place and examine their impact on marsupials, it is particularly instructive to review the settlement history of New South Wales. The development of this state has been well studied and patterns of change in its marsupial fauna are reasonably well documented through the efforts of several indefatigable naturalists, animal collectors and explorers.

The bare facts are that New South Wales occupies a vast area of 809,500 square kilometres (312,500 square miles), contains a broad range of coastal, arid and alpine habitats, and encompasses the greatest topographic range in Australia. European settlement began at Sydney Cove in 1788 but expanded rapidly along the coast and then to the inland plains after 1813 when routes were discovered over the Blue Mountains. Once home to 65 species of marsupials, the state boundaries now surround only 48 species, 23 of which are listed as threatened under state legislation. The historical record shows that marsupials disappeared earliest from the state's far west, hung on for longer in the central plains and fared best in the coastal forests. The scale of loss follows a similar trend: 16 of the 39 species of marsupials originally in the far west have gone, as have 10 of the 51 species that once occupied the Great Dividing Range and western slopes, and three of the 36 species from coastal localities. What processes introduced by European settlers could account for these patterns? And do marsupials face similar threats now?

Seven of the 16 marsupials that disappeared from western New South Wales were last recorded there in 1857 or earlier, less than 70 years after first settlement. Despite the influx of pastoralists to the region throughout the 1840s, it seems unlikely that their activities or the arrival of sheep and cattle could have caused such a rapid cascade of extinctions. However, there is some evidence to implicate another imported species in the early extinctions: the feral Cat. Cats were brought to Australia by the first settlers, who valued their ability to control rats and mice. Except for the Crescent Nailtail Wallaby, which weighed around 3.5 kilograms (7.7 pounds), all six other marsupials that made their last appearance by 1857 averaged less than 1.5 kilograms (3.3 pounds), a size that could be killed readily by Cats. In addition to the marsupials, eight species of native rodents also were last recorded in western New South Wales in 1857 or earlier; all weighed 200 grams (7 ounces) or less. Some of these species would have been conspicuous to prowling cats because of their bounding or hopping movements; and their use of open habitats would have also made them easy to see and hunt.

Some authors have speculated that feral Cats exerted their effects not solely by predation but also perhaps by competition or transmission of disease too. Competition is usually most intense between ecologically similar species because they have similar requirements for food, shelter and other resources. Western Quolls (*Dasyurus geoffroii*) would thus have been most likely to compete with Cats owing to their meat-based diet. However, there is little to suggest that quolls were out-competed or even that food was in short supply in the presence of Cats, so the potential importance of competition must remain an open question.

Evidence for the importance of disease is also equivocal, with most debate focusing on the possible effects of a Cat-vectored pathogen called *Toxoplasma gondii*. This parasitic organism can only complete the reproductive phase of its life cycle in the cat, so it must have been absent from Australia before felines were introduced. Marsupials, other mammals and birds can be infected by the parasite and may show symptoms such as lethargy, poor coordination, blindness and even death. However, it is not clear that the disease could have caused the mass extinction event observed in the early 1800s. Tests on free-living marsupials show that many species carry antibodies to *T. gondii* and have thus been exposed to the parasite but they show no clinical symptoms of disease. Toxoplasmosis is also more prevalent in cool temperate than in arid areas, making the hot, dry west of New South Wales an unlikely place for the disease to gain hold. Despite these arguments, it would still be premature to dismiss epidemic disease as an extinction process for marsupials. A detailed recent study by Ian Abbott of Western Australia's Department of Environment and Conservation identified an unspecified disease as the cause of a dramatic collapse in that state's mammals between 1875 and 1925. And presently, numbers of the Tasmanian Devil (*Sarcophilus harrisii*) are crashing quickly to alarming levels due to the ravages of a disease that causes animals to develop crippling facial tumors.

The arguments above suggest that, if cats caused early extinctions of marsupials, they did so more by predation than by other processes. Before leaving this topic, we must note an important incongruity in the historical record: Cats were very seldom mentioned in the diaries and journals of early European explorers. Was this because Cats were so common as to warrant no mention in diary entries or because they were present but too scarce to be noticed? The most thorough research, again by Ian Abbott in Western Australia, has confirmed that Cats were encountered very rarely in remote areas and probably did not in fact move far from settlement until the late 19th century. Thus if Cats were responsible for early losses of marsupials in western New South Wales, it may be that they were widely ranging strays that maintained some association with humans. Whatever the correct interpretation, there is little doubt that feral Cats can wreak havoc on naïve native prey. They have been shown to deplete the bird and mammal faunas of small islands soon after their introduction, and to depress the survival of small and medium-sized marsupials in reintroduction programs. If Cats are experimentally removed, populations of small marsupials can be expected to increase several fold.

The period between 1857 and 1880 was one of consolidation and expanded settlement in western New South Wales, with few records or reports being made of marsupials. However, this quiescence was shattered over the next 20 years, with startling and broadscale changes being made to the landscape that triggered a holocaust for the marsupials and other native fauna. The scale and magnitude of damage to the environment over these two decades remains unprecedented in Australian history. The changes, initiated and driven largely by the pastoral industry, saw the rapid demise of some nine species of marsupials from the region, including the Bilby, Numbat (*Myrmecobius fasciatus*), and five species of bettongs and wallabies. To some degree the writing had been on the wall for native species for many years. Although pastoralism did not initially have extensive effects in the far west of the state, the impacts of the industry were nonetheless dramatic in some localities. This is clear from the visionary writings of Gerard Krefft, who spent nine months collecting and observing native vertebrates in the Murray–Darling junction region in 1857–1858. In attempting to capture specimens of the Pig-footed Bandicoot (*Chaeropus ecaudatus*), Krefft

commented that 'The large flocks of sheep and herds of cattle occupying the country will soon disperse those individuals which are still to be found in the so-called settled districts, and it will become more and more difficult to procure specimens for our national collection'. He wrote elsewhere how the low lands of the Murray River 'swarmed' with cattle and sheep.

From the 1840s, increasing numbers of pioneers and their stock moved to western New South Wales, lured by visions of expansive green pastures on the banks and floodplains of the major rivers. The introduction of wire fences and new technology for sinking bores and wells in the 1860s allowed waterless areas of the hinterlands to be opened up, so that by 1878 most of western New South Wales was occupied by stock. Clearing of trees intensified so that more pasture could be grown, especially in the 20 years before 1900, and marsupials were actively persecuted to reduce their supposed competition with the new herbivores. A long run of good seasons saw stock numbers increase up to the turn of the century, when the sheep population in the state's far west exceeded 10 million. As if these changes were not enough for the beleaguered native fauna to cope with, populations of the European Rabbit (*Oryctolagus cuniculus*) peaked in western New South Wales between 1885 and 1892, taking advantage of the improved pastures; the Red Fox crossed the Murray River in 1893; snares, traps and poison baits were set to control the numbers of these and other pests; and the fire regimes that had been imposed by Aboriginal people were eliminated.

Let us look at the effects of these calamities. In the first instance the clearing of trees would have removed the habitats of arboreal marsupials. Clearing was concentrated on the eastern and southern parts of the western region in the 1880s and 1890s, and this led to declines of Koalas, Common Brushtail Possums (*Trichosurus vulpecula*) and Common Ringtail Possums (*Pseudocheirus peregrinus*). The Red-tailed Phascogale was probably an early victim of land clearing; smaller species such as Feathertail Gliders (*Acrobates pygmaeus*) would have also been affected but records are sparse. The removal of trees causes other, more subtle, effects on the environment that may take years to become obvious. These include the depletion of organic material

Black Wallaroo *Macropus bernardus* Female
ARNHEM LAND, NORTHERN TERRITORY

in the soil, increased soil hardness and water runoff, and consequent erosion, and increased salt concentration at the soil surface.

Land clearing is still a serious problem in some Australian states, and unfortunately one that is officially sanctioned. In 2006 the New South Wales Auditor-General confirmed that 639,930 hectares (1.6 million acres) had been approved for clearing by the state government over the previous eight years. A subsequent report commissioned by the World Wide Fund for Nature (WWF), to which I was able to contribute, estimated that this would kill almost 600,000 marsupials a year, including some 4000 Koalas and 314,000 possums and gliders. Between 1997 and 1999 the government of Queensland allowed 446 square kilometres (172 square miles) to be cleared annually. Another WWF report in 2003 estimated that this extraordinary orgy of destruction killed about 100 million native vertebrates each year, with 1.87 million marsupials contributing to this stunning and disgraceful total. Clearing still proceeds in parts of the Northern Territory, Victoria and Tasmania, although not at the levels that are allowed in Queensland and New South Wales.

Pastoralism is the major land use of western New South Wales and much of inland Australia, and was historically the main cause of land clearing in these regions. Sheep and cattle have three potentially deleterious effects on marsupials. In the first instance, they remove long grass and shrubs, and thus deplete the amount of available shelter. This would have left cover-dependent marsupials exposed to predators and extremes of weather. Perhaps not coincidentally, the Brush-tailed Bettong (*Bettongia penicillata*), Northern Bettong (*B. tropica*), Eastern Hare-wallaby (*Lagorchestes leporides*) and Bridled Nailtail Wallaby (*Onychogalea fraenata*) made their last appearance in western New South Wales when sheep numbers were at their peak; all shelter (or sheltered, in the case of the now-extinct Eastern Hare-wallaby) in scrapes under vegetation. Secondly, livestock eat grass and forbs, and would thus have competed with grazers such as the Bridled and Crescent nailtail wallabies during tough times such as drought. Finally, by removing surface vegetation and powdering the soil surface with their hard hooves, livestock would have damaged burrows and increased local erosion of soil, in places down to the underlying bedrock. Conspicuous burrowing

species such as the Northern Hairy-nosed Wombat and Burrowing Bettong (*Bettongia lesueur*) disappeared when sheep flocks reached their greatest extent.

The adverse effects of overgrazing can be deduced by differences in how long marsupials persisted in different regions. For example, the Bilby, Bridled Nailtail Wallaby, Brush-tailed Bettong and Northern Hairy-nosed Wombat all disappeared during the pastoral holocaust of the wild west but persisted until the early years of the 20th century in more easterly districts where stocking was less intense. All species survive today in regions without stock. Despite these observations, it is not clear that livestock are inevitably associated with marsupial declines. The industry appears to have been damaging in Queensland and South Australia, but in Western Australia marsupials and other native mammals have declined in a similar manner in both pastoral and non-pastoral areas. Other factors for their declines must therefore be invoked.

The European Rabbit is a problem for agricultural production and perhaps for native species over much of the southern two-thirds of Australia. Arriving in western New South Wales in 1879, the Rabbit probably competed with native species for food and would have acted in concert with livestock to destroy the cover of native vegetation. As its populations erupted during the 1880s and 1890s, and food became scarce, local newspapers carried reports of starving Rabbits defoliating shrubs and even killing mature trees by stripping their bark. Anecdotal reports also suggested that Rabbits usurped the burrows of Bilbies and Burrowing Bettongs. Certainly, these species and several other marsupials that were potentially susceptible to competition, disappeared from western New South Wales during the Rabbit plagues. Recent studies have failed to confirm that Rabbits are strong competitors but it remains possible that they exert their effects only when at high density. With the introduction of myxoma virus in 1950 and calicivirus in 1995, Rabbit populations no longer reach the astonishing numbers that overwhelmed observers in the late 19th and first half of the 20th centuries. Although unwelcome pests, Rabbits now occur mostly in highly modified agricultural areas and their direct effects on marsupials are probably limited.

Another beneficiary of pastoral expansion is the Red Fox. I have to take a deep breath when writing about this species. It is, on the one hand, a beautiful, smart, adaptable and ecologically successful animal. It has the largest geographical range outside Australia (175 million square kilometres; 68 million square miles) of any of the 23 species of mammals that have established in the great south land. It is, on the other hand, a highly efficient and flexible predator, and an enormous threat to the persistence of many marsupials and other native species. It arrived and spread in western New South Wales in the wake of the Rabbit and thus was probably too late to be the main executioner of the marsupials that disappeared. But it now poses a major threat to many that remain, especially small and medium-sized species weighing more than 35 grams (1.2 ounces). The Red Fox covers most of Australia except the northern arid and tropical regions, and penetrates the central deserts when conditions permit. In what I believe is one of the worst acts of bio-terrorism perpetrated in Australia, the Red Fox was deliberately imported to Tasmania in 1999 or 2000. Despite investigations, we still do not know who the offender was or why this crime was committed.

Why is the Red Fox such a threat? As the species spread from points of release in Victoria in 1871, its appearance in any new locality coincided with the demise – often in the same year – of marsupials, birds and other native species. Letters written to newspapers at the time in South Australia and Western Australia lamented these losses and attributed them to the Red Fox, as did articles written in naturalists' magazines and scientific journals. Writing perceptively in his *Mammals of South Australia* in 1925, the great zoologist Frederic Wood Jones considered the Red Fox to be '… the most baneful disturbing influence brought about by the human folly of introducing animals into a new country'. Its attacks on poultry and livestock brought the Red Fox formal pest status in Victoria only 23 years after its release; it was declared a noxious animal in southern parts of New South Wales in 1903. Foxes are still making slow advances into parts of the Kimberley region of north-western Australia and Cape York Peninsula, and leaving diminished communities of native mammals as they invade.

While working on a confined population of the Southern Brown Bandicoot (*Isoodon obesulus*) near Perth in the mid-1980s, I was able to witness the breathtaking efficiency of fox predation at first hand. A single Red Fox breached the exterior fence of the study area and announced its presence by killing bandicoots and strewing their carcases on the ground. Some were eaten, especially small females in open habitats, but many were killed and simply left. This ostensibly odd behavior is expressed when Foxes find themselves amid abundant prey and are unable to turn off the kill switch. Called 'surplus killing', this destructive behavior is seen commonly when Foxes encounter poultry, young livestock or other prey in confined circumstances. At the end of 6 months the rogue Fox had killed about 90% of the bandicoots I had so laboriously captured and ear-tagged, despite my best efforts to shoot it, catch it and poison it. I suspect it died of gluttony or old age. Pedro Borges, a PhD student at the University of Sydney, had a similar experience of a single Fox wiping out a confined population of Long-nosed Bandicoots (*Perameles nasuta*) on the central coast of New South Wales.

Devastating but anecdotal accounts of this kind abound but there is now a lot of well-documented experimental evidence that quantifies the destructive effects of Red Fox predation. The first such study, by Jack Kinnear and colleagues of the Western Australian Department of Conservation and Land Management (now Environment and Conservation), experimentally removed Red Foxes from outcrops with small populations of the Black-footed Rock-wallaby (*Petrogale lateralis*) and found that wallaby numbers increased by 4–5 fold after several years. Control sites, where Foxes were not suppressed, showed no increases. Similar removal experiments have documented increases in a broad range of other medium-sized marsupials, including Southern Brown Bandicoots, Common Brushtail Possums, Brush-tailed Bettongs, Long-nosed Potoroos (*Potorous tridactylus*) and Rothschild's Rock-wallaby (*Petrogale rothschildi*). Direct observations suggest that animals as large as Whiptail Wallabies (*Macropus parryi*) and even Eastern Grey Kangaroos (*Macropus giganteus*) can be affected by Foxes, probably through losses of newly independent young. Small marsupials do not seem to be consistently oppressed by foxes but often avoid areas bearing signs of Foxes, such as their droppings.

The most dramatic impacts of Foxes are caused by direct predation but competition or transmission of disease may cause more subtle effects. Foxes, like feral Cats, probably compete with quolls for food and den sites, and carry roundworms, tapeworms, mites and other parasites that cause disease in many species of marsupials. A particularly nasty example is the sarcoptic mite, *Sarcoptes scabei*. In the Common Wombat (*Vombatus ursinus*) this mite causes severe itching, excoriation and inflammation of the skin; if it progresses, it causes an excruciating death. Even more insidious are so-called indirect effects from the presence of other species. For example, in areas where European Rabbits are common they support dense populations of Foxes that can, in turn, have very damaging effects on scarce native species. Foxes also eat and disperse the seeds of exotic and invasive plant species such as Blackberry (*Rubus fruticosus*), Bitou Bush (*Chrysanthemoides monilifera*) and Olive (*Olea europea*), which replace native vegetation and reduce the suitability of habitat for marsupials and other species. All up, the Red Fox is a serious problem in its adopted environment in Australia.

While Cats, then Rabbits and Foxes, and were cutting a swathe through the marsupials of western New South Wales, other introduced species were probably exerting additional, if more limited, effects. These include the House Mouse (*Mus domesticus*), Brown Hare (*Lepus capensis*), Pig (*Sus scrofa*) and Goat (*Capra hircus*). Horses (*Equus caballus*) were also present in large numbers, usually as part of the pastoral enterprise but also as feral populations in many areas. The effects of these species are difficult

Long-nosed Bandicoot *Perameles nasuta*
Female with young aged
approximately 2 months
Atherton Tableland, Queensland

to deduce. However, Brown Hares increased with great rapidity on the improved pastures that were expanding in the second half of the 19th century and were considered such a problem for agriculture that thousands could be shot on a property in a single day. In some districts hundreds of thousands of bounties were paid annually on Hares. Goats too became a problem, with feral herds competing for herbage and being helped through tough times in the rangelands by the water provided for livestock. Historical records show that Brush-tailed Rock-wallabies (*Petrogale penicillata*) declined sharply in some areas on the western slopes of the Great Dividing Range around the turn of the 20th century in the presence of Goats, and the Yellow-footed Rock-wallaby (*Petrogale xanthopus*) currently is under threat from this species and from the Red Fox in the far north-west of New South Wales. Pigs and mice are very damaging to the livestock and wheat industries, respectively, but their effects on marsupials have not been studied.

With so many introduced mammals establishing themselves as serious pests to producers, counter measures were soon forthcoming. We have seen already that the large kangaroos and other marsupials were killed in their millions from the 1880s until the early part of the 20th century as part of extensive (and expensive) bounty schemes. Bounties were also placed on the heads of Rabbits, Hares, Foxes and wild dogs. In 1883 the New South Wales government passed the *Rabbit Nuisance Act*, and this allowed for capitation fees to be paid on scalps. The gin traps, other steel jaw devices, snares and nets that were set for Rabbits inevitably captured many native mammals. After seven years the escalating costs of capitation fees led the *Rabbit Nuisance Act* to be repealed, and it was replaced in 1890 with the *Rabbit Act*. This gave responsibility for Rabbit control to landholders. In response, trapping abated, but strychnine, arsenic and phosphorus were laid in baits or used to poison water supplies over large areas of pastoral land. Eric Rolls documented the art of the poisoner in his 1969 classic *They all ran wild*, and also described the horrific toll that indiscriminate poisoning took on native birds, bandicoots, rat-kangaroos and other marsupials. Broad-scale poisoning was rampant until the middle of the 20th century.

At the same time that pasture pests were being poisoned, the Dingo (*Canis lupus dingo*) and other wild dogs were being targeted to reduce their direct attacks on livestock. The Dingo stirs very deep passions in the bush. Although there is sometimes grudging respect for this wily predator, Dingoes are universally hated wherever livestock are run and have been subjected to every conceivable method of control. The first legislation aimed at dog control was enacted in 1830 to reduce the numbers of strays in settled areas near Sydney, and expanded to include Dingoes specifically in 1852 in *An Act to Facilitate and Encourage the Destruction of Native Dogs*. Eric Rolls again provides a graphic account of the time, resources and ingenuity expended on Dingo control – including the celebrated dog-fence – and hints also at the terrible toll exacted on marsupials that were caught innocently in the control effort. Dingoes are still subject to rigorous culling and it is not uncommon to see Dingo scalps or entire carcases hanging from fences and sign posts in pastoral areas. In New South Wales, wild dogs generally are a declared pest and must be continuously suppressed and destroyed on public land unless an exemption is negotiated.

The dog-fence runs for about 5000 kilometres (3100 miles) from north-eastern Queensland to the Southern Ocean in South Australia, with some 585 kilometres (363 miles) running along the border of New South Wales with both South Australia and Queensland. It is checked regularly by boundary riders and keeps the sheep rangelands substantially free of Dingoes and other wild dogs. This releases livestock from predation and also allows populations of kangaroos and Emus to achieve dramatically higher densities inside than outside the fence. Paradoxically, Dingo control may not be good for smaller native mammals. Accumulating evidence suggests that Dingoes suppress the numbers of Cats and, especially, Red Foxes; if so, they are likely to have net beneficial effects on small and medium-sized marsupials that would otherwise succumb to the impacts of the invasive predators.

The concerted attempts to subjugate the land and the native biota had one further, important effect: they changed the fire regimes. Before Europeans arrived, Aboriginal people of western New South Wales practised a 'fire-stick' method of farming in which numerous small-scale fires were lit throughout the year. These frequent fires produced a diverse mosaic of vegetation patches of different age. On the one hand this patchy environment would have provided a range of different resources for different marsupial species, thus enhancing species

Southern Brown Bandicoot
Isoodon obesulus Male
Shown with Firewood *Ixodia achilleoides*
and leaves of Tufted Grass-tree
Xanthorrhoea semiplana
KUITPO FOREST, SOUTH AUSTRALIA

diversity at local and district scales. On the other, it also maintained fuel loads at low levels. As Aborigines were removed from their lands, pastoralists initially continued to burn the land on an annual basis but, from the 1880s, sought progressively to eliminate fire as a management tool. This inadvertently allowed intense and destructive wildfires to take hold; several uncontrolled fires have swept western New South Wales over the last century. The combined effects of the changed fire regime and grazing have allowed the proliferation of native and exotic shrubs (woody weeds) in recent decades. Most likely, as wildfires removed vegetation cover over large areas, marsupials that survived the flames would have fallen prey to Foxes and Cats in the newly opened areas. We have documented just such effects over several thousands of square kilometres in the Simpson Desert following extensive wildfires there over the summer of 2001.

It is difficult to distinguish the relative importance of all these impacts on marsupials more than a hundred years after the event but it seems most likely that combinations of threats pushed many species to, and then over, the brink. A drought at the turn of the 20th century sharply reduced stock numbers and affected settlers so severely that the government initiated a Royal Commission to examine what had happened. *The Royal Commission to Inquire into the Condition of the Crown Tenants* of the Western Division in New South Wales was completed in 1901 and paints a comprehensive but chilling picture of how the land looked at that time. The Royal Commission found that Rabbits and inedible species of shrubs had spread throughout most of the western lands, and also that extensive areas of native vegetation and fragile topsoil had been destroyed. Large areas of shrub and tree cover had been removed, allowing winds to create sand storms that led properties to be described as 'wind-swept barren wastes'. The most salutary effect of the Royal Commission was the recognition that the land could not tolerate overstocking and that drought is a regular occurrence in inland regions. Crucially too, the Royal Commission recommended that stocking controls be imposed; together with the now-reduced productivity of the land, this has kept stock numbers well below the peak of just before 1900.

Many of the threats to marsupials that were set in train by pastoral expansion in New South Wales are still threats today, with continued clearing, overgrazing and predation from the Red Fox probably topping the list.

Poisons such as arsenic and strychnine are now little-used, Rabbits never achieve the densities seen before the introduction of myxoma virus, and new models allow us to predict and control outbreaks of pests such as the House Mouse. But, other threats have become apparent. Of the 23 marsupials that have managed to survive in western New South Wales, 10 are listed as vulnerable or endangered under the state's conservation legislation. And of the 13 that are not listed, all except the four large species of kangaroos are vanishingly rare or have strongholds further east. New threats include the spread of irrigated crops such as cotton and rice in productive riparian and floodplain areas, use of dry lake beds for crops, widespread use of pesticides to enhance crop productivity and, perhaps, small population size. When large populations become confined to remnants of habitat, they become vulnerable to threats such as predation, disease, drought or single catastrophic events such as wildfire. With long-term confinement additional threats may arise from inbreeding and loss of genetic integrity. The Yellow-footed Rock-wallaby and Southern Hairy-nosed Wombat may be in this predicament.

Moving east from the arid and semi-arid plains of New South Wales, the story becomes progressively brighter. Nine of the 10 marsupials that disappeared from the western slopes and Great Dividing Range were species that had their strongholds further west, and they almost certainly succumbed to the same suite of threats noted above. The only forest-dweller to disappear entirely from this rugged and more heavily timbered part of the State was the Eastern Quoll (*Dasyurus viverrinus*). Populations of this species began to decline around the turn of the 20th century, along with other carnivorous marsupials such as the larger Spotted-tailed Quoll (*Dasyurus maculatus*) and the Thylacine and Tasmanian Devil in Tasmania. Several researchers have argued that the declines of these carnivores were caused by an outbreak of disease, with some identifying toxoplasmosis as the most likely candidate. I am sceptical of this because Cats – the vectors of the disease-causing *Toxoplasma* parasite – had been already present for at least 100 years in Tasmania and on the mainland. If there had been a delay in the parasite infecting native fauna, it is hard to see why outbreaks would erupt simultaneously in both geographical regions. I think, instead, that broad-scale poisoning for Rabbits took its toll. Fleets of purpose-built poison carts began plying the paddocks in the late 19th and early 20th centuries, dropping thousands of tons of strychnine baits over vast tracts of land. The Eastern Quoll would have been particularly susceptible to poisoning as it foraged preferentially in areas of mixed pasture and woodland, and would have been affected secondarily by eating poisoned Rabbit carcases. The Spotted-tailed Quoll, which is dependent on heavier forest cover and is more likely to kill its own food than to scavenge, would have been less affected.

Marsupials have fared best in eastern New South Wales, especially along the coast and immediate hinterland. This is counterintuitive. If you pick up a newspaper any day of the week there will be a report on yet another land release for housing, another remnant of bush to be cleared for a shopping centre, a controversial plan for a new road or freeway, or a protest about wood-chips being given away for export at bargain-basement prices from our fast-diminishing old-growth forests. Local media often carry reports of dog attacks on Koalas or deaths of these and other large marsupials on the roads. And yet, 33 of the 36 marsupial species that lived along the coast when Europeans first arrived are still there. Only the Eastern Quoll, Tasmanian Bettong and Brush-tailed Bettong have gone. The two bettongs persisted until the early years of the 20th century, perhaps succumbing to the rampant poisoning campaigns of that time, the quoll until 1964 in the (un-poisoned) eastern suburbs of Sydney. The situation is not entirely rosy, though. Seventeen of the 33 survivors, including the Eastern Quoll, are listed as threatened under state legislation, and some are scarce indeed. You would be very lucky to encounter a Koala or a Long-footed Potoroo (*Potorous longipes*) on the south coast of New South Wales, for example, or a Southern Brown Bandicoot anywhere at all. But these species still persist.

Except where forest has been cleared for settlement or agriculture, the coastal regions retain varying degrees of tree cover and provide large protected areas for many species of marsupials. These areas retain food and shelter resources and provide sufficient cover to blunt the hunting activities of feral Cats and Foxes. Forested areas provide little sanctuary for such pests as Rabbits, House Mice and Goats, are not sprayed with pesticides and herbicides and – with one important exception, to which we return below – are no longer repositories for vast quantities of poison. Forestry activities are tightly regulated with respect to conserving threatened species and maintaining continuity of forest habitat, and aim to provide a mosaic of patches of different age since harvesting. These allow species with different tolerances of disturbance to be retained.

I have dwelt on New South Wales because the history of its marsupial fauna has been much-studied and the key threats to address are well-known. A similar story can be told for much of the southern half of Australia, with marsupials hanging on in productive regions with higher rainfall and faring most poorly in areas given over to broadscale agriculture. However, there are some exceptions. Tasmania has suffered the loss of just one marsupial, the Thylacine, and populations of most other species are still healthy (the Tasmanian Devil is the only species that is ailing). The south-west of Western Australia also retains a suite of medium-sized marsupials that have disappeared elsewhere. Unless

White-tailed Dunnart
Sminthopsis granulipes Male
SOUTH–WEST OF WESTERN AUSTRALIA

the native vegetation has been cleared, much of the original fauna of this region remains protected by the presence of certain shrubs in the legume family that contain high concentrations of poison. This poison – sodium monofluoroacetate or 1080 – is deadly to Foxes, Cats and other introduced mammals, and limits their incursions into areas where the plants are dominant. Because of their long association with the legumes, marsupials have evolved considerable tolerance to the poison and persist where their enemies cannot. Synthetic baits containing 1080 are currently used to protect marsupials from Fox predation in several states. I return to this issue below.

Parts of northern Australia also retain their original marsupial faunas but it is unclear how long this fortunate situation will remain. Vast areas of tropical savanna are burnt to grow fresh pasture each year and result in a greatly simplified environment. Although there has been limited study, I suspect that such newly opened habitats can be exploited readily by feral Cats. There is some evidence that Cats prey upon small dasyurids such as planigales, and Red-cheeked and Stripe-faced dunnarts (*Sminthopsis virginiae* and *S. macroura* respectively) in the Top End and Kimberley; populations of these marsupials are in local decline. The recent arrival of the Cane Toad (*Chaunus marinus*) is also causing concern. This pest is highly poisonous to native predators such as the Northern Quoll (*Dasyurus hallucatus*) and its voracious appetite probably depletes invertebrate populations on which smaller marsupials depend. Anecdotal reports suggest that many marsupials are becoming scarce as this adaptable amphibian continues its westward march across the Top End.

Taken together, these observations confirm what is obvious but also seldom reported: retention of habitat is critically important for the conservation of marsupials. This is exactly the same conclusion that would be reached if we had reviewed native rodents, birds or any other group that has been subject to any study. Where habitat is comprehensively destroyed, as it is for broad-scale agriculture, much of the native fauna is destroyed with it. The lessons for conservation and management are obvious.

Recovery planning

What happens after evaluations have been made? The published IUCN lists are a wonderful resource for conservation scientists who wish to identify trends in species status and patterns of species decline or recovery at different scales. Because the IUCN also publishes the reasons why species have been allocated to a particular category, the general causes of declines and recovery can also be reviewed. However, IUCN lists do not require action. As an international organization, the IUCN has no jurisdiction over state or national governments and thus cannot compel them to take the lists seriously. In some parts of the world the IUCN lists nonetheless provide the only guide to species' status but in Australia there are two further layers of review that dictate how threatened species should be identified and conserved. The first layer is set by the federal government, and the second by state and territory governments. At first sight the imposition of these further levels of review by government might appear to represent unnecessary duplication of the effort expended under the IUCN process or even a means of avoiding assessments that are controversial. However, the IUCN definitions and listing process have had a powerful influence on conservation legislation enacted by the federal and state governments in Australia. And listing of species under either level of government has important consequences for conservation.

At the federal level, the *Environment Protection and Biodiversity Conservation Act 1999* is the key legislative tool for conserving and managing Australia's living resources. It defines Extinct, Extinct in the Wild and the same three categories for Threatened species as the IUCN but differs in specifying the further category of Conservation Dependent for species that require ongoing management to conserve them. The federal Act also has no provision for listing species that are Near Threatened or of Least Concern, although all native species are protected. In much the same way that the IUCN specialist groups evaluate species for listing, a Threatened Species Scientific Committee does the same job for Australia. This committee uses parallel criteria to those specified by the IUCN, the main difference being that recommendations for listing must be confirmed by the Minister for the Environment and Water Resources. Anyone can nominate a species for listing, including

the committee itself. Not surprisingly, because of the influence of IUCN procedures on the federal legislation, the lists of threatened species recognized by both organizations are similar.

Legislation at the state and territory level allows further assessment of species' status to be made. Categories and criteria for assessment vary between jurisdictions but in general state and territory legislation operates in a similar manner to the *Environment Protection and*

Red-cheeked Dunnart
Sminthopsis virginiae Male
NEAR DARWIN, NORTHERN TERRITORY

Biodiversity Conservation Act. Panels of scientists are responsible for keeping lists of threatened species under review and for advising the relevant Minister for the Environment when changes should be made. Except in New South Wales, where that state's scientific committee finalizes all determinations itself, the Minister is then responsible for making changes to the species' lists.

If a species is listed as, say, Endangered at the state level, the federal government is notified and considers whether it should enter the species on its lists too. Conversely, a listing at the national level triggers an equivalent review by the states or territories where the species occurs. On the face of it, this reciprocal procedure could be expected to result in both levels of government producing similar lists. In fact, there is considerable disparity. On the one hand, this is understandable because states and territories list only those species that occur within their borders. There would be little point in Tasmania listing the marsupial moles as species of concern because they have never occurred there; the Thylacine does not appear on the threatened species lists of mainland states for the same reason. On the other hand, many species listed by the states and territories are considered to be at no risk by the federal government. Consider the Black-striped Wallaby (*Macropus dorsalis*). This species occurs in small numbers in the north-eastern corner of New South Wales and is consequently listed as endangered under the NSW *Threatened Species Conservation Act 1995*. It has its distributional stronghold in Queensland and is so abundant in places there that it is considered a local pest. It is therefore not listed as threatened by either the Queensland or federal governments. To illustrate the magnitude of the disparity between the two levels of government, the federal government has listed 30% of Australia's species of marsupials as extinct or threatened, compared with 19% (Queensland) to 62% (New South Wales) for the individual states.

Disagreements between levels of government are nothing new, and so it is tempting to dismiss the differences in species lists here as being of little consequence. But, this would be a mistake. Once listed as threatened, species gain considerable legislative recognition that aims to protect both individuals and their habitats from harm. More importantly, listed species become subject to recovery planning that is designed to protect and restore their populations to reduce the risks they face to their survival. Species listed under both federal and state or territory legislation are usually given priority in recovery planning, as are species that are assessed as being critically endangered or endangered. How does the recovery process work?

In the first instance, information on threatened species is gathered from all sources that are available, threats to the species are identified and ways to neutralize them are considered. The information is incorporated into a recovery plan, and the actions specified by this document are then implemented by teams that include scientists, managers and others concerned to ensure that the species recover. Federal, state or territory nature conservation agencies are responsible for publishing the plans; they usually also coordinate and fund the necessary actions and monitor how well the plan is working. Plans have life spans of 3–10 years and are reworked as needed when they conclude. Recovery plans and the actions they specify are used throughout Australia, and are proving to be an effective means of conserving species that are at risk. Let us consider the Western Quoll as an example.

Now restricted to the south-west of Western Australia, this charismatic quoll once ranged over the southern two-thirds of the continent, from the west coast to the western slopes of the Great Dividing Range in the east. It disappeared from the eastern fringe of this huge area soon after the turn of the 20th century, from South Australia by 1931 and from the central arid zone in the 1950s. It was still present near Perth in the 1950s but by the end of the 1980s had become confined largely to the jarrah forest south and inland of the state capital. The population probably numbered less than 6000 animals at this time. The Western Quoll was listed as a Threatened species in Western Australia in 1983 and Endangered by the federal government in 1991 but later reassessed to be Vulnerable. A recovery plan was prepared in 1994 with the explicit aim of removing the Western Quoll from all lists of threatened species.

The plan aimed firstly to increase population size in the jarrah forest, and then to establish populations in other habitats where the species had previously occurred. Dedicated work by Keith Morris and colleagues at the

Western Australian Department of Conservation and Land Management (now Environment and Conservation) showed that quolls handle current forestry practices quite well and even benefit when the jarrah forest is subjected to prescribed fires. But experiments confirmed that quolls benefit most when Foxes are controlled. Broad-scale poisoning of this destructive carnivore has since allowed quoll populations to increase at least five-fold at monitoring sites in the forest. To establish additional populations, captive-bred quolls have been released into five Fox-controlled areas since 1992. These new populations have been largely successful, including one in semi-arid habitat at Lake Magenta Nature Reserve where quolls had earlier died out. The success of these various recovery actions has led to a dramatic and significant improvement in the status of the Western Quoll.

Recovery tools

Recovery plans provide guidance on how to conserve marsupials but do not always spell out exactly what to do. How should Foxes be controlled, for example, or genetic diversity be enhanced or habitat improved to ensure that a target species increases? There are many tools in the recovery toolkit, some tried and trusted, and others that are in the early stages of being tested.

Ensuring that species have sufficient habitat to maintain viable populations is perhaps the most important component in a recovery plan, and can be achieved by reserving or protecting suitable land. This has been done effectively for the critically endangered Northern Hairy-nosed Wombat, which now occurs only in Epping Forest National Park (gazetted specifically for the wombat in 1971) and for several species of marsupials on islands off the coast. Many more species have their largest populations in national parks and reserves, with smaller and less secure populations on agricultural leases or private land. Increasingly, too, non-government organizations such as Bush Heritage Australia and the Australian Wildlife Conservancy are purchasing land dedicated for conservation. Bush Heritage has the magnificent vision for 2025 of protecting 1% of Australia, conserving land of high value for conservation. The Australian Wildlife Conservancy already owns some 917,000 hectares (3540 square miles). It maintains predator-proof fences around some of its properties, actively reintroducing threatened species back into these protected areas to restore the diversity of the original inhabitants.

When good habitat is available, the next step is often to control threatening processes. Feral predators, for example, do not respect the boundaries of protected areas unless they are fenced and can wreak havoc by moving in from areas that are not controlled. For many marsupials, 1080 poison is the sharpest tool in the toolkit because it is so effective at killing their major nemesis, the Red Fox. Baits containing 1080 are currently used in all states to control Foxes, with some 3.5 million hectares (13,500 square miles) of conservation land being protected regularly in Western Australia alone. Feral Cats are less inclined to eat baits, so research now is focusing on placing cat-specific toxins into foods that fussy felines will eat. Other threats are often addressed quite simply. For example, specific methods can be applied to control many introduced pest species, the risk of wildfire can be reduced by the judicious use of control burns, and genetic diversity can be maintained by introducing new individuals into small or isolated populations.

If threats are unknown or too severe to manage in the wild, threatened species can be taken into captivity and maintained until suitable habitat has been secured. The California Condor (*Gymnogyps californianus*) and Przewalski's Horse (*Equus caballus przewalski*) are famous recipients of such intensive care, and would probably now be extinct without it. Some of the last surviving South Australian Tammar Wallabies, mainland Rufous Hare-wallabies and Eastern Barred Bandicoots were moved into captivity for similar reasons. Many zoos and wildlife parks maintain small but important populations of threatened marsupials; they manage genetic diversity through studbooks, and provide source animals for translocations to islands, fenced reserves or other protected areas.

Another very important set of recovery tools is found under the label 'monitoring'. If we are worried about the small size of a population, the rate of its downward slide or the effectiveness of measures that we have put in place to conserve it, monitoring is essential. This can often be done by regularly trapping and counting the target species, by searching for animals by day or by spotlighting at night. But some species are shy and cryptic and, at low densities, very hard to detect. Some, such as the Eastern Pygmy-possum (*Cercartetus nanus*) and Leadbeater's Possum can be censused by checking their visits to nest boxes, others by identifying the hairs they leave on specially designed traps. In the case of the Northern Hairy-nosed Wombat, for example, loose hairs are trapped on sticky tape placed outside burrow entrances and DNA fingerprinting is used to assess the numbers of different individuals. Wombats are hard to trap and sulk for days after being handled, so such unobtrusive methods of monitoring are very important. In future it should be possible to apply DNA monitoring to faecal pellets.

Other high-tech methods are coming on line to assist threatened species to recover. The most promising target reluctant breeders, and aim to increase their rates of reproduction. These include chemical stimulants to induce oestrus and thus extend the breeding season, artificial insemination of females to improve fertility, and even cross-fostering of young between species. The last technique offers much promise. Female Northern Bettongs, for example, can be persuaded to produce

Stripe-faced Dunnart *Sminthopsis macroura*
Male (bottom) and female
SOUTH-WESTERN QUEENSLAND

about twice as many young if some are removed and placed in the pouch of Brush-tailed Bettongs. Survival and development rates of young are similar in both the actual and foster mothers. Attempts are now being made to cross-foster the young of such threatened species as Gilbert's Potoroos and Northern Hairy-nosed Wombats to their commoner relatives. The utility of other new techniques, such as cloning and freezing eggs and sperm for later use, remains uncertain. Still, the recovery of so many threatened marsupials remains a great challenge and the more tools at our disposal the more options – and success – we are likely to have.

The future

From humble beginnings in the distant Cretaceous period, marsupials have followed a path that has taken them from worldwide ascendancy to their now-significant presence in South America and dominance in the Australian region. They have survived shifting continents, changing climates and seen off challenges from an array of erstwhile competitors. Their tenure in Australia extends to at least 60 million years. But times are changing again. The rapid and recent expansion of human populations prompts the question of how our flagship mammals are likely to fare in future.

Black-striped Wallaby
Macropus dorsalis Female
Clarence River, New South Wales

A pessimistic view might say that the story is over. Recent research suggests that old or phylogenetically distinct species are more prone to extinction than recently evolved models, perhaps because they breed slowly or become too specialized to cope with the demands of changing environments. The marsupial moles, rainforest possums and the Numbat most obviously fit this category but there are other candidates too. Then there is the looming spectre of climate change. Cold-loving species such as the Mountain Pygmy-possum, Leadbeater's Possum and Mountain Brushtail Possum will have little habitat left in the hothouse to come and so walk the world as zombies until their time is up. Longer droughts, flooding rains and more intense wildfires will put other species at risk. If we take the pessimistic view to its extreme and assume that the 34 species of living but threatened marsupials in Australia (60 species if we include those that are Near Threatened) will fall by the wayside in the future, the continent will be impoverished indeed.

A moderate, and more realistic view, is that the status quo will be maintained. We still clear the land and view some marsupials as pests, and as a nation do not accord the environment with enough respect. But we also conserve marsupials in small amounts of land that we set aside, recognize them under legislation, and actively protect them in some areas from the worst and most pervasive threats that afflict them. Organizations such as the Australian Mammal Society and the Australasian Wildlife Management Society are flourishing, and have replaced the advocates for acclimatization and for marsupial destruction. Many marsupials remain under threat but we know about them and a lot of people care. With diligence and action, we should be able to maintain the fragile balance between judicious conservation and rampant development in future.

The most optimistic view is that more Australians will learn to appreciate the environment, and overcome their ambivalence about the fate of the marsupials and other unique animals and plants that it sustains. This could happen if we teach our children a sense of place, if we show more clearly why marsupials and their habitats

are valuable and important, and if we understand the terrible consequences and finality of extinction. It could happen also if we better inform different levels of government so that political decisions about the environment are not made in ignorance of marsupials and other natural resources but with due regard for their interests against the competing demands of developers and economic rationalists. I think there are signs of progress and am confident that we will balance our needs with those of our fellow mammals more robustly in future. Of course, I have just spent 30 years working on our magnificent marsupials and nothing but a positive perspective is possible after that!

In reality, dear reader, this story remains unfinished. Its conclusion – the future of Australia's marsupials – is now up to you.

Tammar Wallaby *Macropus eugenii* Male
KANGAROO ISLAND, SOUTH AUSTRALIA

Agile Wallaby *Macropus agilis*
Female with pouch joey
ARNHEM LAND, NORTHERN TERRITORY

SPECIES ACCOUNTS

Compiled by Adele Haythornthwaite

This section provides a brief description and account of all currently recognized Australian marsupials, including the living species and those that have recently become extinct. The classification and taxonomy of marsupials is under constant review and several approaches can be taken. We have generally followed the recommendations of Don Wilson and DeeAnn Reeder in their comprehensive overview *Mammal species of the world: A taxonomic and geographic reference*, published in 2005, but have also taken into account more recently published research. Species are listed firstly by order, then by family and genus. Some genera have several species and these have been ordered alphabetically by their common names.

Species names The common names used here generally follow those recommended by the Australian Mammal Society and those adopted by reference texts such as Van Dyck and Strahan's *The mammals of Australia*, published in 2007. If two or more common names are available the first listed is the 'official' one and is used throughout the text. There has been a lot of debate about whether to use Aboriginal names as common names. Some, such as the Dibbler, Monjon and Quokka are already in common usage; others, such as the Nyoongar (Common Ringtail Possum) and Quenda (Southern Brown Bandicoot), are used in parts of the species' ranges. We have not used them extensively here. Many of the smaller marsupials were given the same common name by Aboriginal people and so cannot be used to differentiate between species. Conversely, more conspicuous and broadly distributed marsupials often had different names in different Aboriginal countries and it is thus not clear which name should have priority.

The scientific names follow the Linnean arrangement of the genus + species name, as explained in the chapter, What is a marsupial? After the scientific name we have also given the authority name and a date, that is, the person who first described the species and the year in which their account was published. If the authority name is in brackets it means that the species is still recognized but has been transferred at some later date to another genus. As in law, this recognizes the importance of precedence and allows taxonomists to refer back to original descriptions.

Distributions The maps depict the present ranges of marsupials and make no attempt to compare these with species' historical distributions. Maps are therefore not given for extinct species. They were compiled from several sources including specimen localities, atlases and records held by state and territory conservation departments, and from consulting field guides and reference texts such as *The mammals of Australia*. We also had access to maps being compiled by the World Conservation Union (IUCN) as part of their Global Mammal Assessment; these were kindly made available by Mike Hoffmann and Wes Sechrest of that organization. Written descriptions confirm the broad distributional areas occupied but also list islands within species' ranges and give the names of towns or settlements that serve to locate species with very restricted distributions. Only large or important islands have been included. The descriptions also identify species with distributions that extend naturally into Papua New Guinea or, after deliberate introduction by people, to other parts of the world.

Descriptions and biology The external appearance of each marsupial is described briefly using both field experience and reviews of published work, and key features are noted that distinguish similar species. Body weights represent the minima and maxima that have been reported, rather than averages, and information on behaviour, diet, habitat and reproduction is taken from the primary literature or secondary reviews. The information is necessarily selective because of space limitations but aims to capture unusual or quirky aspects of the biology of each species.

Status If a species is listed formally by the IUCN as Extinct, Critically Endangered, Endangered or Vulnerable, this is stated at the beginning of the comments on status. Qualifying statements follow if the species is known to be more common or more threatened in particular parts of its range. Marsupials that are not listed under one of the four IUCN categories are generally described less formally as rare, restricted, sparse, common, threatened or unknown. Species can be rare or even apparently threatened over parts of their range yet lack formal listing. Reviews of conservation status are ongoing; the results of the last major review by the IUCN, in 2005, should be available soon after this book has appeared.

THYLACINIDAE

THYLACINUS

Thylacine, Tasmanian Tiger
Thylacinus cynocephalus (HARRIS, 1808)

DISTRIBUTION
Previously Tasmania, now extinct

By the time of European settlement, the Thylacine was already extinct in mainland Australia, possibly owing to competition with dingoes. In Tasmania, it was hunted heavily by farmers and bounty-hunters to reduce predation on sheep, and the population decreased rapidly. In 1936, the last known Thylacine died in Hobart Zoo. The largest of the marsupial carnivores (15-35 kg; 33-77 lb), this distinctive animal had a wolf-like head and fore-quarters and prominent striping across the back. Its extraordinarily wide gape enabled it to grasp large prey. A mostly nocturnal predator, it hunted kangaroos and wallabies in forest and open woodland, usually exhausting prey by pursuit over long distances. It rested during the day in a lair, occasionally emerging to bask in the sun. Animals sometimes hunted in pairs but were more often solitary. Breeding occurred throughout the year, with 2 or 3 pups carried in the backward-opening pouch. Once young were too large to remain in the pouch, they were left in the lair until weaned.

STATUS *Extinct*

DASYURIDAE

SARCOPHILUS

Tasmanian Devil
Sarcophilus harrisii (BOITARD, 1841)

DISTRIBUTION *Tasmania*

This distinctive animal is common throughout Tasmania and was present on mainland Australia until about 430 years ago. Its future is uncertain as many animals are currently succumbing to an easily transmitted but fatal facial tumor disease. Devils are found in many habitat types but are most common in mixed forest and open grassland agricultural areas. Its build is compact and robust, and males (5.5-11.8 kg; 12-26 lb) are larger than females (4.1-8.1 kg; 9-17.8 lb). Its coat is black with a white band across the chest, and sometimes across the rump. The skull and

teeth are incredibly strong and it has massive jaw closure muscles for crunching bones. Animals feed mostly at night, scavenging on carrion and hunting small and medium-sized mammals, and eating all parts of the carcase including skin and bone. Animals are mostly solitary although many individuals can converge on large carcases and compete aggressively for the food. Females can carry up to 4 young in a backward-opening pouch for 13-15 weeks and then wean them at 8-9 months of age.

STATUS *Common in disease-free areas; common to vulnerable elsewhere*

DASYURUS

Eastern Quoll
Dasyurus viverrinus (SHAW, 1800)

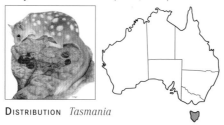

DISTRIBUTION *Tasmania*

Present in south-eastern mainland Australia until the mid-20th century, this species is now restricted to Tasmania. It is found in forest, woodland, scrub, heath and cultivated habitats but is most common in areas where forest or woodland adjoins open grassy areas. This small quoll (0.7-2.0 kg; 1.5-4.4 lb) is unusual in having 2 color phases. Its fur is either fawn-brown (75% animals) or black (25%) and covered with bold white spots; a single litter can contain pups of both color phases. The tail is the same color as the body but not spotted. The hind-foot lacks the 'big toe' of all other quolls. It is a nocturnal predator, eating mostly large invertebrates but also small mammals, ground-dwelling birds and lizards, grass and fruits. Animals are solitary and den during the day in hollow logs, rock piles or burrows. Breeding pairs may share a den for short periods during the breeding season (May-June). Up to 30 young may be born in a single litter but only those that attach to 1 of the 6 pouch teats will survive. After 8 weeks, the young are left in the nest until weaning at 4-5 months of age.

STATUS *Extinct on mainland Australia; common in Tasmania*

Northern Quoll
Dasyurus hallucatus GOULD (1842)

DISTRIBUTION *Four main areas in northern Australia: Hamersley Range, northern Kimberley, western and northern Top End, and north-eastern Queensland*

Previously widespread across northern Australia, this animal's range has contracted significantly since European settlement. It is found in rocky country and open eucalypt forest, often near water, usually within 200 km of the coast. An aggressive hunter, this small (300-1000 g; 10.5-35 oz) quoll eats a range of foods including small mammals, reptiles, frogs, invertebrates and fruits. It has a grey-brown coat with large white spots, and a brown sparsely furred tail. The muzzle is pointed, and the eyes and ears are large. Both sexes have a strong, pungent odor. Mostly nocturnal, it dens in tree hollows and rocky shelters during the day. In rocky habitats animals survive for 2-3 years but in grassland areas nearly all males die after breeding in June. Young are born in July, most in litters of 6 or fewer. Pups attach to teats within the circular pouch area for 8-10 weeks before being left in the nest until weaned at 4-5 months of age.

STATUS *Common within a limited range in some areas; declining or vulnerable in others*

Spotted-tailed Quoll, Spot-tailed Quoll, Tiger Quoll
Dasyurus maculatus (KERR, 1792)

DISTRIBUTION
North-eastern Queensland, south-eastern Australia, Tasmania

The largest of the quolls (1.5-7.0 kg; 3.3-15.4 lb), this predator is easily recognized by its long, spotted tail. It is found in a diverse range of habitats, such as rainforest, open forest, heathland and woodland, from low to high altitudes. Habitat loss, competition with introduced predators and hunting pressure have all contributed to a substantial contraction in its range. It is a mainly nocturnal predator that hunts larger prey (possums, bandicoots and rosellas) than other quolls, as suits its larger size. It also eats smaller prey such as invertebrates, and scavenges carrion from carcases left by foxes and wild dogs. Animals have a thick, coarse reddish-brown coat, covered with prominent, large white spots. The spotted tail is long and coarsely furred. Animals are solitary and den individually during the day in tree hollows, logs or rock crevices. Mating from April to July produces an average of 5 young born after a gestation period of 21 days. The young attach to a teat in the pouch for around 7 weeks, and are weaned at 4-5 months of age.

STATUS *Vulnerable; endangered in north-eastern Queensland*

Western Quoll, Chuditch
Dasyurus geoffroii Gould (1841)

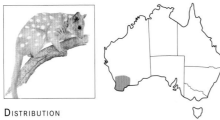

DISTRIBUTION
South-western Western Australia

Formerly distributed throughout arid and semi-arid Australia, including NSW, Queensland, Northern Territory and South Australia, this quoll is now confined to south-western forest and woodland. A medium-sized quoll (0.6–2.2 kg; 1.3–4.8 lb), it has a brown coat with bold white spots and a sparsely furred brown tail. It hunts at night for small vertebrates, reptiles, frogs, birds, freshwater crustaceans and other invertebrates. It is an agile climber, catching some of its prey in trees. During the day, these solitary animals den in hollow logs or burrows. Animals have large home ranges and cohabit only when breeding. Most young are born in the winter (June–July), in litters of up to 6 young that remain in the pouch for 9 weeks, and are weaned at 5–6 months.

STATUS *Vulnerable*

DASYUROIDES

Kowari
Dasyuroides byrnei Spencer (1896)

DISTRIBUTION
Eastern central arid zone of Australia, in the channel country of south-western Queensland and north-eastern South Australia

This species is known only from harsh, sparsely vegetated gibber (stone and clay) deserts. A small (70–140 g; 2.5–5 oz) but voracious predator, it feeds on large invertebrates, small mammals and reptiles. Its tail ends in a distinctive tuft of black hairs that covers half the length and is used to communicate with other Kowaris. On the upper-parts, its fur is a light fawn-grey; the underparts are paler. The head is broad and triangular shaped, with big eyes and large, erect, rounded ears. Animals hunt at night, and rest during the day in burrows, sometimes emerging for short periods to bask in the sun. They excavate most burrows after rainfall, when the soil is softer, which is also when they breed. Apart from breeding time, animals are solitary and territorial. Young are born in the second half of the year. Females raise litters of up to 6 or 7 young which attach to teats in a pouch formed from folds of skin and remain there for 7–8 weeks before being transferred to a nest until weaned at around 15 weeks.

STATUS *Vulnerable; sparsely distributed*

DASYCERCUS

Brush-tailed Mulgara
Dasycercus blythi (Waite, 1904)

DISTRIBUTION
Arid western and central Australia

This species is similar in appearance and distribution to, and difficult to distinguish from, the Crest-Tailed Mulgara. As its name suggests, this animal has a brush-shaped tuft of dark hairs at the tip of its tail; the crest-tailed species has a crest on the upper side of the tail only. This animal has a sandy-brown coat above, with paler belly and chest, large eyes and a thick tail. It is a robust animal (50–140 g; 1.8–5 oz) that inhabits the spinifex deserts of central Australia. It is usually found in sandy habitats, where it is easier to dig and maintain burrows. Animals are solitary, and live in sometimes complex burrow systems that provide insulation from extreme temperatures. Animals emerge at night to hunt for invertebrates and small vertebrates, such as dunnarts, planigales, mice and reptiles. Breeding occurs once a year in winter. The females have 6 teats (the Crest-tailed Mulgara has 8), which limits the number of young in the litter (1 young per teat).

STATUS *Vulnerable; some populations endangered*

Crest-tailed Mulgara
Dasycercus cristicauda (Krefft, 1867)

DISTRIBUTION
Arid central Australia

This attractive animal occupies sand dunes and the fringes of salt lakes, and probably occurs further west than shown here. It is an efficient predator of small mammals and large invertebrates. Males (80–160 g; 2.8–5.6 oz) are slightly larger than females (60–120 g; 2.1–4.2 oz) but are otherwise similar in appearance. The coat is sandy brown above and whitish below. The stout tail is colored orange at the base and tipped with a crest of black hairs. It has a broad, triangular head with large eyes and erect rounded ears. Primarily active at night, it does emerge from its burrow to sunbathe in colder weather. It is well adapted to its harsh habitat, conserving water by producing concentrated urine, storing fat in its tail when food is abundant, and digging deep burrows to escape extreme temperatures. Animals are solitary and maintain a constant home range or territory. Breeding in winter is followed by births 5–6 weeks later of up to 8 young. They attach to a teat in the females pouch for 7–8 weeks and are then left in the nest for increasing periods.

STATUS *Vulnerable; some populations endangered*

PHASCOGALE

Brush-tailed Phascogale, Tuan
Phascogale tapoatafa (Meyer, 1793)

DISTRIBUTION *Northern and south-western Western Australia, northern Cape York, south-eastern Queensland and Wellesley Island (Queensland), north-eastern New South Wales (NSW), eastern and southern Victoria*

An agile climber, this rat-sized (105–310 g; 3.7–11 oz) animal spends most of its life above ground in forest and woodland. It hunts for invertebrates beneath tree bark, and nests on fur, feathers and leaves in tree hollows. It has a distinctive black 'bottle brush' tail which can be raised when it is excited. The coat is grey above and whitish below. Its muzzle is pointed, and both eyes and ears are prominent. At night, when it is active, it moves rapidly up and down tree trunks and along branches, aided by flexible ankle joints capable of rotating 180°. After breeding between May and July, all males die. Young are born after 1 month in litters of 3–8. At 7 weeks of age the young are transferred from the pouch to the nest and at 20 weeks are weaned. Hollow-bearing (old) trees are becoming increasingly scarce in its range and much of its preferred habitat of dry sclerophyll forest with sparse understorey has been cleared for agriculture.

STATUS *Rare and threatened over most of range; more common in south-western Western Australia*

Northern Brush-tailed Phascogale
Phascogale pirata Thomas (1904)

DISTRIBUTION
Top End of the Northern Territory, including Melville Island and Sir Edward Pellew Group

Recent research has shown this northern phascogale to be a species in its own right, not a disjunct population of the widely distributed Brush-tailed Phascogale. It is about the same size as its relative (105–300 g; 3.7–10.6 oz) but differs genetically and in having longer ears, white fur on the belly and pure white surfaces on the hind-feet. A voracious and largely arboreal predator, it hunts invertebrates on trees, descending to the ground infrequently. It builds nests in the hollows of large old trees, preferring open forest and woodland but turning up occasionally in suburban backyards. The species has suffered a large decline since it was first encountered by Europeans in the late 19th century, and few details of its life history are known. There is some evidence that males die after breeding in the dry season, and that females carry litters of up to 8 young, weaning them 5 months later at the start of the summer monsoons.

STATUS *Rare, patchily distributed*

Red-tailed Phascogale
Phascogale calura GOULD (1844)

DISTRIBUTION
South-western Western Australia

This phascogale, formerly distributed over western central Australia and arid regions of NSW, Victoria and South Australia, now occupies remnants of Wandoo and Rock Oak woodlands in the semi-arid wheatbelt. Much smaller (38–68 g; 1.3–2.4 oz) than the similar Brush-Tailed Phascogale, it too has a (smaller) black brush towards the tip of the tail but the brushless lower half of the tail is distinctively colored red. It has a grey coat with whitish underparts, a pointed muzzle, and large eyes and ears. It hunts at night mostly for invertebrate prey but also includes small mammals and birds in its diet. These agile climbers are able to move quickly and easily through the forest canopy but they also forage extensively on the ground. Mating occurs in a tightly synchronized 3-week period in July, after which all males die. Litters of up to 13 young are born after 4 weeks but only 8 can attach to a teat in the mother's pouch; some litters are much smaller. Young are weaned at 3–4 months of age.

STATUS *Endangered*

ANTECHINUS

Agile Antechinus
Antechinus agilis DICKMAN, PARNABY,
CROWTHER AND KING (1998)

DISTRIBUTION
*South-eastern Australia,
in south-eastern NSW,
and central, eastern
and southern Victoria*

Only recently differentiated from the Brown Antechinus, this small insectivorous animal is found in the forest, heathland and woodland of south-eastern Australia. Although it occurs in many different vegetation types and at altitudes up to 2000 m (6560 ft), it requires habitat that has dense groundcover and abundant logs. It is smaller (16–44 g; 0.56–1.5 oz) than the Brown Antechinus and very similar in appearance. However, its coat is more grey-brown than brown, as are the tops of its feet and tail. As its name suggests, it is an agile animal that quickly and easily climbs trees and branches in search of food. It is equally adept at foraging on the ground, searching through leaf litter for invertebrates, and also taking small vertebrates and fruits when available. Animals build nests of leaves in tree hollows, where they nest communally outside of the breeding season. Breeding occurs over a short 2-week period that does not coincide with that of the Brown Antechinus, and all males die soon after. Between 6–10 young are born after 4 weeks gestation and carried in the pouch until 5 weeks of age.

STATUS *Common*

Atherton Antechinus
Antechinus godmani (THOMAS, 1923)

DISTRIBUTION
*Southern edge of Atherton
Tableland, north-eastern
Queensland*

This animal is found only in 2000 km² (770 sq miles) of damp, misty tropical montane rainforest where annual rainfall exceeds 2700 mm (106"). Large (55–125 g; 1.9–4.4 oz) when compared with most antechinuses, this animal has a dull brown coat with paler underparts, and orange to ginger cheeks. It is small-eyed, and probably detects prey more by smell and hearing than by sight. The tail is almost naked, having a crest of sparse hairs on the underside. Mostly nocturnal, animals forage in tree canopies and among vegetation on the forest floor for invertebrates, frogs and lizards. It is likely that adults also eat small mammals such as rats and mice. Animals are generally secretive, and shelter during the day in nests of leaves which they build in epiphytic plants and tree hollows. After breeding in July, all males die. A maximum of 6 young are born in August, and are carried in the pouch until 5 weeks of age.

STATUS *Rare in restricted range*

Brown Antechinus
Antechinus stuartii MACLEAY (1841)

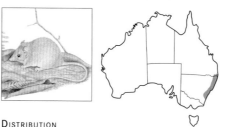

DISTRIBUTION
*Eastern Australia, from south-eastern
Queensland to Kioloa in southern NSW*

Although it is common in urban bushland in and around several large towns and cities, most people have neither seen nor heard of this small animal. It is found in a wide range of moist, wooded habitats, preferably those with dense groundcover, from coastal to subalpine altitudes. It is small (18–60 g; 0.6–2.1 oz) and males are up to twice as large as females. Its coat is uniformly brown above and pale brown below. Its feet are brownish, with no yellow or grey tones. An active, nocturnal hunter, it forages on the ground and climbs trees to find invertebrate prey. Animals nest communally in tree hollows until a month or 2 before the breeding season. They then live solitary lives until the 2-week breeding period in August–September, dependent on location. After breeding, all the males die from stress-related illness. Between 6–10 young are born after 31 days and carried in a shallow circular pouch until 5 weeks of age when they are transferred to the nest and weaned at 3–4 months.

STATUS *Common*

Cinnamon Antechinus
Antechinus leo VAN DYCK (1980)

DISTRIBUTION
*East coast of Cape York
Peninsula*

Found only in semi-deciduous rainforest between the McIlwraith and Iron ranges, this antechinus is the only dasyurid endemic to Cape York Peninsula. It is an agile climber that moves easily and rapidly up trees trunks, along branches and through the canopy. When on the ground, it tends to run along buttresses and logs rather than through the leaf litter. Males (65–120 g; 2.3–4.2 oz) are almost twice as large as females (32–74 g; 1.1–2.6 oz) but are otherwise similar in appearance. The coat is a cinnamon (reddish-grey) color above, and cinnamon to yellowish below. The forehead and front of face has a darkish stripe, and the eyes and ears are relatively large. It forages at night for invertebrates but probably also takes small vertebrates opportunistically. Animals rest during the day in nests in hollow trees and logs. Mating, in mid-September, is soon followed by the death of all males. Litters of up to 10 young are born a month later and are carried in the mother's rudimentary pouch for around 6 weeks. They are then transferred to the nest.

STATUS *Common within restricted habitat*

Dusky Antechinus
Antechinus swainsonii (WATERHOUSE, 1840)

DISTRIBUTION
*South-eastern Australia, from south-eastern
Queensland to south-eastern Victoria; Tasmania*

The largest of the antechinuses (males 43–178 g, 1.5–6.2 oz; females 37–100 g, 1.3–3.5 oz), this robust animal is found in wet forest and heathland. It is more common in mountainous regions but is also found in coastal woodland and heath. It requires a dense understorey of vegetation and debris to provide food and protection from predators. The coat is dark brown with grey below but animals are both darker and smaller at high altitudes. The eyes and ears are small. Animals hunt invertebrates and small vertebrates, which they dig and glean from leaf litter and soil. Mostly terrestrial, animals are rarely observed climbing. During winter, females build spherical nests of leaves in burrows excavated in creek banks or below the soil surface. Mating occurs over a 2-week period in May–September, with animals at high altitudes breeding later than those in coastal areas. All males die soon after mating. Young are born a month after mating, in litters of 6–10 pups, and remain in the open pouch for 8 weeks.

STATUS *Common in suitable habitat*

Fawn Antechinus
Antechinus bellus (Thomas, 1904)

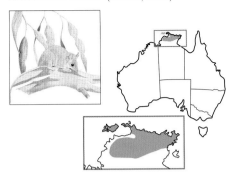

Distribution *Top End, Northern Territory*

The only antechinus found in the Northern Territory, this animal inhabits the eucalypt woodland and open forest of the monsoonal northern interior, in Kakadu and Arnhemland. Males (42-66 g; 1.5-2.3 oz) are larger than females (26-41 g; 0.9-1.4 oz); both sexes have pale to medium grey upperparts with paler chest and belly. The eyes have prominent rings of pale fur, and eyes and ears are large. It is an agile hunter that emerges at night in search of invertebrate prey. Foraging in trees and on the ground, it is likely that it also includes small vertebrates in its diet. Animals rest during the day in nests in tree hollows and logs. Breeding during a 2-week period in August, is followed by a complete male die-off in September. After 4 weeks gestation, litters of up to 10 young are born and attach to a teat in the mother's pouch. They remain there for up to 5 weeks, and are weaned in January at around 4 months of age.

Status *Locally common in patchy distribution*

Rusty Antechinus
Antechinus adustus (Thomas, 1923)

Distribution
North-eastern Queensland, from Paluma to Mount Spurgeon (near Mossman)

Now recognized as a separate species and not a subspecies of the Brown Antechinus, this animal is found only in dense, tropical vine forest at altitudes above 800 m (2625 ft). It is small, and males (30-42 g; 1.1-1.5 oz) are only slightly larger than females (21-34 g; 0.7-1.2 oz). Its coat is dark brown above, and reddish brown below and on the flanks. Its feet are brownish, and the muzzle is a little shorter than in other antechinuses. Its tail is dark brown and shorter than head-and-body length. It is nocturnal, and forages both on the forest floor and in trees for its invertebrate prey. During the day, it rests in a nest constructed from leaves in a hollow log or an epiphytic fern. Mating occurs earlier than in the southern species, in June-July, and by the first week of August all males have died. Litters of 6 young are born from early August onwards. The young remain in the pouch for 5 weeks and are weaned 3 months later, in November.

Status *Restricted to small area*

Subtropical Antechinus
Antechinus subtropicus Van Dyck and Crowther (2000)

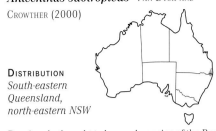

Distribution
South-eastern Queensland, north-eastern NSW

Previously thought to be a subspecies of the Brown Antechinus, this animal is found in the moist, dense, subtropical vine forest of Queensland and NSW. It is slightly larger than the Brown Antechinus, and males (52-67 g; 1.8-2.4 oz) are up to twice as large as females (24-32 g; 0.8-1.1 oz). Its fur is mid-brown on the upperparts and slightly darker on the flanks, and pale brown below. The lightly furred ears and tops of the feet are mid to pale brown and usually lack any pinkish coloration. The tail is a little shorter than the head-and-body length and is colored mid to olive brown. Animals are nocturnal, and forage for food in trees and on the forest floor. Its diet consists of invertebrates such as beetles, spiders and cockroaches, and small vertebrates (mice, lizards) and fruits when available. Animals breed in a short, synchronized 3-week period in September, after which all males die. The young are born around 4 weeks later in litters of up to 8 pups, and carried in the pouch for 5 weeks. Weaning occurs at around 3-4 months of age.

Status *Common in vine forest habitat; rare elsewhere*

Swamp Antechinus
Antechinus minimus (Geoffroy, 1803)

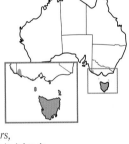

Distribution
Coastal Victoria and South Australia; Tasmania and Flinders, King and Great Glennie islands

As its name suggests, this animal is found in wet, marshy habitats such as tussock sedgeland and heathland where vegetation is dense; it avoids open areas. One of the larger antechinuses (males 30-103 g, 1.1-3.6 oz; females 24-65 g, 0.8-2.3 oz), it has a mid-brown coat which becomes yellow-brown over the flanks and rump, while the underparts are yellow-grey. It has distinctive pale eye rings, and both the eyes and ears are small. The tail is short. Animals forage by day and night for invertebrates by digging in soil and leaf litter, and are entirely terrestrial. Animals rest in grass-lined nests that are usually built into the bases of grass tussocks. After mating over a tightly synchronized period from May to September, all males die. Females give birth a month later; between 6 (Tasmania) and 8 (mainland populations) young are reared in the pouch for around 8 weeks before being left in the nest.

Status *Rare in mainland populations, less so elsewhere*

Yellow-footed Antechinus
Antechinus flavipes (Waterhouse, 1838)

Distribution
South-western Western Australia, north-eastern Queensland, eastern and south-eastern Australia, from central Queensland coast to eastern South Australia

This most widely distributed antechinus species has distinctively yellowish fur that covers its feet, legs and flanks. It is small (21-79 g; 0.7-2.8 oz), and lives in a variety of wet and dry wooded habitats in tropical and temperate regions, including tropical vine forest, coastal heath and semi-arid mulga woodland. Its coat color is variable: animals in the north-east have reddish upperparts; those further south and west are grey-brown or brown. The eyes are ringed with whitish fur and the tail is tipped black. Animals mostly forage for food at night but are frequently active during the day. They are agile climbers and active hunters, eating mostly invertebrates; they also opportunistically include small mammals, birds, flowers and nectar in their diet. Foraging animals make short, rapid, darting movements in trees, on the ground and among rocks. After the 2-week breeding period in late winter or spring, depending on location, all males die. Litters of up to 12 young are born around 4 weeks later and are carried in the pouch until 5 weeks of age.

Status *Common*

Dibbler
Parantechinus apicalis (Gray, 1842)

Distribution
South-western Western Australia, and Boullanger and Whitlock islands, Western Australia

This rarely caught animal was re-discovered in 1967 having been presumed extinct for the previous 83 years. It requires dense cover and inhabits long-unburnt dense coastal heathland and shrubland, as well as *Banksia* woodland and forest if suitable understorey is present. Easily distinguished by pronounced white rings around the eyes, this small (40-120 g; 1.4-4.2 oz) yet robust animal has greyish-brown fur which is flecked with white, giving it a grizzled appearance. The underparts are white and tinged with yellow. Its densely furred tail is stout and tapers towards the tip. It is a solitary, elusive animal that is mostly terrestrial but can climb 2-3 m (6.6-9.8 ft) up into trees, foraging at night and during the day for invertebrates, lizards, birds, small mammals and nectar from *Banksia* flowers. Animals shelter in burrows that they excavate in the sandy soil, or sometimes use those dug by seabirds. Breeding occurs in March–April, and litters of 8 young remain in the pouch for up to 5 weeks. In island populations, all males die after breeding in some years.

STATUS *Endangered*

PSEUDANTECHINUS

Carpentarian False Antechinus
Pseudantechinus mimulus (THOMAS, 1906)

DISTRIBUTION
Alexandria Station, Northern Territory, near Mount Isa, Queensland; North, Central and South West islands of Sir Edward Pellew Group, Gulf of Carpentaria

Rarely seen on the mainland since its initial capture in 1905, this animal is better known from its island populations. It is the smallest of the false antechinuses (15–18 g; 0.5–0.6 oz), and is pale brown above and greyish below, with rufous patches behind the ears. When food is abundant, the tail becomes carrot-shaped from fat deposits; when food becomes more scarce, this fat store is metabolized and the tail becomes much thinner. It is found in rocky areas vegetated with woodland that has a shrubby understorey and spinifex groundcover. Animals are nocturnal and forage for invertebrates in rocky crevices; by day, they retreat to shelters between rocks. Little else is known of the species.

STATUS *Vulnerable; common on islands but rare on mainland*

Fat-tailed False Antechinus
Pseudantechinus macdonnellensis
(SPENCER, 1895)

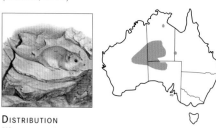

DISTRIBUTION
Western and central arid Australia, from central Western Australia to far western Queensland

This stocky, small animal (18–33 g; 0.6–1.2 oz) is found in rugged areas where crevices and cracks in rocks provide shelter in extreme temperatures. Occasionally it is also found in large termite mounds away from rocks. As its name describes, this animal has a short carrot-shaped tail that becomes distended with fat deposits when food is abundant. Its coat is greyish brown above and

whitish below, with rufous patches behind the ears. The muzzle is quite short, and both the eyes and ears are large. Mostly nocturnal, animals forage among rocks for invertebrates. They sometimes emerge from their rocky shelters to sunbathe during the day. Breeding in winter sees litters of up to 6 young born from July to October, depending on location. After 6–7 weeks in the pouch, the young are left in grass-lined nests in crevices while the mother forages for food. They are weaned at 14 weeks. Both males and females usually survive to breed for 2 years.

STATUS *Common to uncommon depending on location*

Ningbing False Antechinus
Pseudantechinus ningbing KITCHENER (1988)

DISTRIBUTION
Kimberley region, north-eastern Western Australia; north-western Northern Territory

This small animal (15–33 g; 0.5–1.2 oz) is found in both sandstone and limestone rocky habitat in a variety of different vegetation types. Its coat is greyish brown above and paler grey below. The muzzle is long and pointed, and both the eyes and the lightly furred ears are large. Fat is stored in the tail, at times giving it a distinctive carrot shape. The tail is the same length as the head–body length and is lightly furred. Little is known of the behavior of this animal, which was only recently designated a separate species. Presumably animals are mostly nocturnal and have a diet consisting of invertebrates and occasional small vertebrates. Young are born 7–8 weeks after mating in June and are weaned at around 3 months of age. Females have only 4 teats, so litters are small.

STATUS *Locally common*

Sandstone False Antechinus
Pseudantechinus bilarni (JOHNSON, 1954)

DISTRIBUTION
Northern Territory in north-western Top End and west coast of Gulf of Carpentaria

This small animal (21–35 g; 0.7–1.2 oz) inhabits rocky sandstone country. It prefers areas where perennial grasses and open eucalypt forest provide some cover but it is found in many other vegetation types. It has a long tail which is perhaps a useful balancing aide when bounding over rocks. Its coat is greyish brown above and whitish below, with reddish-brown patches behind the ears. The muzzle is proportionately longer than in similar species, and both ears and eyes are large. Animals are solitary and mostly nocturnal,

although they are active at times during the day. Their diet consists mostly of invertebrates but it is likely that small vertebrates are also eaten. After breeding from May to July, litters of up to 6 (but usually 4 or 5) young appear in August–September. About a quarter of the males survive to breed for a second year.

STATUS *Locally common in patchy distribution*

Tan False Antechinus, Rory's False Antechinus
Pseudantechinus roryi COOPER, APLIN AND ADAMS (2000)

DISTRIBUTION
Western and central Western Australia including the Pilbara uplands and parts of the Gibson and Great Sandy deserts

Rocky breakaways and outcrops are the habitat of this newly described animal. One of the smallest of the false antechinuses (16–20 g; 0.6–0.7 oz), it has a reddish-brown coat above, with whitish coloration below, including the tail. The tops of the feet are white also. There is a bright patch of orange-red fur behind the ears. The tail becomes swollen with fat deposits when food is abundant. Presumably, as in other false antechinuses, animals are nocturnal and eat invertebrates; little else is known of this species.

STATUS *Unknown, likely to be sparsely distributed*

Woolley's False Antechinus
Pseudantechinus woolleyae KITCHENER AND CAPUTI (1988)

DISTRIBUTION
Western and central Western Australia

This newly described animal is the largest of the false antechinuses (18–43 g; 0.6–1.5 oz). It is found only in western and central Western Australia in arid rocky areas where the predominant vegetation is *Acacia* shrubland, and tussock and hummock (spinifex) grassland. The coat is colored mid-brown above and buff with a pink tinge below. Both eyes and ears are large, and there is reddish fur behind the ears. The tail becomes swollen with fat deposits when food is plentiful. Little is known of its life history or behavior. Litters of 6 young have been recorded in October, and sexual maturity arrives at 10 months of age. Both males and females have survived to breed in 2 consecutive years.

STATUS *Sparsely distributed, perhaps rare*

DASYKALUTA

Kaluta, Little Red Kaluta
Dasykaluta rosamondae (RIDE, 1964)

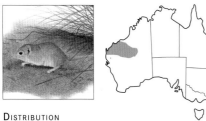

DISTRIBUTION
Western arid areas of Western Australia, including the Pilbara and part of the Great Sandy Desert

This easily recognized animal, found in spinifex hummocklands, has a distinctive rounded body shape, is quite small (20-40 g; 0.7-1.4 oz) but robust, and has a shaggy coat and fat tail. Its coat is copper brown above and paler brown below, and its long fur gives it an unkempt appearance. The muzzle is short, and the eyes are closer to the snout than to the ears. The ears themselves are short and furred and do not protrude above the head. The short, thick tail is often distended with fat deposits and is sparsely furred. It is a nocturnal forager that feeds on invertebrates and small vertebrates. After breeding in September, all males die. Litters of up to 8 young are born in November, and are weaned at 14-16 weeks of age. Females may survive to breed for 2 consecutive years.

STATUS *Locally common*

SMINTHOPSIS

Butler's Dunnart
Sminthopsis butleri ARCHER (1979)

DISTRIBUTION
Northern Kimberley, Western Australia; Bathurst and Melville islands, Northern Territory

Very few of these animals have ever been collected and, consequently, little is known of their distribution, breeding biology and behavior. All animals collected have been found in coastal forest, some with grassy understorey, on sandy substrates. Butler's Dunnart is a small animal (10-20 g; 0.35-0.7 oz) with a mouse-grey coat above and whitish belly, chest and feet. Its muzzle is pointed, and the eyes and ears are large. The tail is bi-colored and thinly furred, and is the same length as the head-and-body. Animals are probably nocturnal and insectivorous, and there is some evidence that they avoid recently burnt vegetation. Little is known of their breeding habits; however, a female with 7 pouch young has been caught in December.

STATUS *Vulerable; rare*

Chestnut Dunnart
Sminthopsis archeri VAN DYCK (1986)

DISTRIBUTION
Northern Cape York and Blackbraes National Park, Queensland; also in south-western Papua New Guinea

This little known and poorly studied species is described from less than 24 museum specimens, and much of its behavior and life history is unknown. It is a small animal, around 16 g (0.56 oz) in weight, with a greyish-tan coat above and paler underparts. Its rounded muzzle gives it a distinctive 'roman nose'. The eyes are large and dark but the ears are small and triangular. The thin tail is dark and sparsely haired. It is presumably insectivorous and is most likely nocturnal. All Australian animals have been caught in tall stringybark woodland growing on red earth soils. Breeding is in the dry season, from July to October, and litters of up to 6 or 8 young are born.

STATUS *Unknown, probably limited*

Common Dunnart
Sminthopsis murina (WATERHOUSE, 1838)

DISTRIBUTION
Isolated population on the Atherton Tableland in north-eastern Queensland; south-eastern Queensland to eastern South Australia but absent in coastal areas south of mid-NSW

Despite its name, this animal is only common over a small part of its range. It is found in woodland, open forest and heathland habitats. The small, mouse-sized animal (12-28 g; 0.4-1.0 oz) has a greyish coat above and is white below, including its feet. Its muzzle is pointed, and the eyes and ears are large. The tail is always thin and is lightly furred. A nocturnal forager, its diet includes beetles, cockroaches and spiders. It rests during the day in a nest of dried grasses and leaves which it builds in a hollow log or grass clump. Animals are never found in high densities and seem to prefer habitats where vegetation is burnt periodically. Up to 2 litters of 10 young are born between August and March, after a gestation period of only 12 days. The young are weaned at around 9 weeks of age and are fully mature at 21 weeks of age.

STATUS *Locally common in south-eastern Queensland and north-eastern NSW; uncommon to rare elsewhere*

Fat-tailed Dunnart
Sminthopsis crassicaudata (GOULD, 1844)

DISTRIBUTION
Southern three-quarters of mainland Australia, excluding far south-western and eastern coast

Possibly the most common and widely distributed dasyurid species, the fat-tailed dunnart occupies a variety of habitat types across most of southern and central Australia. This small (10-20 g; 0.35-0.7 oz) nocturnal predator forages widely for food, preying mostly on invertebrates but also occasionally taking small vertebrates such as lizards. In favorable conditions when food is abundant, its tail becomes quite swollen with fat deposits which are used as an extra energy store when food becomes scarce. Its coat is sandy to fawn brown above, and the chest and belly are white. The eyes and ears are large, and it holds its ears upright. During the day, animals shelter under logs or rocks, or in burrows excavated by lizards or rodents. Animals are mostly solitary but can be found huddling together in small groups in cold weather. In low temperatures they may become torpid for several hours. Females give birth after a short gestation period (12 days) to litters of 8-10 young, which are weaned 10 weeks later.

STATUS *Common*

Gilbert's Dunnart
Sminthopsis gilberti KITCHENER, STODDART AND HENRY (1984)

DISTRIBUTION
South-western Western Australia, southern border of Western Australia and South Australia

The distribution of this species lies within that of the Little Long-tailed Dunnart but it is distinguished by its shorter tail and white patches behind the ears. It is also slightly larger (14-25 g; 0.5-0.9 oz). Its coat is mid-grey to dark grey above and whitish below. The dark rings around the large eyes are quite prominent and the large ears are naked. The grey colored tail is never fat. It is found in temperate coastal areas, preferring heathland or woodland with a heathy understorey. It is active at night, when it forages for invertebrates and possibly small vertebrates. It rests during the day in a nest that is usually constructed in a hollow log or in dense vegetation. Breeding in spring and summer produces litters of up to 8 young in October-November.

STATUS *Common within limited distribution*

Grey-bellied Dunnart

Sminthopsis griseoventer KITCHENER, STODDART AND HENRY (1984)

DISTRIBUTION
Southern and south-western coastal, and Boullanger Island, Western Australia, isolated population on Eyre Peninsula, South Australia

This small animal (14-24 g; 0.5-0.8 oz) is found in a diverse range of habitats, including open woodland, dense heathland and swampland. Populations are most concentrated in coastal heathland on sandy soils, where dense vegetation has had ample time to regenerate after fire. Its coat is mid-grey to dark grey above and, true to its name, its chest and belly are pale or olive grey. The tail is always thin and the same length as the head-body. Animals are nocturnal and actively forage in leaf litter for food. Invertebrates form the main part of the diet but small vertebrates and fruit are consumed opportunistically. Animals breed in winter and spring but those on Boullanger Island breed 2 months earlier than mainland animals. Up to 8 young are born in a single litter and remain in the pouch for 4-5 weeks. They are then left in a leaf-lined nest just beneath the soil surface until they are weaned at 10 weeks of age.

STATUS *Common in suitable habitat*

Hairy-footed Dunnart

Sminthopsis hirtipes THOMAS (1898)

DISTRIBUTION
Western and central arid and semi-arid regions, from coastal Western Australia to south-western Queensland

This animal is found in sandy areas inhabiting a wide range of vegetation types including spinifex hummockland, woodland and shrubland. As its name suggests, this small dunnart (13-24 g; 0.45-0.8 oz) has noticeably hairy feet, with fine hairs covering the pads of the hind-feet and longer hairs fringing the soles. Presumably the hairs provide traction on loose sand. Its coat is a sandy brown above and white below, including the feet and cheeks. It has large eyes with bold dark rings and large ears. The base of the whitish tail can become plump with fat deposits when food is abundant. Animals are nocturnal and solitary, and hunt for invertebrates and small lizards. During the day, they shelter in burrows dug by other animals, such as dragon lizards, rodents, spiders or scorpions. After breeding in late winter, pouch young appear in the spring and early summer.

STATUS *Locally common but usually in low densities*

Julia Creek Dunnart

Sminthopsis douglasi ARCHER (1979)

DISTRIBUTION
Between Julia Creek and Moorrinya National Park, central Queensland

This animal is found only in Mitchell grass downs country where the cracking black-brown soils provide daytime shelter and nesting sites. The largest of the dunnarts (40-70 g; 1.4-2.5 oz), it has a greyish-brown coat above, with whitish underparts. There is a distinct head stripe of darker fur, and the eyes are large and dark. The tail is stout and lightly furred. This animal is distinguished from other dunnarts by both its size and its long, white hind-feet (> 20 mm; 0.8"). Little is known of this species but it is nocturnal and probably feeds on large invertebrates, small mammals and lizards. Litters of up to 8 young are born throughout the year. In captivity, females reach sexual maturity (17-27 weeks) sooner than males (28-31 weeks).

STATUS *Endangered*

Kakadu Dunnart

Sminthopsis bindi VAN DYCK, WOINARSKI AND PRESS (1994)

DISTRIBUTION
Top End, Northern Territory

Recorded for the first time in 1980, this small dunnart (12-14 g; 0.4-0.5 oz) has been captured only in hilly eucalypt woodland that has a gravel substrate; it does not seem to inhabit areas with sandy or clay soils. Its coat is pale grey above, with a whitish chest, belly and feet. The eyes are large and dark and ringed with dark fur. The tail is always thin and has a sparse covering of grey fur. Little is known of its behavior and life history. It is mostly nocturnal and its diet probably consists of invertebrates. During the day, it finds shelter to rest in and has been recorded using a burrow. The dry season appears to be breeding time with pouch young appearing in October.

STATUS *Probably common within a limited distribution*

Kangaroo Island Dunnart, Sooty Dunnart

Sminthopsis aitkeni KITCHENER, STODDART AND HENRY (1984)

DISTRIBUTION
Kangaroo Island, South Australia

This small animal (10-25 g; 0.35-0.9 oz), found only on Kangaroo Island, is the island's only endemic mammal. It was originally collected in 1969 and recognized as a new species in 1984 but since then few specimens of this distinctive animal have been captured. It appears to inhabit mallee heathland habitats on laterite and sandy soils but little is known of its habitat requirements and preferences. Unusually for a dunnart, it has a sooty grey coat with pale grey underparts. The tail is always thin and is slightly longer than the head-body length. Presumably it is both nocturnal and insectivorous, and is inactive during the day.

STATUS *Endangered; rare*

Lesser Hairy-footed Dunnart

Sminthopsis youngsoni McKENZIE AND ARCHER (1982)

 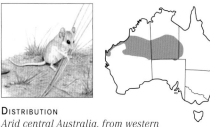

DISTRIBUTION
Arid central Australia, from western Western Australia to western Queensland

One of the smallest dunnarts (7-12 g; 0.25-0.4 oz), this attractive animal is found in sandy habitats. It reaches its highest densities in spinifex hummockland but is also found in desert shrubland and sparsely wooded areas. The sandy-brown coat is flecked with dark grey fur above; it is white below. The large eyes are ringed in dark fur, with a faint stripe extending to the muzzle and a paler stripe above the eyes. The ears are large and thinly furred. In good seasons, the base of the tail thickens with fat deposits but becomes quite skinny in lean times. The soles of the hind-feet are finely haired. Animals are solitary and forage at night for invertebrate prey. They regularly venture into exposed areas far from cover in search of food but can rapidly move to shelter in spinifex hummocks when they sense danger. During the day, animals shelter in burrows excavated by dragon lizards, spiders and scorpions. Breeding from July to September sees litters of up to 6 young appearing in the spring. In good years, animals can breed a second time during the summer.

STATUS *Common in suitable habitat but usually in low densities*

Little Long-tailed Dunnart
Sminthopsis dolichura KITCHENER, STODDART AND HENRY (1984)

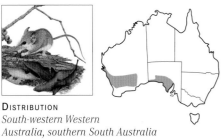

DISTRIBUTION
South-western Western Australia, southern South Australia

This small animal (10-21 g; 0.35-0.7 oz) is found in semi-arid and arid woodland, shrubland and hummock grassland in 2 distinct regions. It is very similar in appearance to the common dunnart but is perhaps most easily recognized by its long tail, which is approximately 1.2 times longer than its head-body length. Its fur is mid-grey above and paler grey or whitish below, with brownish patches behind the long, naked ears and thin dark rings around the large eyes. The tail is always thin and colored like the body. A nocturnal hunter, it actively forages for invertebrate and small vertebrate prey. During the day, it rests in a nest of dry leaves and grasses built in hollow logs, bases of grass hummocks or burrows. Populations reach their highest densities in regenerating areas 3-4 years after fires, when presumably the structure of the understorey vegetation makes conditions favorable for the species. Animals breed between August and March, and a single litter of up to 8 young is born. In some years, a second litter may be produced.

STATUS *Common in suitable habitat*

Long-tailed Dunnart
Sminthopsis longicaudata SPENCER (1909)

DISTRIBUTION
Central Western Australia, from Pilbara to Gibson Desert, and into southern Northern Territory

This attractive but rarely seen animal can be difficult to find but is easy to identify by its impressively long tail. It is found on rocky plateaus, boulders and scree slopes in arid western and central regions where spinifex hummocks are sparsely sprinkled with trees and shrubs. It is small (15-20 g; 0.5-0.7 oz) with a greyish-brown coat above and is white or pale grey below. Its tail, which is over twice as long as its head-and-body length, is tipped with a thin tuft of long hairs. It is probably used as a counterbalance when this agile animal is bounding and climbing over rocks. When running, the animal holds the tail straight out behind its body. As with all dunnarts, it hunts mostly for invertebrates at night but probably includes small vertebrates in its diet when the opportunity arises. Mating occurs in the spring and litters consist of up to 6 young. Little else is known of this species.

STATUS *May be common at times in suitable habitat*

Ooldea Dunnart
Sminthopsis ooldea TROUGHTON (1965)

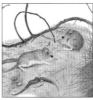

DISTRIBUTION
Central and southern arid Australia, including eastern deserts of Western Australia, Tanami Desert in the Northern Territory and western South Australia

Found mostly in sandy areas, this small animal (8-17 g; 0.3-0.6 oz) inhabits desert woodland, shrubland and hummock grassland. Only recently confirmed as a separate species, it is perhaps more common and widely distributed than is currently thought. Its coat is colored yellowish grey above and whitish below. In common with other dunnarts, it has a pointed muzzle and large eyes and ears; unlike them, it has ears with a triangular shape. The tail becomes plump with fat in good seasons but not as plump as those of the Stripe-Faced Dunnart and Fat-Tailed Dunnart. At night, animals forage for invertebrate prey; they rest during the day in shallow burrows. Litters of up to 8 young are born September-November; a second litter may be born in some years. Young are weaned at 10 weeks of age.

STATUS *Probably common but yet to be defined*

Red-cheeked Dunnart
Sminthopsis virginiae (TARRAGON, 1847)

DISTRIBUTION
Eastern Kimberley, Western Australia, Top End of Northern Territory, and coastal north-eastern Queensland including Cape York; also in Papua New Guinea

The preferred habitat of this species is tropical savanna grassland and it is often found near wet areas, such as swamps and rainforest fringes. Few dunnarts have distinctive markings but this animal has unmistakably reddish-orange cheeks and the coloring sometimes extends to the shoulders and fore-limbs. It is a large dunnart (18-58 g; 0.6-2.0 oz) with a mostly greyish-brown, flecked coat above, and whitish underparts and feet. A dark stripe down the centre of the face extends over the forehead. A nocturnal forager, the dunnart's diet consists mostly of invertebrates. Animals nest in dense vegetation on the ground and are probably adversely affected if frequent fires destroy groundcover. Breeding occurs between October and March, with up to 8 young born in a single litter. The young are weaned at around 10 weeks of age and are sexually mature by 4-6 months of age.

STATUS *Common in preferred habitat; uncommon elsewhere*

Sandhill Dunnart
Sminthopsis psammophila SPENCER (1895)

DISTRIBUTION
Isolated populations in arid Western Australia and South Australia; probably extinct in Northern Territory

This little known animal is one of the larger dunnarts (25-44 g; 0.9-1.6 oz). It is found only in arid, sandy habitats, including many types of desert woodland and shrubland, with a lower storey and groundcover of hummock grasses, usually spinifex. The hummocks provide shelter and effective protection from many predators. Once thought to be extinct, the species was 'rediscovered' in 1969 after a 75 year absence from the record. Its coat is greyish brown above and whitish below, including on the tops of the feet. It has a pointed muzzle and large eyes and ears. The eyes are ringed with dark fur and there is often a dark patch on the forehead. The tail is comparatively long and darker on the underside, the reverse of most other species. This nocturnal animal feeds on invertebrates, and probably on small mammals and reptiles. Its life history remains unstudied but females with up to 5 young have been recorded between October and January.

STATUS *Endangered; sparsely and patchily distributed throughout its range*

Stripe-faced Dunnart
Sminthopsis macroura (GOULD, 1845)

DISTRIBUTION
Arid and semi-arid Australia, from west coast of Western Australia to central NSW

A prominent dark stripe extending from the nose to the top of the head, makes this animal easy to recognize. The size of a large mouse (15-25 g; 0.5-0.9 oz), it has grey-brown upperparts and white underparts. The eyes are large and dark; the ears are large and lightly furred. In common with other arid zone dunnarts, its tail becomes distended with fat deposits when food is abundant and may become quite skinny in poor seasons. It occupies many different habitat types, including shrubland, and tussock and hummock grassland on sandy, stony, rocky and clay soils. At night, it forages for invertebrates in open spaces close to cover. During the day, it rests in soil cracks or under rocks and logs. In the breeding season from July to February, females can raise up to 2 litters of 8 young. The gestation period is very short, only 11 days. The young stay in the pouch for 4 weeks before being weaned at 10 weeks.

STATUS *Locally common*

White-footed Dunnart
Sminthopsis leucopus (GRAY, 1842)

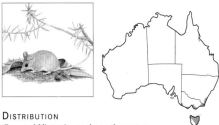

DISTRIBUTION
*Coastal Victoria and south-eastern
NSW, Tasmania and offshore islands;
1 outlying population near Paluma in
north-eastern Queensland*

This animal is generally found in drier habitats, preferring forest, woodland, heath and coastal dune grassland with sparse groundcover. It is absent from areas that have been logged or burned, apparently owing to its dislike for the dense regrowth that these disturbances cause. The White-footed Dunnart is only visibly different from the Common Dunnart by its shorter muzzle and striations on its foot pads. This small animal (19-26 g; 0.7-0.9 oz) is a nocturnal forager, hunting for invertebrates and small skinks on the ground or in leaf litter. It rests during the day in nests built from bark and leaves under fallen logs or in dense litter. In Tasmania, nests have been recorded above-ground in trees. Breeding in late July and August produces a single litter of up to 10 young born from mid-August to mid-September.

STATUS *Unknown; uncommon to rare, depending on location*

White-tailed Dunnart
Sminthopsis granulipes TROUGHTON (1932)

DISTRIBUTION
*Western goldfields and coast,
south-western Western Australia*

The shrublands and woodlands that this small animal (18-37 g; 0.6-1.3 oz) inhabits have a mid-dense understorey of low shrubs. The dunnart's coat is pale brown above and the white of the underparts extends under the chin and to the tops of the feet. The eyes and ears are large. The tail, which is white except for a light brown thin mid-line stripe, becomes fat when food is abundant. Although little is known about this species, it is probably nocturnal and feeds on a very wide range of invertebrates and perhaps occasional vertebrates. Mating appears to occur in autumn and winter, with litters of young born shortly after. Females have been recorded with up to 12 teats, the most for any species of dunnart.

STATUS *Common within limited distribution*

Kultarr
Antechinomys laniger (GOULD, 1856)

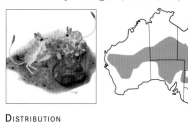

DISTRIBUTION
*Southern and central arid Australia, from central
Western Australia to central western NSW*

With its long hind-feet and brush-tipped tail, the elegant Kultarr is easily recognizable. This elusive animal is small (20-30 g; 0.7-1.06 oz), yet its distinctive, bounding quadrupedal gait gives it great speed and maneuverability. It is strictly nocturnal, and its large, dark eyes and prominent, upright ears are used to detect invertebrate prey at night. Animals rest during the day beneath logs, in the base of grass hummocks or in burrows. Kultarrs are found in a variety of arid habitats but prefer areas where there is little groundcover, which may make them more vulnerable to predation by introduced pests such as foxes and cats. Mating once a year in winter and spring is followed by 12 days' gestation and the birth of a litter of 6-8 young. The young are weaned at 3 months.

STATUS *Unknown; very patchy distribution*

Common Planigale
Planigale maculata (GOULD, 1851)

DISTRIBUTION
*Northern Australia, from the Pilbara in Western
Australia to eastern Arnhemland, including Groote
Eylandt, east coast from northern Cape York to
central NSW*

This animal (6-24 g; 0.2-0.8 oz) is the largest planigale in Australia. It inhabits a wide range of wet and dry habitats, including sand dunes, marshes, forests and rocky areas. Its coat is reddish brown above, sometimes with small whitish spots, and is greyish on the underparts. It has a flattened head, although not as extreme as that of the Long-tailed Planigale. The muzzle is pointed, and the ears are prominent and naked. A nocturnal predator, it actively hunts in leaf litter and in small crevices for insects. Like all planigales, it is renowned for aggressively tackling prey sometimes as large as itself. It jumps on the back of the moth, grasshopper or centipede, and repeatedly bites its head and body until it is overcome. Breeding differs between northern and eastern populations: in the north, breeding throughout the year results in larger litters (8-12 young); in the south, litters of 5-10 young are born from October to January. Future research might divide these populations into 2 distinct species.

STATUS *Common*

Giles' Planigale
Planigale gilesi AITKEN (1972)

DISTRIBUTION
Arid and semi-arid central and eastern Australia

This medium-sized planigale, weighing 6-15 g (0.2-0.5 oz), is found in habitats with cracking clay soils, often vegetated quite densely with grasses. In common with other planigale species, it has a flattened head-and-body shape for squeezing through cracks and crevices in the soil. Its coat is grey above and paler below, and its claws are always blackish. When food is abundant, the tail may become plump with fat deposits but is usually quite skinny. An aggressive hunter, it is nocturnal and actively seeks out its large invertebrate prey in soil cracks, leaf litter and low vegetation. Occasionally it includes small reptiles and mammals in its diet. Normally animals are protected from extreme temperatures when resting in soil crevices but they may emerge to bask in the sun when cold; they can also become torpid for short periods. 1 or 2 litters of up to 6-8 young are born each year, and become independent at around 10 weeks of age.

STATUS *Sparsely distributed*

Long-tailed Planigale
Planigale ingrami (THOMAS, 1906)

DISTRIBUTION
*Northern Australia, from the Kimberley in Western
Australia to Rockhampton, Queensland, with
isolated populations in arid Australia*

The smallest marsupial, this tiny (4-6 g; 0.14-0.21 oz) animal lives and hunts for food in cracks and crevices in the soil. It is common in habitats with cracking soils, such as savanna woodland and grassland. When much of this habitat floods in the wet season, animals move to higher ground until the waters subside. It has a very flat, wedge-shaped head (only 4 mm (0.16") high) which helps it maneuver through soil crevices. Its coat is variably brown or grey above with paler underparts. Its eyes are small and the naked ears are large. The tail is thin and long. These nocturnal animals are voracious predators of large invertebrates and small vertebrates, sometimes as large as themselves. Litters of 4-8 young are born throughout the year but mostly between September and March. The young are carried in the mother's backward-opening pouch for around 6 weeks, and are weaned at 3 months of age. This species, like the Common Planigale, is under taxonomic review and may turn out to be a complex of 2 or more species.

STATUS *Common in suitable habitat*

Narrow-nosed Planigale
Planigale tenuirostris Troughton (1928)

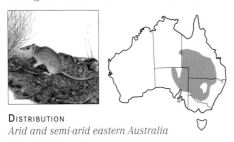

Distribution
Arid and semi-arid eastern Australia

This tiny animal (4-9 g; 0.14-0.3 oz) lives in soil cracks and crevices in various habitats with cracking soils such as mallee shrubland, hummock grassland, Mitchell grassland and on the fringes of watercourses. Its whole body and head is flattened to aid movement through soil cracks. Its coat is pale brown to mid-brown above, and pale grey brown or whitish below. It has light brown claws that serve to distinguish it in the field from Giles' Planigale. Active at night, and frequently hunting in soil cracks rather than on the surface, it is a fearless predator of large invertebrates such as grasshoppers, moths and centipedes. During the day, it rests in soil cracks where it is insulated from extreme temperatures. In winter, animals may become torpid for short periods and can be found nesting together to conserve warmth. Breeding usually occurs once a year in winter or spring. Litters of up to 12 (but usually 6) young are born after a gestation period of 19 days and are carried in the pouch until around 6 weeks of age. Young are weaned at 13 weeks.

Status *Sparsely distributed over a wide range*

NINGAUI

Pilbara Ningaui
Ningaui timealeyi Archer (1975)

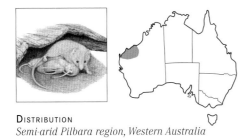

Distribution
Semi-arid Pilbara region, Western Australia

This tiny animal (5-9 g; 0.18-0.3 oz) has a little more red in the coat but is otherwise difficult to distinguish from the Wongai Ningaui. It is found in areas vegetated with dense clumps of spinifex, which may also have a sparse overstorey of emergent trees and shrubs. An agile climber, it moves nimbly up and through spinifex clumps in search of invertebrate prey, and also forages extensively on the ground. It shelters during the day in burrows or in spinifex clumps where dense spiky stems give protection from predators. The breeding season begins in September. Pouch young are still evident in March suggesting that, in some years, 2 litters may be produced. Litters consist of 4-6 young.

Status *Locally common in some areas*

Southern Ningaui, Mallee Ningaui
Ningaui yvonneae Kitchener, Stoddart and Henry (1983)

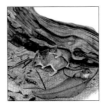

Distribution
Patchily distributed through semi-arid southern Australia

The Southern Ningaui is found in sandy habitats, such as mallee scrub, heath and hummock grassland but only where spinifex grows. This small animal (5-10 g; 0.18-0.35 oz) has a shaggy coat, colored grey brown above and greyish below, with white under the chin. The ears are flattened against the side of the head, and the eyes are small and close-set. It is nocturnal and actively forages for invertebrate prey in and around spinifex hummocks, climbing up through spinifex stems to catch spiders, grasshoppers and moths. It also eats small lizards. Animals shelter during the day in dense spinifex clumps or other vegetation, or in burrows or hollow logs. The breeding season starts in late winter (late August) and the single litter of 5-7 young remains in the pouch for around 4 weeks. Young are weaned and independent at 10-12 weeks of age.

Status *Sparsely distributed*

Wongai Ningaui
Ningaui ridei Archer (1975)

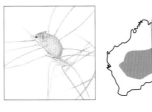

Distribution
Arid central Australia, from central Western Australia to central Queensland

This small (6-10 g; 0.2-0.35 oz), shaggy-haired animal can be found throughout the spinifex deserts of central Australia. Although widespread, it never occurs in large numbers and leads a solitary life outside of the breeding season. It hunts voraciously at night for invertebrate prey, favoring beetles, crickets and spiders up to 10 mm (0.39") in length. It is well adapted for hunting in dense clumps of spinifex: it has a small body, a cone-shaped head, flattened ears for squeezing through small spaces, and a thin, semi-prehensile tail for climbing spinifex stems in search of prey. During the day, animals shelter in lizard or spider burrows or in the base of spinifex hummocks. Females breed once or twice a year, giving birth to at least 10 young. Of these, 6-7 will survive if they attach to teats in the mother's open pouch, where they stay for 6 weeks. The young leave the maternal nest when they are weaned at 12-13 weeks. There is some evidence that both sexes disperse up to several hundred metres at this time.

Status *Sparsely distributed but common*

MYRMECOBIIDAE
MYRMECOBIUS

Numbat
Myrmecobius fasciatus Waterhouse (1836)

Distribution
South-western Western Australia

Widely distributed across central and southern Australia until European settlement, this unique and easily recognized marsupial is now restricted to the remnant eucalypt forests of south-western Australia. It weighs 300-720 g (10.5-25.2 oz), has orange-brown body fur which becomes darker towards the rump and is striped with 6-7 conspicuous white bars across the back. The head is narrow, with a dark cheek stripe. Its long tail is stout and bushy. It has the most specialized diet of any marsupial, eating only termites which it mines from galleries in logs, trees and the ground. Its daily routine depends on termite activity, so it spends the day searching for food and rests at night. In summer, both termites and numbats are inactive during the hottest part of the day; in winter, both prey and predator are active during the warmer hours. Animals are mostly solitary and mate from February to April. Litters of 2-4 young are born after a 14 day gestation period. Young are carried in the pouch for around 4 months, are weaned at 6 months and become independent at 8-9 months.

Status *Vulnerable due to habitat destruction and predation by cats and foxes*

Northern Marsupial Mole
Notoryctes caurinus THOMAS (1920)

Southern Marsupial Mole, Itjaritjari
Notoryctes typhlops (STIRLING, 1889)

Northern   Southern

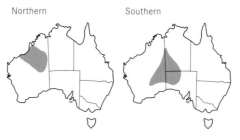

DISTRIBUTION
Sandy deserts: the northern species in western Western Australia, the southern species in south-ern Western Australia, Northern Territory and western South Australia

These 2 enigmatic species are unlike any other marsupials and are almost identical in appearance. They inhabit sandy desert environments where they spend most of their lives burrowing through sand. They are supremely adapted to a burrowing way of life. The body is strongly built and elongated (40–60 g; 1.4–2.1 oz), covered in fine silky fur colored off-white to red-brown depending on the amount of staining from red sands. The eyes are reduced to vestigial lenses, the external ears are small holes covered by hair, and the snout is protected by a tough, horny shield. The fore-limbs are short and powerful with 2 spade-like claws. In the female, the pouch faces backwards to protect the young from sand; in the male, the testes are situated beneath the skin in the abdomen. Moles 'swim' through sand 'diving' to depths of 2.5 m (8 ft) beneath the surface, the tunnels being back-filled behind them. Occasionally animals are observed on the surface, usually after rain, where they leave characteristic tracks in the sand. Moles appear to be insectivorous but may also eat subterranean plant matter. Females produce 2 young in a litter.

STATUS *Endangered, but difficult to assess size of populations*

Desert Bandicoot
Perameles eremiana SPENCER (1897)

DISTRIBUTION
Formerly central and west-ern deserts

Last collected in 1943 and presumed extinct since the 1960s, this small bandicoot was similar in appearance to the Western Barred Bandicoot except for the orange or reddish coats of animals in some areas. In its absence it is unlikely that its status as a separate species or a subspecies of the Western Barred Bandicoot will ever be fully resolved. It inhabited sandy deserts vegetated with spinifex or tussock grasses. It excavated shallow scrapes beneath grass hummocks or shrubs and rested there during the day in nests made from dried grasses, leaves and twigs, creating a leafy tunnel to shelter from predators and extreme temperatures. At night, animals emerged from their nests to forage for invertebrates and possibly plant material also. Litters probably consisted of 2 young, perhaps more (females have been recorded with 8 teats). Extinction of this animal was probably caused by unfavorable changes in vegetation structure from altered fire regimes, and predation by foxes and cats.

STATUS *Extinct*

Eastern Barred Bandicoot
Perameles gunnii GRAY (1838)

DISTRIBUTION
South-western Victoria, northern and eastern Tasmania

Almost extinct on the Australian mainland, this distinctive animal is most common in Tasmania, where it inhabits pastoral grassland with stands of denser vegetation for shelter. It is easily identified by its 3–4 bold whitish bars across its hind-quar-ters, which contrast with the darker brownish-grey coat. The chest and underparts are pale grey. It has a compact body (500–1450 g; 1.1–3.2 lb), and a long snout, large eyes and erect, tall ears up to 5.5 cm (2.2") long. It uses its powerful claws to dig for food and leaves tell-tale conical pits up to 30 cm (12") deep in the soil. An omnivore, it feeds opportunistically on foods such as insects, worms, lizards, small mammals, bird eggs, bulbs, roots and tubers. Animals are nocturnal, and rest dur-ing the day in grass lined nests built in shallow depressions beneath bushes in dense vegetation. They are solitary except when mating. Litters of 2–3 young are born after a gestation period of only 12.5 days and are weaned at 2 months. Females can produce new litters consecutively.

STATUS *Vulnerable; endangered on mainland; sparse to common in Tasmania*

Long-nosed Bandicoot
Perameles nasuta GEOFFROY (1804)

DISTRIBUTION
Eastern coast of Australia, from north-eastern Queensland to southern Victoria

This animal makes itself known to many human residents of eastern Australia by digging in their gardens. Its natural habitat includes wet and dry forest and woodland, and heathland. It is a small animal (850–1100 g; 1.9–2.4 lb) with greyish-brown coarse spiky fur above, and smoother, whitish fur on its underside. It has a long snout, which it probes into the ground when searching for food, and large, erect ears. The claws are long and used for digging, and the tail is short. Animals forage at night, digging holes and sifting through leaf litter for invertebrates, small reptiles, roots and tubers. During the day, animals rest in leafy nests built in shallow scrapes beneath bushes. Breeding occurs throughout the year at low latitudes (it ceases in winter in the south),with litters of 1–5 young, usually 2–3, born after a short gestation period (12.5 days) and carried in a backward-opening pouch. Females can produce a litter every 70 days, so pups are weaned when they are still very young.

STATUS *Locally common; uncommon where fox density is high*

Western Barred Bandicoot
Perameles bougainville QUOY AND GAIMARD (1824)

DISTRIBUTION
Bernier and Dorre islands, Western Australia

Two offshore islands in Shark Bay are now the refuge of this previously widespread species. Until European settlement, it had inhabited a range of semi-arid shrubland habitats across southern and western Australia where dense vegetation provided shelter from predators. The animal is small (175–285 g; 6.1–10.0 oz) and finely boned. Its coat is grey brown with darker and lighter bars across the rump, back and flanks, and the underparts are greyish to cream. It has a long, narrowly pointed muzzle and large erect ears. The tail is short and bi-colored; it has long claws on all feet. Active at night, it feeds on invertebrates and roots, herbs and seeds that it finds by digging and sifting through sand, soil and leaf litter. During the day it shelters in a scrape excavated beneath low shrubs, in a nest lined with leaves and grasses. Animals are solitary except when breeding. Litters of usually 2 young (but up to 4) are born between April and October.

STATUS *Endangered; currently included in re-introduction programs on the mainland where feral predators can be controlled*

Golden Bandicoot
Isoodon auratus (RAMSAY, 1887)

DISTRIBUTION
Isolated populations in north-western Kimberley region, and Barrow, Middle, Augustus and Uwins islands, Western Australia and Marchinbar Island, Northern Territory

The Golden Bandicoot now inhabits only a fraction of its former range of northern, western and central Australia. Changes in land use after European settlement probably caused the extreme contraction of its range. Its coat is colored a warm golden brown, through which coarse, spiky black guard hairs protrude; the underparts are pale brown. The muzzle is pointed, the eyes are small and the ears are short. It has strong claws on all feet for digging, and a short tail. It has been found in a variety of habitats, including spinifex hummockland, semi-arid woodland with tussock grass understorey, rocky sandstone country and rainforest patches. It is nocturnal and forages for invertebrates, small reptiles, mammals and plant matter on the ground and by digging conical pits in the soil. Animals may breed throughout the year, producing litters of usually 2, sometimes 3, young. Recent genetic work suggests that this species may represent the inland form of the Southern Brown Bandicoot.

STATUS *Vulnerable; common on offshore islands; vulnerable in Kimberley region*

Northern Brown Bandicoot
Isoodon macrourus (GOULD, 1842)

DISTRIBUTION
Northern and eastern coast of Australia, from Kimberley region, Western Australia to mid-NSW; also in Papua New Guinea

The largest bandicoot (500–3100 g; 18 oz–7 lb), this animal inhabits open forest, woodland and grassland of eastern and northern Australia, preferring habitats where groundcover is sparse but understorey vegetation, such as shrubs and grasses, provides shelter. It has coarse, brown-black fur above with whitish underparts, a long pointed muzzle, and small eyes and ears. It uses its strong claws to dig for food while foraging at night, unearthing invertebrates, roots, tubers, grass seeds and berries. Animals rest during the day in a scrape covered with leaf litter and twigs, beneath grass tussocks or in hollow logs. Animals are mostly solitary. Breeding times differ depending on location: animals in south-eastern Queensland breed year round; those in other regions breed seasonally. Litters of 2–4 young are

born after a gestation period of 12.5 days, remain in the backward-opening pouch for 7 weeks and are weaned soon after. Females can produce a litter every 2 months.

STATUS *Common where habitat remains intact; uncommon or extinct elsewhere*

Southern Brown Bandicoot
Isoodon obesulus (SHAW, 1797)

DISTRIBUTION
Fragmented range across southern coastal Australia, from south-western Western Australia to south-eastern NSW, Tasmania and offshore islands; isolated population in northern Cape York Peninsula, Queensland

Smaller at 400–1600 g (14 oz–3.5 lb) than the similar Northern Brown Bandicoot, this animal is found in sandy habitats vegetated with heath or scrub. It is most common in areas that are periodically burnt, where regenerating vegetation provides ample shelter and food resources. Changes in fire frequency and land use since European settlement have caused the range of this animal to contract substantially. Its coarse fur is a yellowish brown or grey, with cream underparts. It has small eyes and ears, and a long pointed muzzle. The tail is short and colored dark brown. Animals are nocturnal foragers, digging through the topsoil to find invertebrates such as insects and earthworms, fungi and plant material. During the day, animals rest in nests of grass and leaves hidden in dense vegetation or leaf litter. Breeding from winter through to late summer produces 2–3 litters of 2–4 young in a single season.

STATUS *Locally common in some areas; uncommon elsewhere; extinct in many parts of former range*

Pig-footed Bandicoot
Chaeropus ecaudatus (OGILBY, 1838)

DISTRIBUTION
Formerly western, central and southern arid Australia

Last recorded officially in 1901, this small (200 g; 7 oz), delicately built animal is now extinct, probably as a result of changes in land use. It inhabited spinifex and tussock grassland in arid sandy habitats; in semi-arid areas, it preferred open woodland habitats with a grassy understorey. Its

---

ECHYMIPERA

Rufous Spiny Bandicoot, Long-nosed Echymipera
Echymipera rufescens (PETERS AND DORIA, 1875)

DISTRIBUTION *Eastern Cape York, northern Queensland; also in Papua New Guinea*

This large bandicoot (600–2000 g; 1.3–4.4 lb) inhabits rainforest, gallery forest, grassy woodland and dense coastal heath, and is frequently found near disturbed areas. It is an extremely variable species in color and size: its coat is dark grey to blackish in some areas, dark reddish brown in others. Its fur is very coarse and spiny. The head is elongated with a long, conical snout, and small upright ears; it has a very short, almost naked tail. Males are usually 25–30% bigger than females. Animals are inactive during the day, resting in burrows excavated in sheltered sites. At night, they emerge to forage in soil and leaf litter for insects. Small vertebrates and plant material are probably eaten too. Litters of 1–3 young have been recorded. Although relatively common within its range, little is known of this species.

STATUS *Locally common within a limited range*

coat was orange brown above and fawn below, with a dark crest towards the tip of its tail. It had large ears, a finely pointed muzzle and long slender legs. The front feet were structured so that only the 2nd and 3rd toes were used for locomotion, giving the appearance of a cloven hoof of a pig or deer, hence the common name. On the hindfeet, only the 4th toe was used for locomotion. The presumably nocturnal animal probably had an omnivorous diet of invertebrates and plant matter. It sheltered during the day in a grass-lined nest either under dense vegetation or perhaps at the end of a short, straight burrow. According to Aborigines, animals could move at high speed, sometimes finding shelter in hollow logs, and probably persisted in Western Australia until the 1950s. Little is known of their breeding habits, except that litters of no more than 2 pouch young have been recorded.

Status *Extinct*

THYLACOMYIDAE

MACROTIS

Bilby, Greater Bilby
Macrotis lagotis (REID, 1837)

DISTRIBUTION
Arid Western Australia and Northern Territory with isolated populations in southern Northern Territory and south-western Queensland

With its long, rabbit-like ears, silky blue-grey fur and distinctive black-and-white tail, this large bandicoot (800–2500 g; 1.8–5.5 lb) is immediately recognizable. After European settlement, changes in land use and vegetation, and predation by foxes and cats, caused the sudden and widespread contraction of its range. It inhabits spinifex grass-land on sand or clay soils, open stony plains and *Acacia* shrubland. A powerful digger, it excavates burrows up to 2 m (6.5 ft) deep and 3 m (10 ft) long in which it shelters during the day. At night, it emerges to forage for insects, small lizards, fungi, fruit and seeds, even those that are buried, using its acute senses of hearing and smell. Animals live alone or in pairs, sometimes with newly independent young. Litters of 2–4 young are mostly born in March–May or throughout the year if food is plentiful. Young are carried in the pouch for 11 weeks and are weaned shortly afterward.

Status *Vulnerable; patchy distribution*

Lesser Bilby
Macrotis leucura (THOMAS, 1887)

DISTRIBUTION
Formerly deserts of central Australia

Not reported alive since 1931, this now extinct small bandicoot (310–435 g; 11–15 oz) inhabited the sandy deserts of central Australia. It occupied spinifex sand dunes and sand plains, mulga, woodland and tussock grassland of Western Australia, Northern Territory and South Australia, where, according to local Aboriginal people, it may have persisted until the 1960s. It had a grey-brown coat with pale greyish underparts. The ears were very long, the muzzle pointed and the eyes were small. The tail was mostly white, with a greyish underside. Often found in the same habitat as the larger Bilby, this animal always burrowed in sandy soils, often in the sides of dunes, whereas the Bilby prefers burrowing in loam soils on flat ground. The Lesser Bilby apparently filled in the mouth of its burrow while inside, leaving a faint depression in the sand. Records of this nocturnal forager eating rodents show it to have been partially carnivorous. It also included some plant material, such as seeds, in its diet. Females had 8 teats but litters consisted of 2 young.

Status *Extinct*

DIPROTODONTIA

PHALANGERIDAE

TRICHOSURUS

Common Brushtail Possum
Trichosurus vulpecula (KERR, 1792)

DISTRIBUTION
Northern Australia, central Australia, south-western Western Australia, eastern Australia from Queensland to South Australia, Tasmania; introduced to New Zealand

This large nocturnal possum (1.2–4.5 kg; 2.6–9.9 lb), familiar to many Australians, lives in trees in habitats as varied as forest, woodland, farmland and urban areas. It varies in color and size throughout its range: in the north, animals are smaller and have reddish coats whereas in the southern regions, animals are heavier, with darker, longer-haired coats. In wooded habitats, it feeds on a wide variety of fruit, flowers and leaves but sometimes scavenges on carrion. It dens in hollow logs and crevices, is territorial and solitary outside of the breeding season. It has adapted superbly to life in urban habitats: animals frequently den in roofs and forage in gardens and garbage bins, often to the annoyance of other residents. Females can breed from 1 year of age, with a single young usually born in autumn and sometimes another in spring. After 4–5 months in the pouch, the young is carried on the mother's back until weaning at 5–7 months of age.

Status *Common in south-east and south-west; declining and endangered in central Australia*

Mountain Brushtail Possum
Trichosurus cunninghami LINDENMAYER,

DUBACH AND VIGGERS (2002)

DISTRIBUTION
Coastal plains and ranges of south-eastern NSW, and central and north-eastern Victoria

This large possum (2.0–4.5 kg; 4.4–9.9 lb) is distinguished from the similar Common Brushtail Possum and Short-eared Possum by its stocky build, dark color, thick fur and shortish ears that protrude above the top of its head. It is found in coastal and mountain forests where mature trees provide hollows for dens. A nocturnal forager, it feeds on a wide variety of leafy vegetation and plant material on the forest floor. Animals are territorial and use clear secretions from glands on the chest and under the tail to mark occupied areas with scent. Animals can be long lived, with females surviving for 17 years, and males 12. Breeding animals form long-term monogamous partnerships; females seldom breed until they are 3 years old. A single young is born after only 15–17 days gestation, and remains in the pouch for 5–6 months. After weaning, at 7–12 months, it stays in the same area for up to 3 years before dispersing to a new territory.

Status *Common*

Short-eared Possum
Trichosurus caninus (Ogilby, 1836)

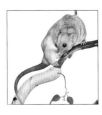

DISTRIBUTION
Coastal plains and ranges of central eastern Queensland to central eastern NSW

This large possum (2.0–4.5 kg; 4.4–9.9 lb) of both dry and wet forests is similar in appearance to the more southerly occurring Mountain Brushtail Possum but has quite short, rounded ears and a long, brushy tail. Both males and females are of similar size and have a stocky build with powerful claws for climbing trees. The dense fur is colored grey brown above, usually with rich cream-colored underparts. Animals forage at night on a wide range of plant material gleaned from the forest floor and understorey vegetation. They build nests in hollows in mature trees or fallen logs and retreat to them to rest during the day. Animals form enduring partnerships. A single young is born after 15–17 days gestation, usually between March and June. It remains in the pouch for around 6 months and is weaned at 8–11 months of age. Longevity and other aspects of life history are very similiar to those of the Mountain Brushtail Possum.

STATUS *Common*

WYULDA

Scaly-tailed Possum
Wyulda squamicaudata Alexander (1919)

DISTRIBUTION
Kimberley coastal region, northern Western Australia

As its name suggests, this mid-sized possum (1.3–2.0 kg; 2.9–4.4 lb) has a distinctive scaly, prehensile tail which it uses for climbing. Its uppermost coat is mostly grey with a dark dorsal stripe, although the thickly furred tail base is reddish; the belly is a pale cream. It inhabits rainforest, woodland and vine thickets that persist in rugged, rocky areas. Animals shelter in rock piles or crevices during the day, emerging at night to feed on blossoms, leaves and fruits. If animals are disturbed when feeding, they will descend from the tree to return to the rock shelter below. Animals are solitary except in the breeding season (March–August). The single young is retained in the pouch for 6 months and weaned at about 8 months.

STATUS *Rare to common in a limited area*

Common Spotted Cuscus
Spilocuscus maculatus (Desmarest, 1818)

DISTRIBUTION
Northern Cape York Peninsula

A distinctive animal with a pink, bare-skinned face, small ears and forward-looking reddish eyes, this tree-dwelling species (1.5–4.5 kg; 3.3–9.9 lb) lives in the tropical rainforest and mangroves of Cape York. Its fur is dense and variably colored, with upperparts grey with whitish cheeks and shoulders, and underparts whitish. The backs of males are often mottled with white blotches. The bumps covering the underside of the naked prehensile tail help with climbing. Although agile, it is a relatively slow climber that moves through the tree canopy at night in search of fruits, leaves, flowers, insects and possibly even small vertebrates. Animals usually spend the day sleeping in the tree canopy in temporary nests of twigs and leaves. They are usually solitary, except during the extended breeding season. Up to 3 young have been recorded in a single litter but it is likely that only 1 will survive to maturity.

STATUS *Sparsely distributed in Cape York; a different subspecies is common in Papua New Guinea*

Southern Common Cuscus
Phalanger mimicus Thomas (1922)

DISTRIBUTION
Eastern side of Cape York Peninsula

Smaller (1.5–2.2 kg; 3.3–4.8 lb) and more secretive than the Common Spotted Cuscus, this species is restricted to the rainforest of a small part of the Cape York Peninsula. Its coat is mostly grey with pale underparts and a distinctive, narrow, dark-brown stripe that extends down the back from the head to the tail. The tail itself is furred at the base, then dark and scaly to the tip, and is used extensively for climbing through the tree canopy. Animals are vegetarian, eating a variety of fruits, leaves and blossoms from trees at night. Tree hollows or dens are used for sleeping during the day. Animals are solitary except during the breeding season. Females have 4 teats and give birth to 1–2 young but little is known of the species' reproductive biology and breeding behavior.

STATUS *Sparse over a limited distribution*

Eastern Pygmy-possum
Cercartetus nanus (Desmarest, 1818)

DISTRIBUTION
South-eastern Australia and Tasmania

Wet and dry eucalypt forests, temperate rainforests, heathlands and subalpine woodlands are the preferred habitats of this small possum (15–40 g; 0.5–1.4 oz). Its coat is grey-brown above, and pale grey and white below. The almost-naked prehensile tail is often distended with fat deposits. Animals are nocturnal, feeding mostly on nectar and pollen which are collected from flowers by means of a brush-tipped tongue. It may also eat arthropods but usually when flowers are not available. Nests, which are made from strips of bark, seem to be built by mothers with young only, and are usually situated in clefts and crannies in trees. Breeding can occur throughout the year on the mainland but stops in summer in Tasmania. Up to 3 litters a year are produced, each of 4 young, which remain in the pouch for about 4 weeks and are weaned at 7–9 weeks of age.

STATUS *Sparse to locally common*

Little Pygmy-possum
Cercartetus lepidus (Thomas, 1888)

DISTRIBUTION
South-eastern South Australia and Kangaroo Island, south-western Victoria, Tasmania

The smallest of the possums, this tiny (6–10 g; 0.2–0.35 oz) animal is found in the mallee shrubland and spinifex deserts of South Australia and Victoria, and in wet and dry sclerophyll forests of Tasmania. Its coat is grey above and paler on the underparts, its ears are large and pinkish, and its sparsely haired prehensile tail is often swollen with fat. Unlike other pygmy-possums, its diet includes many arthropods and even small vertebrates; it also eats nectar. It is nocturnal and arboreal but prefers to hunt and nest in thick vegetation close to the ground instead of the tree canopy, probably to avoid predators such as owls. Nests are found in hollow logs and stumps, and in spinifex hummocks. Animals can breed all year round on the mainland and in spring in Tasmania. Litters comprise 4 young, which become independent at about 12 weeks of age.

STATUS *Uncommon in all habitats*

Long-tailed Pygmy-possum
Cercartetus caudatus (Milne-Edwards, 1877)

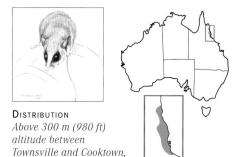

Distribution
Above 300 m (980 ft) altitude between Townsville and Cooktown, north-eastern Queensland; also in Papua New Guinea

This small (25–40 g; 0.9–1.4 oz), agile possum is easily recognized by the large black patches around the eyes and muzzle, and the long prehensile tail. Its coat is grey to brown on upperparts, and pale underneath; the skin of the naked tail is dark. This possum is nocturnal and arboreal, and jumps quickly from branch to branch when foraging for arthropods and nectar from flowers in its rainforest habitat. It builds almost spherical nests from leaves in shelters such as tree hollows. It sleeps during the day in the nest where it leaves young that have left the pouch. Adults are sometimes found nesting together, although they are often solitary. Females give birth to 1–4 young in up to 2 breeding seasons a year. Young remain in the pouch for 6–7 weeks and are weaned at around 10–12 weeks.

Status *Locally common in a restricted distribution*

Western Pygmy-possum
Cercartetus concinnus (Gould, 1845)

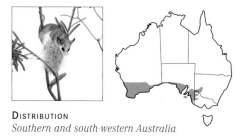

Distribution
Southern and south-western Australia

The mouse-sized Western Pygmy-possum (8–20 g; 0.3–0.7 oz) is distinguished from other pygmy-possums by the uniformly white fur on its underparts, and reddish-brown fur on the upperparts. It has a dark ring around the large eye, long rounded dark ears and a sparsely furred tail that is often coiled. It inhabits mostly dry woodland and heath that has a dense undergrowth of shrubs. A nocturnal forager, it climbs through trees and shrubs to feed on flower nectar and arthropods. It uses leaves to construct spherical nests in tree hollows and grass-tree stumps, and sleeps in them during the day. However, some animals have been observed sheltering in disused birds' nests or under tree stumps on the ground. Adults breed up to 3 times a year, when females give birth to litters of up to 6 young which remain in the pouch for around 4 weeks. Weaning occurs at 50 days but sexual maturity is not achieved until 1 year.

Status *Widespread in southern Western Australia; uncommon in dry areas*

Mountain Pygmy-possum
Burramys parvus Broom (1896)

Distribution
Alpine areas above 1400 m (4600 ft) in NSW Snowy Mountains and Victorian Alps

Originally described from a fossil specimen, this largest of the pygmy-possums (35–80 g; 1.2–2.8 oz) is restricted to a small alpine area. Its coat is grey above and pale below, with reddish flanks in the breeding season. The tail is long and almost naked. It is the only Australian mammal that hibernates, before which it is capable of doubling its body weight. It gathers its diet of arthropods, fruits and seeds by foraging in and around boulder patches. The nooks and crannies between boulders are insulated from extreme temperatures, and provide good shelter and hibernation sites. Animals can be long-lived (up to 12 years old). Adults hibernate for 7 months each year (April–October), juveniles for 5–6 months. Breeding is in a tightly restricted season, the alpine spring, with 4 young carried in the pouch for 4–5 weeks, and weaned after another 4–6 weeks. All young are independent by the end of summer.

Status *Endangered; locally common in a tiny area*

Feathertail Glider
Acrobates pygmaeus (Shaw, 1794)

Distribution
Forests of eastern and south-eastern Australia to south-eastern South Australia

The smallest of the gliders (10–14 g; 0.35–0.5 oz), the Feathertail Glider is easily recognized by its distinctive tail, which is 'feathered' with a fringe of stiff hairs. Its coat is grey above and pale below, with prominent dark eye markings; the gliding membrane extends from the elbows to the knees. It feeds at night on nectar, which it collects with its brush-tipped tongue, and on arthropods. It is an extremely agile and fast climber, aided by ridged feet and sharp claws which grip well on most surfaces. It glides from tree to tree up to a distance of 25 m (80 ft), using the tail to help steer and brake. Animals feed and nest together in groups of 7–20 adults, which are not necessarily related. In tropical areas, animals breed year round; in the south they tend to breed in late

winter and spring. In Victoria, females give birth to 2 litters a year, with 3–4 young in each litter. Young remain in the pouch for about 7 weeks and are weaned at 14 weeks.

Status *Common in old growth and wet forests, less common in younger or dry forests*

Leadbeater's Possum
Gymnobelideus leadbeateri McCoy (1867)

Distribution
Forests of central highlands of Victoria, also an isolated population near Melbourne

Thought to be extinct until rediscovered in 1961, this small possum (100–160 g; 3.5–5.6 oz) inhabits the cool Mountain Ash forest of central Victoria. Its coat is grey above and cream on the underparts and cheeks, with a dark stripe below the cheek and another extending down the back to the tail. The tail is narrow at the base and broader at the tip. A fast-moving, agile climber, it forages at night for arthropods and drinks sap from a wide range of plants. Animals use shredded bark to build communal nests in hollows in mature trees. They live in small colonies of up to 8 individuals consisting of a breeding pair, 1 or more generations of offspring, and unrelated males. They aggressively defend the territory around the nest (1–2 ha; 2.5–4.9 acres) from other colonies. Each year, females produce 1–2 litters of 1–2 young which remain in the pouch for 12–13 weeks and are weaned at around 4 months. When sexually mature (10–12 months) daughters are forced to disperse by being evicted from the colony.

Status *Endangered*

Mahogany Glider
Petaurus gracilis (De Vis, 1883)

Distribution
A 100 km (62 mile) coastal strip in north-eastern Queensland, south of Tully

This endangered species is found only in a narrow strip of swampy woodland. It is similar in appearance to the Sugar Glider but is larger (255–410 g; 9–14.5 oz) and has a longer tail that is narrow at the base. Its coat is variably colored, often mahogany brown above, but may be a smoky-grey color; the underparts are pale but may be buff or apricot.

The bushy tail is 1.5 times the body length, and has a darker tip. A dark mid-line stripe extends from between the eyes to the lower back. It is an agile climber and glides between trees using the membrane of skin that extends from the wrists to the ankles. Animals feed at night on tree sap, nectar and pollen from flowers, lichens and invertebrates. It nests during the day in tree hollows, either singly or in pairs. Little is known of this animal's breeding habits.

Status *Endangered; at risk from habitat destruction by land clearing for agriculture*

Squirrel Glider
Petaurus norfolcensis (Kerr, 1792)

Distribution
Eastern Australia, from north-eastern Queensland to central Victoria

Similar in appearance but twice the size (190–300 g; 6.6–10.5 oz) of the closely related Sugar Glider, the Squirrel Glider is found in dry and wet sclerophyll forests. Its coat is blue grey to brown grey above, with a creamy white belly, and a dark stripe on the head and back. It has a long, bushy prehensile tail, and a gliding membrane, extending from the tip of the 5th finger to the hind-feet, that enables it to glide 50 m (164 ft) or more. It is a highly active, agile, fast-moving animal that forages at night for invertebrates, nectar, pollen, gum, sap and seeds. Animals live in territorial colonies of up to 8 family members, consisting of 1 adult male, 1–3 adult females and young. Territories are vigorously defended, and colony members scent-mark the boundary and chase intruders. Animals begin to breed in winter, with litters of 1–2 young born after 20 days gestation. Young remain in the pouch for 10–12 weeks, and are weaned at 16 weeks.

Status *Locally common in dry forests; habitat destruction in the south is a threat*

Sugar Glider
Petaurus breviceps Waterhouse (1839)

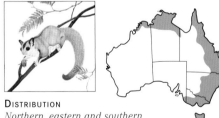

Distribution
Northern, eastern and southern Australia; introduced to Tasmania; also in Papua New Guinea

The most common and widespread of the gliders, this attractive, medium-sized (115–140 g; 4.0–4.9 oz) species inhabits a variety of forest types throughout its range, from tropical rainforest in the north to dry woodland in the south, as well as more marginal areas, such as corridors of remnant vegetation in cleared farmland. Its coat is grey above and creamy white below, with dark patches around the eyes, and blackish stripes below the ears and on the spine. The bushy tail is slightly tapered and often has a white tip. At night, this animal glides from tree to tree in search of sap,

nectar, pollen and invertebrates, before returning to the communal nest to rest during the day. Animals live in social groups of up to 7 adults, and have clearly defined territories. Litters of 1–2 young are born in winter or spring and remain in the pouch for 10 weeks; they are then transferred to the nest for a further 6–7 weeks. At 7–10 months, the independent young begin to disperse to new territories.

Status *Common*

Yellow-bellied Glider, Fluffy Glider
Petaurus australis Shaw (1791)

Distribution
Eastern Australia, isolated population between Mount Windsor and Cardwell, then from Mackay, Queensland, to south-western Victoria

Yellow or cream underparts distinguish this large glider (450–700 g; 1.0–1.5 lb) which is grey to brown black above. Its ears are large, dark and unfurred, and it has a long and bushy tail. The gliding membrane extends from wrists to ankles. It lives in wet sclerophyll forest and feeds at night on tree exudates (sap) that it 'milks' from trunks by biting through the bark; it also eats invertebrates and pollen. Agile and highly mobile, it moves up to 2 km (1.2 miles) in a night, gliding from tree to tree throughout its large territory (around 35 ha; 86 acres), which it marks with scent. The daytime den is built in tree hollows; animals nest in small social groups of an adult male, 1–3 adult females and their offspring. Females usually give birth to a single young, in winter or spring, which is retained in the pouch for 13–14 weeks and weaned at 4–5 months.

Status *Uncommon; patchily distributed throughout its range*

DACTYLOPSILA

Striped Possum, Triok
Dactylopsila trivirgata Gray (1858)

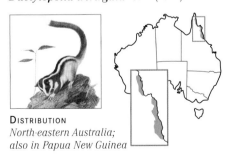

Distribution
North-eastern Australia; also in Papua New Guinea

This mid-sized (250–550 g; 8.8–19.4 oz) possum is distinctively marked with black-and-white stripes on its head and body. The long tail is grey at the base, becoming black, then tipped with white. Unlike other possums, it feeds predominantly on invertebrates which it extracts from cracks in tree bark by using its large incisor teeth, long tongue and elongated, sharply clawed 4th finger. Its diet also includes fruit and honey,

and occasionally small vertebrates and eggs. This nocturnal forager and agile climber is rarely observed on the ground. When threatened, it produces a strong, penetrating skunk-like odor from glands beneath its tail. Animals nest in tree hollows or among epiphytes, often in pairs. Litters of 1–2 young are born shortly after mating from February to August. After weaning, the young appear to disperse to new territory.

Status *Uncommon but widely distributed through many forest types*

PSEUDOCHEIRIDAE

PSEUDOCHIRULUS

Daintree River Ringtail Possum
Pseudochirulus cinereus (Tate, 1945)

Distribution
North-eastern Queensland in 3 populations: Thornton Peak massif, Mount Windsor Tableland, Mount Carbine Tableland

Only found in a small area of montane rainforest (above 400 m; 1280 ft), this mid-sized possum (0.7–1.4 kg; 1.5–3.1 lb) is easily distinguished by its coat of pale fawn or cinnamon brown above and creamy white below. It has large red-brown eyes, a short, furred tail, often tipped with white, and a brown mid-line stripe extending from the forehead to the lower back. Animals are mostly nocturnal but have also been observed during the day. They are mainly arboreal but are slow, cautious climbers that can jump short distances between stout branches. The diet consists of leaves and fruits (particularly figs) of rainforest trees. Animals are mostly solitary and spend most of the day resting in nests built in tree hollows or clumps of epiphytes. Usually litters consist of 2 young that are born during an extended breeding season.

Status *Uncommon; sparsely distributed over a restricted range although local populations may have relatively high densities*

Herbert River Ringtail Possum
Pseudochirulus herbertensis (Collett, 1884)

Distribution
Restricted area of north-eastern Queensland between Ingham and Cairns

This arboreal possum (around 1 kg; 2.2 lb) is only found at altitudes of 350 m (1150 ft) or above and in dense rainforest habitat. Its coat is dark chocolate brown above and white on the chest and belly, although some males may be dark all over. The long prehensile tail is naked on the underside and tipped with white. Animals are slow-moving and forage at night for leaves, fruits and flowers from many different rainforest plants. They are usually solitary except during the breeding season. Most young are born between May and July, with young remaining in the pouch for 15 weeks and carried on the mother's back for a further 2 weeks. The young are then moved to the nest, which is usually in a tree hollow or fern clump, and start to make short journeys independently.

STATUS *Locally common in protected habitat*

and cream or grey below. The tail is long and thin with a white tip, the eyes are ringed with dark fur, and the ears are rounded and held erect. This nocturnal forager is rarely observed on the forest floor, preferring to feed and sleep in the tree canopy. Animals are solitary outside the breeding season, resting singly in tree hollows during the day. Litters of 1 (occasionally 2) young are born in winter, and remain in the pouch for 3 months. They are independent at 6–7 months.

STATUS *Vulnerable owing to predation by foxes and habitat destruction by land clearing*

is darkly striped from the forehead to the middle of the back. It has whitish rings around the eyes and behind the ears. The base half of the prehensile tail is thickly furred, the terminal half is naked. It is the only ringtail possum that does not shelter and sleep in trees; however, it is an agile climber of rocks and trees. Young are born throughout the year, usually singly, and remain in tightly knit family groups with their parents for prolonged periods.

STATUS *Common within restricted habitat*

PSEUDOCHEIRUS

Common Ringtail Possum
Pseudocheirus peregrinus (Boddaert, 1785)

DISTRIBUTION
Eastern Australia, including Tasmania

This widespread and relatively common possum is found in a wide variety of habitats, including eucalypt forest and woodland, rainforest, coastal scrub, and suburban parks and gardens. A medium-sized animal (700–1000 g; 1.5–2.2 lb), it has a grey, reddish or dark-brown coat on upperparts, and a creamy white belly. It has short ears, and a long prehensile tail that is bi-colored dark grey and white, with the last third naked and rough on the underside, to assist in climbing. It is an agile climber that feeds at night on eucalypt leaves and flowers, and often travels along the ground between trees. In the north of its range it uses tree hollows as nests but in the south builds dreys (spherical nests of twigs and bark) on tree limbs or in dense vegetation such as mistletoe. Animals form small social groups of 1 adult male and 1–2 adult females which deliver litters of 2 young between April and November. Young are carried in the pouch for 16 weeks and are weaned at around 28 weeks. Unusually for marsupials, males help to care for the young.

STATUS *Common*

Western Ringtail Possum
Pseudocheirus occidentalis (Thomas, 1888)

DISTRIBUTION
Coastal forests of south-western Western Australia

This once-widespread species is now restricted to a few isolated populations in coastal forest, which nevertheless can reach high densities; animals even visit gardens on town fringes. Most animals inhabit forest with lush, dense vegetation where they feed on the leaves of a few favored tree species. The species is similar in size (0.9–1.1 kg; 2.0–2.4 lb) and appearance to the Common Ringtail Possum; its fur is dark brown or greyish above

---

PSEUDOCHIROPS

Green Ringtail Possum
Pseudochirops archeri (Collett, 1884)

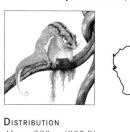 

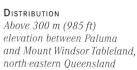

DISTRIBUTION
Above 300 m (985 ft) elevation between Paluma and Mount Windsor Tableland, north-eastern Queensland

The combination of black, yellow and white banding on the fur of this medium to large-sized (800–1300 g; 1.8–2.9 lb) possum gives its coat a very distinctive and unusual green tinge. The underparts are creamy white, as are patches around the eyes and behind the ears. There are 3 dark stripes running down the back and the fully furred prehensile tail is often tipped with white. When foraging at night, the possum remains mostly in trees where it eats leaves from a wide variety of montane rainforest plants. It is the only possum to curl up into a ball on a branch to sleep during the day and not use a nest. Animals are mostly solitary. A single young is born between August and November. When it has left the pouch, its mother carries it on her back until weaning, usually from October to April.

STATUS *Uncommon, sparsely distributed in a restricted range*

PETROPSEUDES

Rock Ringtail Possum
Petropseudes dahli (Collett, 1895)

DISTRIBUTION
Rocky escarpments of northern Australia

A large (1.2–2.0 kg; 2.6–4.4 lb) thick-set possum, this animal shelters in rocky crevices during the day and emerges at night to feed on the leaves, flowers and fruits of a wide range of trees and shrubs. Its coat is grey above and pale below, and

---

HEMIBELIDEUS

Lemuroid Ringtail Possum
Hemibelideus lemuroides (Collett, 1884)

DISTRIBUTION
Above 450 m (1475 ft) altitude south of Cairns, north-eastern Queensland

This medium to large-sized (750–1100 g; 1.65–2.4 lb) animal is the only possum that makes free-fall jumps (of 2–3 m; 6–10 ft) between branches. Its body fur is brown black above, and yellowish below, although some individuals (particularly those from northern populations) are creamy white all over. The prehensile tail is dark and bushy. As its name suggests, its face is similar to that of a lemur, with a short muzzle, large eyes and short, round ears. An agile climber, it lives in the tree canopy of cool, wet rainforest and forages at night on leaves, although it may also eat some flowers and fruits. Both males and females secrete a musky-smelling liquid which can be detected over some distance and may be used to communicate within social groups or to mark territory. Up to 3 animals nest together in dens built in tree hollows from leaves and bark. Young are born singly between August and November, and retained in the pouch for 6–7 weeks before being left in the nest or carried on the mother's back for a further 20 weeks. The young may remain with its mother for many months after weaning.

STATUS *Common within a restricted range*

PETAUROIDES

Greater Glider
Petauroides volans (Kerr, 1792)

DISTRIBUTION *Eastern Australia*

This largest of the gliders (900–1700 g; 2.0–3.7 lb) can glide a distance of 100 m (330 ft). Its coat color is variable: dark grey, dusky brown, light grey or cream on upperparts, and cream below. It has a long furry tail which is not prehensile. The gliding membrane extends from the elbow of the fore-limbs, not the wrist as in other gliders. It has very large, erect, fluffy ears and a short muzzle. It is found in *Eucalyptus*-dominated forest and woodland up to altitudes of 1400 m (4600 ft). A nocturnal forager and agile climber, it feeds exclusively on eucalypt leaves. Animals build nests in tree hollows, which are shared by a male and female during the breeding season. Mating between March and June produces a single young that is carried in the pouch for 16 weeks, then left in the nest or carried on the mother's back for 12-16 weeks. When independent, sons are driven from the nest by the father but daughters remain with the mother for up to another year.

STATUS *Common in forest with old, hollow-bearing trees; rare elsewhere*

TARSIPEDIDAE

TARSIPES

Honey Possum
Tarsipes rostratus GERVAIS AND VERREAUX (1842)

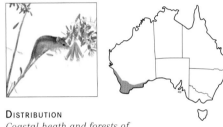

DISTRIBUTION
Coastal heath and forests of south-western Western Australia

This mouse-sized possum (7-16 g; 0.25-0.6 oz) feeds exclusively on flower nectar and pollen, which it collects on its brush-tipped tongue. Its coat is distinctively striped dark brown along the spine, grey brown above and cream below. It has a very long, pointed snout, and a long, tapering, prehensile tail. All toes (except 2nd and 3rd digits) have minute claws and expanded tips which allow it to 'stick' on vertical surfaces. This fast, nimble climber darts rapidly from flower to flower when feeding. It is nocturnal but is also active during the day in cold weather. Disused birds' nests and small hollows are used as nests. Animals are found in small social groups and individuals huddle together. They breed year-round, with 1-4 (usually 2-3) young born in each litter, and up to 4 litters produced each year. Females produce the smallest young of any mammal at birth (5 mg; 0.0002 oz), and males produce the longest sperm of any mammal (>0.3 mm; 0.01"). Mothers carry young in the pouch for 8-9 weeks and wean them 3-4 weeks later.

STATUS *Locally common but requires high diversity of flowering plants*

PHASCOLARCTIDAE

PHASCOLARCTOS

Koala
Phascolarctos cinereus (GOLDFUSS, 1817)

DISTRIBUTION
Eastern Australia but not Tasmania

The easily recognizable Koala is found in eucalypt forest and woodland below 1000 m (3300 ft) altitude. Animals are much larger in southern parts of the range (7-14 kg; 15.4-30.8 lb) than in the north (4-9 kg; 8.8-19.8 lb), and males are larger than females. Its fur is grey to greyish brown on the upperparts and creamy on the chest and belly. It has a compact, powerful body with a relatively large head and a small fluffy tail. Its ears are large with tufts of white hair, and the nose is wide, hairless and dark. It has muscular limbs and strong claws for climbing. Animals feed almost exclusively on *Eucalyptus* leaves, foraging in the tree canopy for around 4 hours each night and resting for the remainder of the day in tree forks. An agile climber, it descends to the ground only when moving between trees. Animals are usually solitary, except when breeding from September to December. A single young is born and remains in the pouch for 6 months; it is then carried on its mother's back for a further 6 months. When independent, sons disperse to new territories.

STATUS *Common in intact forest; less common where forest is unprotected*

VOMBATIDAE

VOMBATUS

Common Wombat
Vombatus ursinus (SHAW, 1800)

DISTRIBUTION
South-eastern Australia, Tasmania (including Flinders Island)

A prolific excavator of burrows, this large (25-40 kg; 55-88.2 lb), stocky animal is found in a variety of habitats throughout its range: in forest, woodland, scrub and heathland from coastal to alpine altitudes. Its coat of coarse, dense hair is uniformly brown grey on the upperparts. The ears and tail are short, and it also has a large hairless nose. The Common Wombat's short, powerful legs, broad feet and stout, flattened claws enable it to rapidly dig burrow complexes over 30 m (98 ft) long. A mostly nocturnal grazer, it eats native grasses, sedges and the roots of shrubs and trees. In cold

conditions, animals also graze during the day; they even forage in snow. Individuals usually inhabit their preferred burrows but appear to visit other animals' burrows occasionally. Adult males defend their territories against intruders. Breeding can occur at any time of the year and females produce 1 young every 2 years. The single young remains in the backward-opening pouch for 6-8 months and is weaned a year later.

STATUS *Common in northern areas; less common in south-western part of range*

LASIORHINUS

Northern Hairy-nosed Wombat
Lasiorhinus krefftii (OWEN, 1872)

DISTRIBUTION
Epping Forest National Park, central Queensland

Restricted to just 1 population of about 100 animals, this critically endangered species is found in dry, grassy woodland. A large (30-35 kg; 66.1-77.0 lb), robust animal, it digs clusters of burrows in the sandy soil. Its appearance is similar to the Common Wombat except that it has a grey coat; erect, pointed ears; a flat, broad muzzle; and a broad nose covered in short whiskers. As with all wombats, its teeth are open-rooted and continue to grow throughout its life. It forages at dawn and dusk, and at night on native grasses, retreating to its cool burrow during the day. Individuals maintain their own burrow systems and appear to be mostly solitary. A single young is born in spring or summer, and remains in the pouch for up to 10 months. Females can produce up to 2 young every 3 years.

STATUS *Critically endangered*

Southern Hairy-nosed Wombat
Lasiorhinus latifrons (OWEN, 1845)

DISTRIBUTION
Southern Australia, in semi-arid regions

Slightly smaller than other wombats (19-32 kg; 41.8-70.4 lb), this animal is found in semi-arid shrubland and grassland, and mallee woodland. It has a greyish-brown coat with whitish patches on the face usually beneath the eye, a broad, whiskered muzzle and pointed, erect ears. Animals emerge from their burrows at night, dawn and dusk to graze on grasses and shoots; they return to the burrow to rest during the day. Individuals excavate burrows in sandy soils that are clustered together to form a warren, which may contain the burrows of 5-10 animals. The social hierarchy within the warren is dominated by males, which defend the territory around it from males living in nearby colonies. Males also show aggressive behavior in the breeding season. A single young is born between September and December, remains in the pouch for 6-9 months and is weaned at over 1 year of age.

Status *Common within a restricted area; uncommon in farmed areas*

Hypsiprymnodontidae

Hypsiprymnodon

Musky Rat-kangaroo
Hypsiprymnodon moschatus RAMSAY (1876)

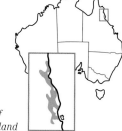

Distribution
Tropical rainforests of north-eastern Queensland

This forest-dwelling animal is the smallest macro-podoid (360-680 g; 12.6-23.8 oz). Its coat is mid to dark chocolate brown above, lighter below. It has a long slender head with large eyes, small rounded ears, and a naked, scaly tail. Unusually, it has 5 toes on the hind-feet with an opposable 'big toe' which is used for climbing over branches. Both males and females have a strong, musky odor. During the day, it forages on the ground for fallen fruits, seeds, fungi and, occasionally, insects. It rests at night in a sheltered nest which it builds from leaves; it uses its coiled semi-prehensile tail for carrying nesting material to the nest. Animals are usually solitary but often feed together in loose groups. Litters of 1-3 young, born between October and April, remain in the pouch for 21 weeks and then stay in the nest until they are weaned.

Status *Common in a limited area; requires extensive patches of rainforest*

Potoroidae

Potorous

Broad-faced Potoroo
Potorous platyops (GOULD, 1844)

Distribution
Previously recorded in coastal areas in south-western Western Australia

Not seen since 1875, this extinct species was probably already rare by the time of European settlement. Museum specimens show that this animal was a small (approximately 800 g; 28.0 oz) potoroo, with a grey-brown, streaked coat and pale grey underparts. As its name indicates, its face was broad and short. It appeared to live to the north and east of the tall forests of the south-

west, in shrubland or low forest. Regrettably, little else is known about this species, although some subfossil remains have been found.

Status *Extinct*

Gilbert's Potoroo
Potorous gilbertii (GOULD, 1841)

Distribution
South-western Western Australia, in 1 restricted area near Albany

Thought to be Australia's rarest mammal, this potoroo had not been sighted for more than 100 years before it was rediscovered in 1994. It is smaller (785-965 g; 1.7-2.1 lb) than the similar Long-nosed Potoroo, with a coat that is reddish brown on upperparts and pale grey below. The face is a little broader, particularly around the cheeks. The potoroo lives in dense undergrowth in coastal heathland and forages almost exclusively for underground fungi. In common with other fungus-eating potoroos, this species probably plays an important role in dispersing fungal spores. Timing of breeding and litter size are probably similar to the Long-nosed Potoroo.

Status *Critically endangered*

Long-footed Potoroo
Potorous longipes SEEBECK AND JOHNSTON (1980)

Distribution
Restricted to small areas in south-eastern NSW and north-eastern Victoria

The largest potoroo (1.6-2.2 kg; 3.5-4.8 lb), this animal is restricted to temperate rainforest and tall open forest in only 3 small areas. Its coat is brownish-grey above and pale grey below. It is distinguished from the smaller Long-nosed Potoroo by its very long, powerful hind-feet; it also has a long, thick, round tail. This nocturnal, terrestrial animal feeds mainly on underground fungi, which it digs up using its strong fore-limbs and long claws; it also eats other plant material and some invertebrates. It favors forest with a dense understorey of ferns, grasses and sedges, which unfortunately provide little protection from predators such as foxes and dingoes. Animals are solitary except when breeding. A single young is born after 38 days gestation and remains in the pouch for around 5 months; it may not be able to breed until 2 years of age. Breeding appears to continue year-round, and females can raise 2 young each year.

Status *Endangered; habitat not secure and animals very prone to predation*

Long-nosed Potoroo
Potorous tridactylus (KERR, 1792)

Distribution
Eastern Australian mainland, Tasmania

Similar in appearance to the larger Long-footed Potoroo, this 660-1640 gram (1.45-3.6 lb) animal lives in coastal woodland and heathland with a dense understorey. Its coat is dark reddish-brown above and pale grey below, and it has large hind-feet with strong claws. The muzzle is long in southern populations but noticeably shorter in Queensland populations. It forages mostly at night by digging up a wide variety of food including fungi, and smaller amounts of roots, tubers, seeds, fruits and invertebrates. Mostly solitary, animals rest by day in individual nests or burrows in dense vegetation. Animals may mate throughout the year and adult females usually breed twice a year. The single young remains in the pouch for 19-20 weeks.

Status *Common in Tasmania; sparse or vulnerable elsewhere*

Bettongia

Brush-tailed Bettong, Woylie
Bettongia penicillata GRAY (1837)

Distribution
Far south-west of Western Australia; reintroduced to islands off South Australia

This animal used to be widely spread across southern and north-western Australia but now inhabits a fraction of its former range. It is found in semi-arid woodland with a dense understorey of grass or shrubs. A mid-sized animal (1.1-1.6 kg; 2.4-3.5 lb), its coat is pale to reddish grey above, pale grey below. It has long hind-feet, a pointed face and large eyes. The semi-prehensile tail has a black crest of hairs and is as long as the head-body length. This bettong forages at night for the underground fungi that make up 90% of its diet, digging them up with its powerful fore-paws and claws. Animals build substantial domed nests from grass and bark, and carry nesting material in their coiled tail. Individuals are solitary, breeding throughout the year. A single young is usually born and remains in the pouch for 12-13 weeks; it is weaned 4 weeks later. Up to 3 young are produced each year.

Status *Locally common where foxes are controlled; rare to absent elsewhere*

Burrowing Bettong
Bettongia lesueur (QUOY AND GAIMARD, 1824)

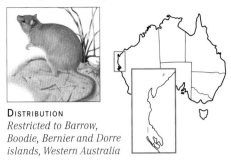

DISTRIBUTION
Restricted to Barrow, Boodie, Bernier and Dorre islands, Western Australia

Once widespread across much of central and western Australia, this species is now presumed extinct on the mainland. Fox predation is believed to be the cause of the rapid decline of this once common species. It is similar in appearance to other bettongs, weighs 900–1600 g (2.0–3.5 lb) is colored yellow grey above and paler below, and has a short muzzle and small ears. The only bettong to consistently den in burrows, it emerges at night to forage for a wide range of different foods, including roots, tubers, fungi, invertebrates, fruits, flowers, seeds and leaves. Animals are gregarious, and dig individual burrows that interconnect to form large warrens that sometimes house up to 100 bettongs. Year-round breeding sees females raising up to 3 young a year, each born singly after a gestation period of only 3 weeks. The young remains in the pouch for about 16 weeks.

STATUS *Vulnerable; abundant on Bernier, Dorre, Barrow and Boodie islands; extinct elsewhere (although programs are currently re-introducing it to mainland Australia)*

Northern Bettong
Bettongia tropica WAKEFIELD (1967)

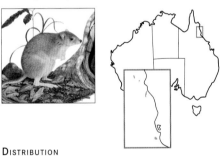

DISTRIBUTION
North-eastern Queensland (Mount Carbine and Mount Windsor tablelands in the Daintree area, Lamb Range near Cairns and Coane Range west of Townsville only)

Restricted to just 4 small areas in north-eastern Queensland, this rare species inhabits woodland and forest adjacent to tropical rainforest, preferring habitat with a dense, grassy understorey. A small bettong (900–1400 g; 2.0–3.1 lb), its coat is dark grey above and pale grey below. The grey tail is long and half-covered with a crest of short black hairs. It forages at night, digging for fungi, and plant roots and tubers, with strong fore-paws and claws; it also eats some grass. This solitary animal dens alone during the day in a sheltered nest built from grass. Breeding is probably year-round, with up to 3 single young produced annually.

STATUS *Endangered; rare to locally common in a restricted range*

Nullarbor Dwarf Bettong
Bettongia pusilla McNAMARA (1997)

DISTRIBUTION
Formerly Nullarbor Plain, South Australia

Known only from jawbones and teeth collected in Koonalda, Old Homestead and Weekes caves in the Nullarbor Plain, this species appears to have become extinct recently, around the time of European settlement. The dentition, and size and shape of the jaw indicate that this bettong was smaller than other bettong species (possibly < 1.6 kg; 3.5 lb) but it probably had a similar omnivorous diet. Little else is known of this species.

STATUS *Extinct*

Tasmanian Bettong
Bettongia gaimardi (DESMAREST, 1822)

DISTRIBUTION
Eastern and central Tasmania

This bettong was also found in south-eastern mainland Australia until the beginning of the 20th century. Paler in color than other bettongs, its coat is brownish grey flecked with white above, and pale grey below; the tail is often tipped white. A relatively large bettong (1.2–2.3 kg; 2.6–5.1 lb), it forages at night for underground fungi but also eats some tubers, roots, seeds and soil invertebrates. In common with other bettongs, it is found in habitats that have dense grassy understoreys, usually in open forest or woodland. Animals build nests from grass in sheltered sites, and usually den alone. Breeding continues year round with females giving birth to a single young 2–3 times a year. Gestation lasts for 21 days only, and young are carried in the pouch for around 15 weeks and weaned 6–8 weeks later.

STATUS *At risk due to unsecured habitat (forestry, agriculture); may be locally common in protected areas*

Rufous Bettong
Aepyprymnus rufescens (GRAY, 1837)

DISTRIBUTION
Eastern Australia, from 300 km (186 miles) inland in northern NSW and Queensland to the coast

Found in a variety of forest and woodland habitats in eastern Australia, this animal forages where the understorey is open or grassy. This large (2.3–3.6 kg; 5.1–7.9 lb) animal's coat is reddish grey

above and whitish below, with a pale stripe at the hip. It has a short, hairy muzzle, large ears and a solid, tapering tail. Strongly nocturnal, it forages for fungi, tubers, invertebrates, grasses and other plant matter, and often chews on animal bones to gain calcium. It builds nests from grasses that it carries in its semi-prehensile tail. Animals are often found in loose social groups of 3 (1 male and 2 females), and have large territories, up to 110 ha (270 acres). Breeding continues throughout the year, with a single young, born after 22–24 days, remaining in the pouch for 16 weeks and being weaned at 5–6 months. Up to 3 young may be born in a year.

STATUS *Locally common in grassy forests, rarer inland; vulnerable to fox predation*

Desert Rat-kangaroo
Caloprymnus campestris (GOULD, 1843)

DISTRIBUTION
Formerly in central Australia, around the South Australia-Queensland border

As the last recorded sighting was in 1936, this little-known species now is presumed extinct. A small (630–1060 g; 1.4–2.3 lb) but extremely fast and agile animal, it was probably never common within its limited range in central Australia. It occupied sparsely vegetated gibber plains and sand-ridge habitat where it presumably fed on vegetation and invertebrates. Surprisingly, in such an exposed environment, this species did not excavate burrows; instead it built distinctive nests beneath bushes or in scrapes in the open, from leaves, grasses and twigs. Its coat was pale reddish brown above and paler below. It had a short muzzle, large ears, and very long feet and tail. Pouch young were recorded in December and August, and it is likely that only single young were born.

STATUS *Extinct*

Banded Hare-wallaby
Lagostrophus fasciatus (PERON AND LESUEUR, 1807)

DISTRIBUTION
Bernier and Dorre islands, Western Australia

This animal was once more widespread in south-western Western Australia. The distinctive grey shaggy coat and horizontally striped back, make this medium-sized (1.3–3.0 kg; 2.9–6.6 lb) hare-wallaby easily recognizable. It is found in dense shrub habitat where it forms runs and shelters beneath the shrub layer, and browses at night on shrub foliage, grasses and spinifex. Animals have distinct territories but are often found in small groups. Males are very aggressive to other males, whereas females are largely unaggressive. A single young is usually born in late summer; however females that do not produce a young at this time breed later in the year. Young remain in the pouch for 6 months, and are usually weaned at 9 months of age.

Status *Vulnerable; common in suitable habitat on islands; extinct on mainland; vulnerable to future introductions of predators*

LAGORCHESTES

Central Hare-wallaby
Lagorchestes asomatus FINLAYSON (1943)

Distribution
Formerly western and central deserts

This species is known only from a single skull collected in 1932, and from accounts by western desert Aborigines. It was perhaps the smallest of the hare-wallabies with long, soft grey fur and thickly furred feet. It was found in spinifex deserts, inhabiting sand plains and dunes. Unlike similar species, it did not flush from cover when disturbed, making it easy prey for Aboriginal hunters. Animals built scrapes or shallow burrows beneath spinifex hummocks, and fed on grasses, seeds and fruits. It is likely that introduced predators and changes in habitat and land use caused this species extinction.

Status *Extinct*

Eastern Hare-wallaby
Lagorchestes leporides (GOULD, 1841)

Distribution
Previously common in semi-arid NSW, Victoria and South Australia

About the size of a European hare, this species astonished early collectors with its speed and agility. It was capable of jumping over 2 m (6.5 ft) in height, and outrunning dogs. John Gould, who first described this species, recorded that one '…bounded clear over my head' while Gould was standing. Its coat was grizzled brown with white-tipped hairs above, pale grey below, with reddish patches around the eyes, and reddish fore-paws and hind-limbs, and a black patch at the elbow. The short, muscular tail was grey above, pale below. It was found in open habitat on grassy plains, and constructed shallow scrapes beneath cover for shelter. It was probably nocturnal, and presumably grazed on grasses and other plants. Little else is known about this once common animal. It was last recorded in 1890.

Status *Extinct*

Rufous Hare-wallaby, Mala
Lagorchestes hirsutus GOULD (1844)

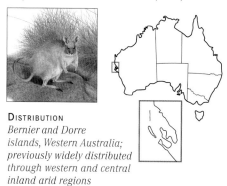

Distribution
Bernier and Dorre islands, Western Australia; previously widely distributed through western and central inland arid regions

This smallest surviving hare-wallaby (800–1600 g; 1.8–3.5 lb) was once common throughout spinifex desert and shrubland. Changes in habitat due to altered fire regimes and predation by introduced cats and foxes have caused its extinction on the mainland. Easily recognized by its shaggy coat, it is colored rufous above and paler below, and has white eye rings, ear fringes and muzzle. A nocturnal forager, it eats forbs, grasses and spinifex, seeds and some invertebrates. It is usually solitary, spending the day in scrapes or shallow burrows beneath shrubs. Breeding occurs throughout the year, with young spending around 17 weeks in the pouch.

Status *Vulnerable; locally abundant on offshore islands; extinct on mainland*

Spectacled Hare-wallaby
Lagorchestes conspicillatus GOULD (1842)

Distribution
Semi-arid and arid northern Australia, absent in far north

Easily distinguished by its orange eye-rings, this animal is the most common of the hare-wallabies. It inhabits tropical grassland in northern Australia, and is also found in open forest and woodland. Its shaggy coat is greyish brown, often with a reddish tinge, and has white belly fur. A robust animal weighing 1.6–4.5 kg (3.5–9.9 lb), it is thickset with a short neck. It feeds by night in open areas, on grasses, herbs, leaves and spinifex tips. Daytime is spent in burrows dug into the base of spinifex hummocks, or beneath dense shrubs. This usually solitary animal will occasionally graze in small groups. Breeding is year-round, with a single young born after a 1-month incubation period. Young remain in the pouch for around 4–5 months. Predation by introduced cats and foxes, and changes in fire regimes and land use, have caused population declines in the western part of its range.

Status *Locally common through central and eastern part of range; declining rapidly in Western Australia and western Northern Territory*

ONYCHOGALEA

Bridled Nailtail Wallaby
Onychogalea fraenata (Gould, 1841)

Distribution
Restricted to 2 populations (1 remnant, 1 re-introduced) in central Queensland; previously widespread in central Queensland, western NSW and north-western Victoria

With its distinctive white bridle markings, this attractive and once-common wallaby now inhabits only a fraction of its former range in grassy woodland. Changes in land use, particularly destruction of groundcover by pastoral activity, and predation by introduced foxes, have caused the extinction of most populations. Animals weigh 3–6 kg (6.6–13.2 lb), and have a yellow-greyish coat and white stripes around the shoulders, back and face, and whitish belly. The long, muscular tail is tipped with a horny spur. Animals graze at dawn, dusk and night on grasses, forbs and shrubs, in open areas close to cover. The day is spent resting in a scrape under bushes. Animals are usually solitary but sometimes graze in loose groups. Breeding continues throughout the year. The single young remains in the pouch for around 4.5 months.

Status *Endangered*

Crescent Nailtail Wallaby
Onychogalea lunata (GOULD, 1841)

Distribution
Formerly western, central and southern semi-arid and arid regions

This previously widespread species declined rapidly in the late 19th and early 20th centuries, and was last sighted in the 1950s. A small wallaby (approximately 3.5 kg; 7.7 lb), it had an ashy-grey coat above with a white belly and a distinctive white crescent-shaped stripe around the shoulder. The coat also had prominent white markings around the cheeks, muzzle, flanks and ear fringes. It emerged from thickets of dense arid woodland and shrubland to feed at night in open areas, probably browsing on grasses and leaves. When disturbed, it would rush for cover, sometimes sheltering inside hollow trees. It rested during the day in shallow scrapes beneath shrubs or a spinifex hummock, sometimes moving to clearings to bask in the sun for a short time. Little else is known about this animal, although reproduction and life history were probably similar to the Bridled Nailtail Wallaby.

Status *Extinct*

Northern Nailtail Wallaby
Onychogalea unguifera (Gould, 1841)

Distribution
Drier parts of northern Australia, from Kimberley to Cape York

The largest and most common of the nailtail wallabies, this species is widely distributed through the drier regions of northern Australia. Weighing 5–9 kg (11.0–19.8 lb), this sandy-colored animal has distinctive white stripes on the cheek and hip. As with other nailtail wallabies, it has a small horny spur at the tip of its long tail. It is found in many habitat types, including grassy woodland, floodplains, swamps, savanna, spinifex and tussock grassland. It forages in open areas at night, feeding on herbs and other succulent vegetation, and retreats to rest in a scrape under dense cover by day. Animals are mostly solitary but often feed together in small groups. Breeding is probably continuous throughout the year.

Status *Locally common*

DENDROLAGUS

Bennett's Tree-kangaroo
Dendrolagus bennettianus De Vis (1887)

Distribution
North-eastern Queensland, between Mount Amos (near Cooktown) and Daintree River, and inland to Mount Windsor Tablelands

The largest tree-dwelling mammal in Australia, with the Koala, this species (8–13.5 kg; 17.6–29.7 lb) is restricted to an area of 3500 km² (1350 sq miles). It is found in lowland vine forest and montane rainforest, where it lives in the trees and feeds at night on rainforest leaves, and fruits when in season. Its coat is greyish brown above and below, with a reddish neck and dark grey face and muzzle. Its fore-limbs are long and muscular, and fore-paws and hind-feet are dark brown black. The tail is very long and tipped with a dark tuft. It is a cryptic species which remains in the tree canopy during the day while resting. Animals are territorial, and males aggressively defend their large (25 ha; 62 acre) territory against other males. Females in captivity breed each year and the single young remains in the pouch for around 9 months. In the wild, most births occur in the early wet season from November to January, and females rear 1 young every 2 years.

Status *Sparsely distributed in a restricted area*

Lumholtz's Tree-kangaroo
Dendrolagus lumholtzi Collett (1884)

Distribution
North-eastern Queensland

A surprisingly agile climber, this muscular, robust animal (6–9.5 kg; 13.2–20.9 lb) spends most of its life in trees. It has long, powerful front limbs, a greyish coat with pale underparts, and a distinctive black face and muzzle. Its non-prehensile tail is very long and ends in a dark tuft. Once common in now-cleared lowland rainforest of north-eastern Queensland, it is restricted to mountainous regions where rainforest is still intact. It is a nocturnal forager that browses on leaves in trees, occasionally descending to the ground to move between trees. During the day it rests hunched on a tree branch. It is capable of jumping up to 15 m (50 ft) between trees, and climbs using all 4 of its powerful limbs and its tail as a counterbalance. Unlike all other kangaroos and wallabies, tree-kangaroos can move their hind-limbs independently of each other, a distinct advantage when climbing. Animals are mostly solitary. They breed year round and rear a single young every 2 years.

Status *Locally common where habitat is protected; absent elsewhere*

PETROGALE

Allied Rock-wallaby
Petrogale assimilis Ramsay (1877)

Distribution
Central to north-eastern coastal Queensland, from Croydon (north-west) to Home Hill (near Townsville), and on Palm and Magnetic islands

Similar in appearance but genetically quite different from other rock-wallabies in north-eastern Queensland, this species occurs in both wet (coastal) and dry (inland) tropical habitats. It eats a wider range of vegetation than species restricted to coastal areas; it feeds at night on leaves and grass shoots. Its weight is around 4.5 kg (9.9 lb). The coat lacks distinctive markings and is greyish brown above; on some animals it can be quite dark. The underparts are paler sandy-brown, and there is a pale stripe on the cheek. The fore-feet are dark grey brown, and the tail becomes darker towards the tip. Animals live in hierarchical colonies, with mature males and females forming long-lasting partnerships. These pairs can be observed grooming each other and grazing together. They breed throughout the year and a single young is suckled in the pouch for 6–7 months before becoming sexually mature at 18 months.

Status *Common but becoming less so in the western part of its range*

Black-footed Rock-wallaby
Petrogale lateralis Gould (1842)

Distribution
Widely scattered through central Australia and on offshore islands of Western Australia and South Australia

This widely distributed species varies greatly in appearance across its range; 3 subspecies are currently described. It is a small–medium sized rock-wallaby (2.3–5.3 kg; 5.1–11.7 lb), that is generally greyish and/or brownish with a paler belly and chest, dark feet and tail, and a white stripe on the cheek and side. A dark stripe also runs from the forehead or top of the head to at least part way down the spine. The coat often turns paler in summer. Animals forage for grass and leafy plants mostly at night, and seldom move far from rocky shelter. During the day, they retreat to cool and safe shelters in crevices and caves, and occasionally bask on sunny rocks in cold weather. Animals are social and live in colonies. This previously common and widespread species has been severely affected by fox predation, particularly in the western and southern parts of its distribution.

Status *Scattered; ongoing fox control required to ensure survival of local populations*

Brush-tailed Rock-wallaby
Petrogale penicillata (Gray, 1825)

Distribution
Great Dividing Range, from south-eastern Queensland to eastern Victoria; very recently extinct in the Grampians, western Victoria

Once both widespread and abundant, this species declined dramatically in the early 20th century. It was perceived to be an agricultural pest that competed with stock for food, and was hunted heavily for its fur. Competition with rabbits and goats for food and predation by foxes caused the extinction of many populations. It prefers rocky areas in eucalypt forest and woodland that have caves, crevices and ledges, especially those with a sunny northerly aspect that can be used for basking. A large rock-wallaby (5–11 kg; 11.0–24.2 lb), it has a dense, shaggy reddish-brown coat with grey shoulders and a paler stripe on the cheek. All 4 feet are dark brown black, and the cylindrical tail is tipped with a dark brown brush. It is an extremely agile climber, capable of moving quickly across steep cliff faces. Mostly nocturnal, it eats grasses and other foliage, and will also feed on seeds, flowers and fruits when available. Animals live in colonies (2–43 animals) which they defend vigorously against intruders. Males mate with 1–3 females, and a single young is born after 30–32 days gestation. Young remain in the pouch for 29 weeks and are weaned after 41 weeks.

Status *Vulnerable; endangered in Victoria and western NSW; common in north-western NSW and Queensland; introduced to Oahu, Hawaii, and Kawau, Motutapu and Rangitoto islands in New Zealand (subsequently eliminated from Motutapu and Rangitoto after reaching pest levels)*

Cape York Rock-wallaby
Petrogale coenensis ELDRIDGE AND CLOSE (1992)

Distribution
Far north Queensland, on eastern Cape York Peninsula from Pascoe River to Musgrave, south-west of Princess Charlotte Bay

The rarest of the north-eastern Queensland rock-wallabies, this little-known animal closely resembles its common relatives further south. It weighs 4.0–5.0 kg (8.8–11.0 lb), and its coat is grey brown above and paler below. It has a pale cheek stripe and a dark stripe extends over the top of the head and down the upper spine. The tail base is also dark but the tail becomes progressively paler towards the tip until it is almost white. Little else is known of this animal; it is likely that its behavior is similar to that of the more southerly species.

Status *Rare within restricted distribution*

Godman's Rock-wallaby
Petrogale godmani THOMAS (1923)

Distribution
North-eastern Queensland, near Cooktown, from Bathurst Head in the north to Mount Carbine

Similar in appearance to other rock-wallabies of north-eastern Queensland, this species is only definitively identified through genetic analysis. It is a medium-sized animal (4.3–5.2 kg; 9.5–11.4 lb). Its coat is brownish grey above and paler below, with a pale cheek stripe and dark feet. The dark tail is often whitish towards the tip, for up to one-third or one-half of its length. It is mostly nocturnal and feeds on grass, and foliage from shrubs and forbs that grow on or around rock outcrops. Animals usually rest by day in rocky shelters, emerging to feed and groom. They are gregarious and live in colonies, and breeding pairs form enduring partnerships. Some animals have bred with Mareeba Rock-wallabies to produce hybrid populations in the south of its range.

Status *Common within restricted distribution but has declined sharply around Black Mountain*

Herbert's Rock-wallaby
Petrogale herberti THOMAS (1926)

Distribution
South-eastern Queensland, up to 200 km (124 miles) inland from coast, from Fitzroy River to 100 km (62 miles) north of Brisbane.

The most widespread and common of 7 similar species of rock-wallaby in eastern Queensland, this animal is also the largest (3.7–6.7 kg; 8.1–14.7 lb). Its coat is greyish brown above and paler below, with a dark stripe mid-line from the forehead to behind the neck. There are dark grey or brown-black patches on the face and armpits, and whitish stripes on the cheeks and flanks. Both fore-feet and hind-feet are dark, and the cylindrical tail is long and becomes darker towards the tip. The species lives in rocky habitat and feeds mostly at night on grass and leafy foliage. It is also active for part of the day, when animals can be observed feeding and grooming. These social animals form long-lasting breeding partnerships.

Status *Widespread and common in suitable habitat*

Mareeba Rock-wallaby
Petrogale mareeba ELDRIDGE AND CLOSE (1992)

Distribution
North-eastern Queensland, west of Cairns, from near Mount Carbine to Mungana in the west and Mount Garnet in the south

This species lacks any physical characteristics that distinguish it from the similar Unadorned, Allied and Sharman's rock-wallabies. Its coat is brownish grey with paler underparts, although the coat color usually matches the rock on which the animal lives. It lacks prominent markings but has a pale cheek stripe, and dark tail and paws. The tip of the tail in some individuals is whitish. Its size is also similar to other north-eastern Queensland species (3.8–4.5 kg; 8.4–9.9 lb). Analysis of genetic material is the only reliable method of identification. Animals are gregarious and form long-lasting pair-bonds between breeding adults. They live in small colonies, sheltering during the day in nooks and caves in rock outcrops. They emerge at night to feed on grasses, shrubs and forbs, and can be sometimes seen grooming and feeding during the day. The species is poorly known but is probably similar in behavior and life history to the Allied Rock-wallaby.

Status *Common within restricted distribution*

Monjon
Petrogale burbidgei KITCHENER AND SANSON (1978)

Distribution
Kimberley coast and some islands in Bonaparte Archipelago, Western Australia

This smallest of the rock-wallabies (960–1430 g; 2.1–3.1 lb) is restricted to remote and inaccessible areas of north-western Australia. Its coat is unusually colored olive above, marbled with brown and black; underparts are whitish yellow. The brownish face has a paler streak along the cheek, and a darkish stripe over the top of the head and neck. It has large eyes but short ears. The tail is brownish and tipped with a prominent dark brush. In the deeply fissured King Leopold sandstone country it finds shelter in caves and crevices. It is an extremely agile and fast-moving animal when disturbed. Mostly nocturnal, it feeds on grasses and ferns in open woodland. Although timid, animals are social and appear to live in large colonies. Breeding is protracted.

Status *Common in restricted range*

Nabarlek
Petrogale concinna GOULD (1842)

Distribution
Northern Australia in the Kimberley, Top End between Victoria and Mary rivers, Arnhem Land, and offshore islands including Groot Eylandt

Only slightly larger (1.0–1.7 kg; 2.2–3.7 lb) than the Monjon, this animal is found in a variety of rocky habitats in northern Australia, ranging from low granite hills and boulder fields to steep, rugged cliffs. Its fur is short and silky, and colored reddish brown above and pale grey below. Many animals have a dark shoulder stripe, and dark face and forehead with a pale cheek stripe. The tail is tipped with a brush of dark brown hairs. Animals mostly feed at night on grasses and nardoo, both of which contain high levels of indigestible silica. A system of continuous replacement of its molars (grinding teeth) allows the species to cope with its abrasive diet, an adaptation unique among marsupials. These solitary animals have been observed travelling several hundred metres from rocky shelters in search of food, particularly nardoo, in the surrounding grassland. Breeding is probably year round, and young remain in the pouch for approximately 6 months.

Status *Locally common in restricted distribution; rarer in Victoria River–Mary River region*

Proserpine Rock-wallaby
Petrogale persephone Maynes (1982)

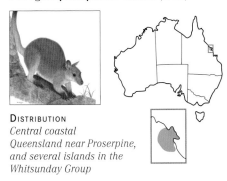

Distribution
*Central coastal
Queensland near Proserpine,
and several islands in the
Whitsunday Group*

Found only on rocky outcrops among the vine forest of a small area of the central Queensland coast, this timid animal has the smallest distribution of all rock-wallabies. The medium-large sized macropod (4.1-8.8 kg; 9.0-19.4 lb) has a grey-brown coat above, with yellow-cream underparts. It has a pale cheek stripe, and dark grey face and feet. The tail is reddish at the base and becomes darker towards the end, although the tip is whitish. Animals are active during the day as well as at night, moving from the shelter of rocky crevices and caves to surrounding woodland to browse on understorey grasses and forbs. This species is particularly vulnerable as its tightly restricted range occupies an area in which the human population is booming. New houses and roads have destroyed rock-wallaby habitat, and animals are often killed on roads. The breeding season is probably protracted or continuous year round.

Status *Endangered*

Purple-necked Rock-wallaby
Petrogale purpureicollis Le Souef (1924)

Distribution
*North-western central Queensland,
from Mount Isa in the west to the Selwyn Ranges
in the east, and near Winton and Lawn Hill*

This rock-wallaby is distinguished by its unusual coloration around the head, face and neck – markings that can range from a pale pinkish wash to a bright purple or reddish color. The rest of its coat is pale grey brown, with darker brown armpit and, sometimes, flank. Facial markings consist of a pale cheek stripe on a grey muzzle. Its tail is long and has a dark tip. A medium-sized animal (4.7-7.0 kg; 10.3-15.4 lb), it is found in rocky country in semi-arid north-western Queensland, and is usually associated with grass hummockland and mulga vegetation types. It is a nocturnal forager that grazes on grasses and other foliage, and rests during the day in crevices and caves in rocky outcrops and fissured limestone pavements. Breeding continues throughout the year. A single young is born after 30 days gestation and remains in the pouch for around 6-7 months.

Status *Locally common in most areas*

Rothschild's Rock-wallaby
Petrogale rothschildi Thomas (1904)

Distribution
*Hamersley and Chichester ranges
in the Pilbara region of Western Australia,
and 5 offshore islands in Dampier Archipelago*

This large rock-wallaby (up to 6.6 kg; 14.5 lb) is found on granite outcrops. Its coat is grey brown above and brown below, with pale cheeks and throat, and dark brown head and ears. The back of the neck and shoulders can sometimes be a purplish grey. The tail is long and slender, becoming darker towards the tip. It shelters during the day in caves and crevices among the rock piles and outcrops, where the temperature is much cooler. At night it emerges to feed on plants (mostly grasses) that grow on the lower slopes of the outcrops and in the surrounding hummock grassland. Breeding is probably continuous, with a peak in the wet season at the end of the year.

Status *Declining on mainland where fox predation is uncontrolled; secure on fox-free islands*

Sharman's Rock-wallaby, Mount Claro Rock-wallaby
Petrogale sharmani Eldridge and Close (1992)

Distribution
*Seaview and Coane
ranges, near Ingham,
Queensland*

This species can be reliably distinguished from other species of rock-wallaby in north-eastern Queensland only by analysis of genetic material. Physically, it closely resembles the Allied and Mareeba rock-wallabies. It tends to be slightly smaller than other species at 4.1-4.4 kg (9.0-9.7 lb). Its coat is grey brown above and paler below, sometimes becoming sandy colored after moulting. The tail darkens towards the tip, and there is sometimes a dark stripe on the top of the head and down the upper spine. It has the pale cheek stripe common to the similar species. This social animal lives in groups and forms long-term breeding partnerships. It is mostly nocturnal and feeds on grasses, foliage from forbs and shrubs growing around rocky outcrops. Breeding probably occurs year round.

Status *Locally common within a restricted area*

Short-eared Rock-wallaby
Petrogale brachyotis (Gould, 1841)

Distribution
*Monsoonal areas of
northern Australia,
from the eastern
Kimberley to the Northern Territory-
Queensland border, including Groote
Eylandt and nearby island groups*

Highly variable in its color and size, this medium-small rock-wallaby (2.2-5.6 kg; 4.8-12.3 lb) is perhaps best recognized by its short ears. It is brownish grey above and paler below, with whitish markings around the face, cheek, flank and thigh sometimes evident. The tail becomes darker towards the end and is tipped with a blackish brush. It is found in rocky areas in tropical savanna grassland. Mostly nocturnal, it forages on grasses and browse in the grassland surrounding the rocks but always remains close to rocky cover. In cooler weather, it often also feeds during the day. Animals are social and live in colonies. Despite its abundance across most of its range, little is known of this species.

Status *Common across most of range; rarer in far west of range*

Unadorned Rock-wallaby
Petrogale inornata Gould (1842)

Distribution
*Central coastal Queensland,
from Home Hill (near
Townsville) to Fitzroy River,
and some islands in
Whitsunday Group*

As its name suggests, this medium-sized rock-wallaby (3.1-5.5 kg; 6.8-12.1 lb) lacks many of the distinctive markings of other species. It is grey brown above, and paler below, although the shade of the coat varies according to the color of the rock in its habitat. The pale cheek stripe and darker stripe on the head and back of the neck are better defined in southern populations, becoming less distinct in the north. The fore-paws are dark, and the tail darkens towards the tip. This social animal lives in colonies and enduring pair-bonds form between mature adults. It is mostly nocturnal; as well as grazing, it browses on leaves from shrubs and forbs. Breeding appears to be year-round, and the single young is born after a short gestation period (30 days). The young remains in the pouch for 6-7 months and is sexually mature by 18 months.

Status *Common throughout most of its range*

Yellow-footed Rock-wallaby
Petrogale xanthopus GRAY (1855)

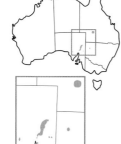

DISTRIBUTION
Adavale Basin,
Queensland; Gap and
Coturaundee Range,
western NSW; Flinders
and Gawler ranges, South Australia

Once greatly in demand for its pelt, this attractive animal is easily recognized by its distinctive markings. As its name suggests, the limbs and feet are a yellow orange color. The coat is generally grey above and whitish on the underparts, with white striping on the cheeks, ears and flanks. The muzzle is dark, and a brown stripe runs from the forehead to the middle of the spine. Uniquely, it has a yellow orange tail ringed with orange stripes that become darker towards the tip. One of the larger rock-wallabies (6–12 kg; 13.2–26.4 lb), it is mostly nocturnal and emerges from rocky shelters at night to forage on grasses, forbs and other foliage. In cooler weather, it can be found basking on warm rock faces. It is a gregarious animal, living in colonies of up to 100 individuals although each animal has its own home range that overlaps with those of other colony members. Breeding occurs year-round, with a single young remaining in the pouch for around 6 months.

STATUS *Appears secure in restricted distribution in Queensland and South Australia; endangered in smaller populations in Queensland and NSW*

Red-legged Pademelon
Thylogale stigmatica (GOULD, 1860)

DISTRIBUTION
Coastal forests of eastern Australia,
from northern Queensland to central NSW;
also in Papua New Guinea

Males of this species (3.7–6.8 kg; 8.1–15.0 lb) are larger than females (2.5–4.2 kg; 5.5–9.2 lb), and have more muscular fore-limbs and chests. Both sexes have similar coloration, with dark brown-grey (in rainforest populations) or light reddish-brown fur (in open forest populations) on upperparts, with paler fur below. The hind-limbs, cheeks and bases of ears are reddish brown, with some paler fur around the muzzle and throat. The tail is shortish, thick and round, and colored grey. Animals are active day and night, feeding on leaves, fruits, grasses and seeds, depending on season and availability. No nest is made, instead animals shelter and rest in shallow depressions excavated in the leaf litter within dense forest. These timid animals are usually solitary but aggregate in

small feeding groups where food is abundant. Breeding, throughout the year, usually results in a single young but occasionally twins are born. The young stay in the pouch for 6–7 months.

STATUS *Common*

Red-necked Pademelon
Thylogale thetis (LESSON, 1827)

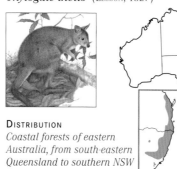

DISTRIBUTION
Coastal forests of eastern Australia, from south-eastern Queensland to southern NSW

Although sparsely distributed through rainforest and open forest, this species is common around forest edges where adjacent pastures and grassy areas provide abundant feeding opportunities. The species has larger males (2.5–9.1 kg; 5.5–20.0 lb) than females (1.8–4.3 kg; 4.0–9.5 lb) but coat markings are similar for both sexes. Its neck and shoulders are colored a rich red, with the rest of the fur being grey brown above, and cream below. The grey-brown face lacks distinctive markings, except for a ring of reddish bare skin around the eye. The tail is relatively short, round and thick, and is held stiffly behind the body when jumping. Animals feed at night on grasses and forbs in open areas within 100 m (330 ft) of cover. They are also active during the day but only where dense vegetation provides protection from predators. Young are born year round, with animals becoming sexually mature at around 12–17 months of age.

STATUS *Common around forest fringes*

Tasmanian Pademelon
Thylogale billardierii (DESMAREST, 1822)

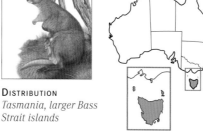

DISTRIBUTION
Tasmania, larger Bass Strait islands

Previously found in south-eastern mainland Australia, this species is now restricted to islands, including Tasmania, where it inhabits dense forest or woodland during the day and emerges at night to feed on grasses and forbs in open areas. Females (2.4–10.0 kg; 5.3–22.0 lb) are substantially smaller than males (3.8–12.0 kg; 8.4–26.4 lb) and have a less muscular build. The short, dense fur is dark grey brown above, buff below and tinged with red on the belly. The ears are large, and the eyes are ringed with reddish fur. The tail is quite short but muscular. Animals do not live in colonies but form feeding groups of up to 10 or more individuals. The fore-limbs are used for holding food, and also for digging through snow to find food in winter. Breeding occurs throughout the year but most young

are born in autumn. The single young stays in the pouch for around 6–7 months, becoming sexually mature at around 15 months.

STATUS *Extinct on mainland; common in Tasmania*

Quokka
Setonix brachyurus (QUOY AND GAIMARD, 1830)

DISTRIBUTION
South of Perth to east of Albany, and Rottnest and Bald islands, Western Australia

This distinctive wallaby is well known to many visitors to Rottnest Island in Western Australia. It has a rounded, compact body shape, and its fur is long, coarse and dense. Its coat is brown with reddish tinges around the ears and cheeks, and has paler belly fur. The ears are unusually short and rounded, and the thick tail is also short. Males (2.7–5.0 kg; 5.9–11.0 lb) are larger than females (2.7–4.0 kg; 5.9–8.8 lb) but otherwise the 2 sexes have a similar appearance. On the mainland, the species is found in dense forest and swamp edges where soil moisture is high. On Rottnest Island it lives in low scrub and open forest where access to water may be seasonally limited. It is mostly nocturnal, and feeds on leaves, grasses and occasionally fruit and fungi. It emerges during the day only if food is in short supply. Animals are social and live in groups which can become quite large (25–150 individuals). Older males are dominant, forming a hierarchy over females and juveniles. Nest sharing is common. Breeding is year round on the mainland but young appear mostly in February–March on Rottnest Island.

STATUS *Vulnerable; common on offshore islands; rare on mainland due to fox predation and habitat loss*

Agile Wallaby
Macropus agilis (GOULD, 1842)

DISTRIBUTION
Northern coastal Australia, from eastern Western Australia to central Queensland; also in Papua New Guinea

This widespread and common wallaby is found in tropical open forest and grassland, often in areas near rivers and streams. In the Northern Territory it is also found in coastal sand dunes and rocky hills further inland. Males (16–27 kg; 35–59 lb) are larger and stockier than females (9-15 kg; 20-33 lb) but share the same overall appearance. The animal's coat is pale brown on the upperparts and whitish below. Each cheek and thigh has a white stripe, and the tail is long and pale. The ears have black margins and there is also a short, dark mid-line stripe between the eyes and ears. Grazing starts in the late afternoon and continues into the night and early morning; animals eat native grasses and shoots, and fallen fruits when available. They rest during the day under shady shrubs in groups of up to 10 individuals, although feeding groups may be much larger. Breeding occurs throughout the year, with the single young remaining in the pouch for 7-8 months. The young become sexually mature after 12-14 months.

STATUS *Common*

Antilopine Wallaroo
Macropus antilopinus (GOULD, 1842)

DISTRIBUTION
Northern coastal Australia, including Kimberley, Top End and central Cape York

This short-haired wallaroo inhabits open woodland in tropical northern Australia. It prefers areas with grassy understorey, generally on plains or rolling country, but can also be observed on rocky hills grazing with the Euro or Black Wallaroo. Males (30-49 kg; 66-108 lb) have reddish-tan coats; females (16-20 kg; 35-44 lb) have greyer heads and shoulders. The underparts and face are pale, and the tail is long and thick. It eats mostly grasses in cooler periods of the day and at night, although it is active throughout the day in the wet season. These gregarious animals graze and rest in small groups. There is a dominance hierarchy among males. Births continue throughout the year and peak towards the end of the wet season. Young remain in the pouch for up to 9 months.

STATUS *Common*

Black-striped Wallaby
Macropus dorsalis (GRAY, 1837)

DISTRIBUTION
Central and eastern Queensland, north-eastern NSW

Common throughout its range, this wallaby is distinguished by the prominent black stripe that extends mid-line from its forehead to the middle of its back. Males (18-20 kg; 39-44 lb) are 2-3 times larger than females (6-7.5 kg; 13-16.5 lb) but have similar markings. The coat is mostly brown with reddish shoulders and fore-limbs.

Underparts are whitish and there are also white patches around the eyes, a white cheek stripe, and a whitish stripe on the thigh. It is found in forested habitats with a dense shrub understorey which provides shelter from predators. A nocturnal forager, it emerges from cover at dusk to graze in open areas at forest margins, rarely straying far from shelter. During the day, animals rest in dense vegetation in social groups of 20 or more individuals. Animals breed year round, with the resulting single young suckled in the pouch for 7-8 months and becoming sexually mature at 14-20 months.

STATUS *Common throughout its range but vulnerable to habitat loss for agricultural development*

Black Wallaroo
Macropus bernardus ROTHSCHILD (1904)

DISTRIBUTION
Western and central Arnhem Land, Northern Territory

The smallest of the wallaroos, this species is found only in open eucalypt woodland in steep, rocky areas. It is small and stocky, and has a long and shaggy coat. Males (19-22 kg; 42-48 lb) are dark brown to black with paler fore-limbs; females (around 13 kg; 29 lb) are grey brown with dark fore-paws and hind-toes. It is a nocturnal animal except in the wet season when it is also active during the day. It feeds on the grasses and forbs that comprise the sparse understorey, and also descends from the escarpment to grassland and waterholes on the plains below. This solitary animal shelters during the day in shade beneath rocks or trees. It is timid and bounds away quickly when disturbed; its extreme agility allows it to readily leap across steep rocky areas.

STATUS *Common within its limited distribution*

Common Wallaroo, Euro, Hill Kangaroo
Macropus robustus GOULD (1841)

DISTRIBUTION
All mainland Australia, except the far south of the continent and western Cape York Peninsula

This widespread species is found in open grassland, shrubland, woodland and forest almost always occupying rocky hills, stony rises and rocky areas that provide caves and shelters. It is well adapted to aridity; it seldom needs free water and survives on low quality food but needs the shelter of cool, moist caves to do so. Its coat is coarse, long and shaggy with its color varying with geographic location from dark grey to reddish brown above; the fur is paler below. It has long, erect, lightly

furred ears and a tapering, stout tail. Animals are of a stocky build with males (20-47 kg; 44-103 lb) larger and more muscular than females (15-25 kg; 33-55 lb). They move from their rocky shelters in late afternoon or early evening to graze on grasses, sedges, herbs and forbs in nearby open areas. The animals feed singly and lead solitary lives except when breeding. After birth, which may be throughout the year, young remain in the pouch for around 9 months.

STATUS *Common*

Eastern Grey Kangaroo
Macropus giganteus SHAW (1790)

DISTRIBUTION
Eastern Australia, including north-eastern Tasmania

A familiar and widespread species, this animal inhabits a range of diverse habitats, including woodland, open forest, farmland with forest remnants, and grassland in tropical, temperate, subalpine and semi-arid regions. It is a large kangaroo, and males (20-66 kg; 44-145 lb) are larger and have a more robust and muscular build than females (15-32 kg; 33-70 lb), particularly around the chest and fore-limbs. The coat color is variable, most commonly being grey brown above and pale grey below. The thick and muscular tail is darker towards the tip. The ears are quite large and colored brown. Animals feed mostly on grass; they emerge at dusk to graze intermittently in open areas until dawn. The day is spent resting in a shady scrape beneath trees or shrubs in wooded or forested areas. Animals are social and aggregate in large feeding mobs. Males form a hierarchy within a group and compete aggressively for females when breeding. Most young are born in autumn but breeding is continuous. The single young leaves the pouch at around 11 months of age.

STATUS *Common on mainland; vulnerable in Tasmania*

Parma Wallaby
Macropus parma WATERHOUSE (1845)

DISTRIBUTION
Forests of the Great Dividing Range in eastern NSW, from Gibraltar Ranges in north to Watagan Mountains in the south; introduced to Kawau Island, New Zealand

Thought to be extinct for 100 years until 'rediscovered' on the Australian mainland in 1967, this timid animal is found in eucalypt forest and rainforest with a dense understorey of shrubs. It is a small wallaby, with males larger (4.1–5.9 kg; 9.0–13.0 lb) and more muscular than females (3.2–4.8 kg; 7.0–10.6 lb). Similar in appearance to pademelons, this cryptic species has reddish-brown or grey-brown fur above, and a whitish throat and underbelly. It has a white stripe along the muzzle and upper cheek, and a dark stripe from the top of the head to the top of the spine. The tail is thinly furred and dark, and is often tipped with white. It feeds at night in grassy clearings close to cover, and eats grasses, herbs and leaves from shrubs. When resting under dense shrubs during the day, an animal stays in a sitting position with the tail between its legs; it seldom lies down. Animals often feed in small groups of 2–3 but are otherwise solitary. Breeding is year round but most young are born in the late summer and autumn. The single young remains in the pouch for around 6–7 months, is weaned at 10 months, and achieves sexual maturity at 16–22 months.

Status *Rare and patchy across current range; range in south of distribution has contracted*

Red Kangaroo
Macropus rufus (Desmarest, 1822)

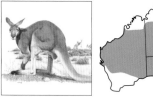

Distribution
All mainland Australia, except northern, central and southern coastal districts

This iconic animal is the largest living marsupial. Throughout the semi-arid and arid regions of mainland Australia it inhabits grassland, desert, woodland and open forest, and is most common in open savanna woodland. Males (25–90 kg; 55–198 lb) reach more than double the body weight of females (17–35 kg; 37–77 lb), and have broad, powerful chests. Males are usually colored orange red and females blue grey, although sometimes the reverse is true. In both sexes, the belly is whitish, the side of the muzzle is marked with a black-and-white patch, and a white cheek stripe extends from muzzle to ear. The long and powerful tail tapers towards the tip. These mostly nocturnal animals primarily graze on grasses but also eat forbs and leaves of shrubs. They spend the day sheltering in a scrape beneath shady shrubs and low trees. Animals form small groups of 2–10 individuals but large aggregations of up to 1500 have been recorded around waterholes during drought. Spring and summer is the time of most births and young remain in the pouch for around 8 months. Weaning takes another 4 months, and sexual maturity 14–20 months in females and 24–36 months in males. Reproduction is very sensitive to environmental conditions: in a drought lasting 2–3 months, up to 50% of pouch young die.

Status *Common*

Red-necked Wallaby
Macropus rufogriseus (Desmarest, 1817)

Distribution
South-eastern Australia (south-eastern Queensland to southern Victoria), including Bass Strait islands and Tasmania; introduced to New Zealand and British Isles

This common species inhabits open and dense forest with a complex understorey which is close to open, grassy areas. Males are larger (15–27 kg; 33–59 lb) and darker than females (11–15.5 kg; 24–34 lb). The coat is fawn grey to reddish above with paler underparts, and both sexes have a reddish nape and shoulder coloration. The cheek stripe is whitish, and the muzzle, fore-paws and hind-toes are black. Animals are darker and shaggier in the southern part of the range, particularly in Tasmania. This grazer primarily eats grasses, herbs and leaves of low-growing shrubs, often in loose aggregations of up to 30 individuals. Mostly nocturnal, animals also graze in the early morning and late afternoon on cloudy days. They rest alone during the day in dense shrubbery. Mothers carry a single pouch young for 8–9 months, before weaning at 1 year of age.

Status *Common*

Tammar Wallaby
Macropus eugenii (Desmarest, 1817)

 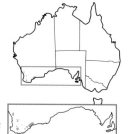

Distribution
Scattered populations through southern Western Australia and South Australia and their offshore islands; introduced to New Zealand

Habitat loss and predation by introduced foxes and cats have left this species sparsely distributed throughout its former wide range across western and southern mainland Australia. Populations on offshore islands are mostly intact, although it is now extinct on Flinders Island. The smallish wallaby has larger males (6–10 kg; 13–22 lb) than females (4–6 kg; 9–13 lb). Its thick coat is greyish brown above and grey below, with reddish fur on the neck, shoulders and flanks. There is a noticeable white cheek stripe, dark patches on the muzzle and around the eye, and a short dark midline from the head to the upper spine. Animals are nocturnal foragers, grazing in grassy areas close to cover. They rest during the day in dense vegetation, usually coastal heath and scrub or dry forest, or mallee with a dense understorey. Animals are usually solitary except when breeding. Most young are born from January to March, and remain in the pouch for 8–9 months.

Status *Common on offshore islands; vulnerable on mainland*

Toolache Wallaby
Macropus greyi Waterhouse (1845)

Distribution
Formerly south-eastern South Australia, probably far western Victoria

This beautiful wallaby was once common in its preferred habitat but hunting pressure – primarily from humans but also from foxes – and habitat loss caused its rapid demise. The last wild individual was recorded in 1927; it was taken into captivity and died in 1939 despite concerted efforts to find it a mate. It inhabited the extensive plains north and east of The Coorong wetlands, where open grassland and sedgeland abuts drier areas of stringybark heath and woodland. A slender, graceful animal, its coat and tail were silvery grey and its underparts whitish. The back of the neck was tinged red, and the mid and lower back were striped with darker grey. The muzzle, ear tips and paws were all black, and its face was striped white. It was noted for its speed and agility, able to outrun and out-manoeuvre all but the fastest hunting dogs by rapidly changing direction and, unusually, altering its stride length. Animals lived in social groups when food was abundant; they were solitary when food was scarce.

Status *Extinct*

Western Brush Wallaby
Macropus irma (Jourdan, 1837)

Distribution
Coastal forests of south-western Western Australia

Unlike most other small wallabies, this animal prefers open forest and woodland habitats with a grassy understorey and avoids dense cover. It is fast and agile over open ground, and is mostly active at dawn and dusk. Males and females are of similar size (7–9 kg; 15.5–20 lb) and appearance, with a grey or dark-grey coat and prominent white facial stripe. The ears are black and white, and it has black 'gloves' on all 4 feet. The grey tail is tipped with a black crest. Some animals have faint stripes across the lower back and thigh. Animals rest during the heat of the day beneath shady shrubs, usually singly or in pairs, then emerge when it is cooler to graze on grasses and forbs. Most young are born in the autumn and spend around 6 months in the pouch.

Status *Uncommon across its range, although numbers increase in response to fox control*

Western Grey Kangaroo
Macropus fuliginosus (Desmarest, 1817)

DISTRIBUTION
*Southern mainland Australia,
and Kangaroo Island, South Australia*

This large, abundant kangaroo occupies a variety of habitats across southern Australia – open forest, woodland, shrubland, heathland and grassland. As with all large kangaroos, the males (20–54 kg; 44–119 lb) are substantially larger than the females (15–28 kg; 33–62 lb) and have powerful, muscular fore-limbs and chest. The fur is thick and coarse, and is colored variably grey brown to chocolate brown above, and paler on the underparts. The long, muscular tail tapers towards the tip. When moving quickly, an animal is bipedal and can travel over 10 m (33 ft) in a single bound. Its diet consists of grasses but it also eats forbs, shrubby browse and herbs if grass is in short supply. Animals are social, grazing in groups in cooler periods of the day and at night, and resting together in the heat of the day in scrapes beneath shrubs and trees. Males fight for dominance and access to females. Most young are born in summer and remain in the pouch for just over 10 months.

STATUS *Common*

Whiptail Wallaby, Pretty-faced Wallaby
Macropus parryi Bennett (1835)

DISTRIBUTION
*Coastal forests of central and eastern
Queensland, and north-eastern NSW*

This 'pretty-face' wallaby has prominent white cheek and ear stripes, and a dark grey brownish face. It also has white hip stripes that blend with the pale sides of the tail to form a distinct 'U' shape when seen from behind. The long, slender tail is pale grey, often with a darker tip. Its body fur is greyish brown above and paler on the underparts. Males (14–26 kg; 31–57 lb) are almost twice as large as females (7–15 kg; 15–33 lb). Its preferred habitat is open forest and woodland with a grassy understorey. Animals rest in the shade during the heat of the day but become active during cooler parts of the day and at night, when they graze on grasses, forbs and ferns. Animals form social groups of up to 50 individuals. Females carry their single pouch young for around 9 months; the young become sexually mature after 18 months (females) or 2 years (males).

STATUS *Common*

Swamp Wallaby
Wallabia bicolor (Desmarest, 1804)

DISTRIBUTION
*Eastern mainland Australia, from
north-eastern Cape York to south-western Victoria
and south-eastern South Australia, at altitudes
below 1200 m (3940 ft)*

This widespread and common wallaby inhabits forest and woodland with a dense understorey of grass, ferns or shrubs; it prefers moist gullies. Although primarily a browser, it eats a broad range of vegetation, including grasses, herbs, shrubs, ferns and fungi, and poisonous plants such as hemlock. Males (12.3–20.5 kg; 27.1–45.1 lb) are slightly larger than females (10.3–15.4 kg; 22.7–33.9 lb). Its fur is very dark brown to black on its upperparts, with reddish-orange underparts. The cheek has a pale stripe and the tail is often tipped white. The species is usually nocturnal but will also forage during the day if food is scarce. It rests by day in a scrape in dense vegetation. Usually solitary, animals will feed together in groups of up to 8–10 individuals at rich food sources. Young are born throughout the year but mostly in winter in southern populations. They remain in the pouch for 8–9 months, and achieve sexual maturity at 15–16 months.

STATUS *Common*

GLOSSARY

Anterior situated towards or pointing towards the front, or head end, of an animal.

Blastocyst an embryo at a very early stage of development, consisting of a hollow ball of cells with no differentiation of organs.

Calcrete (or *hardpan*) a hard layer of calcium-rich material that lies on or just under the soil surface.

Carnassial teeth sharp, elongated, blade-like teeth, specialized for cutting flesh. In marsupials the carnassials are formed from the upper and lower third premolars; in placental carnivores the carnassials are the fourth upper premolar and first lower molar.

Chorio-allantoic placenta an organ system that allows developing mammals to gain nourishment from the blood supply of the mother. The growing young are surrounded by an embryonic sac, or chorion, that is in close contact with the mother's blood supply at the placenta. The union of the chorion with the placenta allows the passage of food materials and gases, thus providing the young with the nutrition needed for growth.

Chorion the outer cellular layer of the embryonic sac of higher vertebrates, which forms the placenta in mammals.

Chorionic villi finger-like projections from the chorion that invade the mother's tissues to form the placenta.

Cloaca terminal part of a gut which also receives the ducts of the urinary and female genital system.

Dasyurid a member of the family Dasyuridae, which comprises mostly meat-eating and insect-eating marsupials in the Australasian region.

Diapause a temporary halt in the growth and development of an embryonic mammal.

Didelphid a member of the family Didelphidae, which makes up the majority of marsupials in the Americas.

Diprotodontian a member of the order Diprotodontia, the largest of the marsupial orders. The order contains possums, kangaroos, wombats and many other species.

Ectomycorrhizal associations integrations of fungi and green plants, in which the fungi live within or on the outside of the plant roots. The associations can be parasitic but most often are mutually beneficial: the fungus gains a host on which to grow and in return takes up nutrients from the soil that the plant can use.

Epipubic bones a pair of forward-pointing bones that are attached to the pubic bones of the pelvis. They are well developed in most marsupials, except in the Thylacine and marsupial moles, as well as in monotremes.

Fossorial pertains to animals that dig burrows and spend some or most of their lives moving and sheltering underground. Fossorial can also relate to adaptations for digging, such as flattened and powerful fore-limbs and broad, stout claws.

Gallery forest a stretch of forest along a river in otherwise open country.

Guild a group of species that uses similar environmental resources, such as food or shelter, in a similar manner.

Lacustrine pertains to animals or plants that live or grow on the shores of lakes.

Marsupial shelf in most marsupials, but not the Koala and Honey Possum, the lower edge of the bottom jaw flares inwards to form an extensive and flattened area of bone which allows insertion of muscles that are important for chewing.

Megafauna very large, usually land, animals such as the living elephants and rhinoceros. The marsupial megafauna were very large herbivorous and carnivorous forms that became extinct in the Pleistocene.

Microbiotheriid a member of the family Microbiotheriidae. The South American Monito del Monte (*Dromiciops gliroides*) is the only living species but fossils of this family have been found also in South America, Australia and possibly Antarctica.

Molecular clock technique a means of measuring the timing of evolutionary events that is based on the assumption that genetic mutations happen at a constant rate. Genetic changes accumulate in animals' genomes, thus the larger the difference between the genomes of two species, the longer it will be since they split apart from each other.

Monotreme member of the order Monotremata, the only mammals that lay eggs. Living members of this order are the Platypus (*Ornithorhynchus anatinus*) and four species of echidnas.

Myrtaceous plants belonging to the family Myrtaceae. Eucalypts and their relatives are the dominant members of this family in the Australasian region; melaleucas and tea-trees are other members.

Occlusion when the jaw is closed, the upper and lower teeth are in contact.

Opposable the situation in which fingers or toes can be flexed inwards and towards each other to allow a grasping action. In some climbing marsupials the big toe opposes the other toes, and the second and third fingers oppose each other to allow grasping of tree branches and other surfaces.

Parapatric distribution abutting but not overlapping geographical distributions of two or more species.

Peramelemorphian a member of the Order Peramelemorphia which contains the bandicoots and bilbies.

Pheromone secretions, often from specialized skin glands, whose odors act as chemical messengers from one animal to another. These odors can convey information about sexual state, physiological condition, territory ownership and other aspects, and often prompt specific responses by animals that receive the chemical message.

Phylogenetic pertains to the evolutionary relationships between groups of animals or plants, often encompassing their evolutionary histories.

Placentation pertains to the process of formation of the placenta or to the way in which the placenta attaches to the uterus.

Plate tectonics a branch of geology that studies continental movements and the plates that carry the continental land masses.

Posterior situated towards or pointing towards the rear end of an animal.

Propleopine a member of the subfamily Propleopinae, a now-extinct group of carnivorous and omnivorous marsupials.

Proteaceous plants belonging to the family Proteaceae. Banksias, grevilleas, hakeas and dryandras are dominant members of the family in the Australasian region.

Pterygoid muscle a thick muscle that attaches the lower jaw to the skull. The sideways movements it produces when it contracts, generate powerful forces for chewing.

Riparian pertains to animals or plants that live or grow on the banks of rivers.

Riverine pertains to a river or stream.

Scincid a member of the lizard family Scincidae, or skinks, which is widely distributed in Australasia and other parts of the world.

Simulation modelling a computer-based means of exploring how different inputs to a problem lead to different solutions. This type of modelling is used in population studies to explore, for example, the contributions of reproduction, lifespan, survival of young or old animals to a population's long-term survival.

Spinifex a type of grass that grows in arid parts of Australia and covers some 25% of the continent. It grows in the shape of a hummock, and is characterized by thin, stiff leaves with sharp points. It is sometimes called 'porcupine grass' or 'pincushion grass' because of its needle-like leaves.

Sporocarp the spore-producing structure of fungi, lichens and some primitive plants. The fungal sporocarp is most familiar as mushrooms, toadstools or truffles.

Squamosal a bone in the skull of mammals that serves as the point of articulation, or hinge, for the lower jaw. This hinge is not present in other vertebrates, and is taken to be a point of demarcation between early reptiles and early mammals.

Sympatric distribution overlapping geographical distributions of two or more species.

Syndactyly a condition where the second and third toes of the hind-foot are partly fused together but separate, with their own claws, at the tips. This condition is seen in many Australasian marsupials, such the bandicoots and diprotodontians.

Taxon (plural *taxa*) a general term that refers to any taxonomic group or level.

Vomeronasal organ a pocket-like opening in the nasal cavity that is important for sensing pheromones and other chemical cues in many mammals. Its function is poorly known in marsupials but is probably similar to that in the placental mammals that have been studied.

Yolksac placenta a type of placenta that provides nourishment for growing young but lacks the chorionic villi that allow the more intimate contact between mother and young in the chorio-allantoic placenta.

BIBLIOGRAPHY

Websites

The Australian Mammal Society has an excellent website that posts news of recent discoveries, conferences, seminars and other forthcoming events. For the latest research on marsupial biology, visit www.australianmammals.org.au

General publications on Australian marsupials

The works listed below are either general references on Australian marsupials, or books that describe the biology of individual species or groups of species. Most are available from public libraries or from bookstores but older works are scarce and likely to be available for reference only at state libraries or museums. Books in this list were used to compile the species accounts.

Archer, M (ed). 1982. *Carnivorous marsupials*, 2 volumes. Royal Zoological Society of New South Wales, Sydney.

Archer, M (ed). 1987. *Possums and opossums: Studies in evolution*, 2 volumes. Surrey Beatty and Sons, and Royal Zoological Society of New South Wales, Sydney.

Archer, M, Flannery, T and Grigg, G. 1985. *The kangaroo*. Weldon, Sydney.

Armati, PJ, Dickman, CR and Hume, ID (eds). 2006. *Marsupials*. Cambridge University Press, Cambridge.

Aslin, HJ and Smith, MJ. 1987. *Marsupials of Australia*. Volume 2: *Carnivorous marsupials and bandicoots*. Illustrated by R Woodford Ganf. Lansdowne Editions, Sydney.

Calaby, JH and Flannery, TF. 2005. *Marsupials of Australia*. Volume 3: *Kangaroos, wallabies and rat-kangaroos*. Illustrated by R Woodford Ganf. Mallon Publishing, Melbourne.

Claridge, A, Seebeck, J and Rose, R. 2007. *Bettongs, potoroos and the Musky Rat-kangaroo*. CSIRO Publishing, Melbourne.

Cole, J and Woinarski, J. 2002. *A field guide to the rodents and dasyurids of the Northern Territory*. Surrey Beatty and Sons, Sydney.

Cronin, L. 1991. *Key guide to Australian mammals*. Reed Books, Sydney.

Curtis, LK. 2005. *Green guide to kangaroos and wallabies of Australia*. New Holland Publishers, Sydney.

Dawson, TJ. 1983. *Monotremes and marsupials: The other mammals*. Edward Arnold, London.

Dawson, TJ. 1995. *Kangaroos: Biology of the largest marsupials*. University of New South Wales Press, Sydney.

Flannery, TF. 1990. *Australia's vanishing mammals*. RD Press, Sydney.

Flannery, TF. 1994. *Possums of the world: A monograph of the Phalangeroidea*. Geo Productions, Sydney.

Flannery, TF, Martin RW and Szalay, A. 1996. *Tree-kangaroos: A curious natural history*. Reed Books, Melbourne.

Fleay, D. 1947. *Gliders of the gum trees*. Bread and Cheese Club, Melbourne.

Frith, HJ and Calaby, JH. 1969. *Kangaroos*. FW Cheshire, Melbourne.

Goldingay, RL and Jackson, SM (eds). 2004. *The biology of Australian possums and gliders*. Surrey Beatty and Sons, Sydney.

Gould, J. 1845–63. *The mammals of Australia*, 3 volumes. The Author, London.

Green, RJ. 1973. *The mammals of Tasmania*. Mary Fisher Bookshop, Launceston.

Grigg, GC, Jarman, P and Hume, ID (eds). 1989. *Kangaroos, wallabies and rat-kangaroos*, 2 volumes. Surrey Beatty and Sons, Sydney.

Guiler, E. 1985. *Thylacine: The tragedy of the Tasmanian Tiger*. Oxford University Press, Melbourne.

Hume, ID. 1982. *Digestive physiology and nutrition of marsupials*. Cambridge University Press, Cambridge.

Hume, ID. 1999. *Marsupial nutrition*. Cambridge University Press, Melbourne.

Hunsaker, D, II (ed). 1977. *The biology of marsupials*. Academic Press, New York.

Hyett, J and Shaw, N. 1980. *Australian mammals: A field guide for New South Wales, Victoria, South Australia and Tasmania*. Thomas Nelson, Melbourne.

Jackson, SM. 2003. *Australian mammals: Biology and captive management*. CSIRO Publishing, Melbourne.

Johnson, C. 2006. *Australia's mammal extinctions: A 50 000 year history*. Cambridge University Press, Cambridge.

Johnson, P. 2003. *Kangaroos of Queensland*. Queensland Museum, Brisbane.

Jones, C and Parish, S. 2006. *Field guide to Australian mammals*. Steve Parish Publishing, Brisbane.

Jones, ME, Dickman, CR and Archer, M (eds). 2003. *Predators with pouches: The biology of carnivorous marsupials*. CSIRO Publishing, Melbourne.

Kerle, JA. 2001. *Possums: The brushtails, ringtails and Greater Glider*. University of New South Wales Press, Sydney.

Krefft, G. 1871. *The mammals of Australia, illustrated by Miss Harriett Scott, and Mrs Helena Forde, for the Council of Education; with a short account of the species hitherto described*. Thomas Richards, Government Printer, Sydney.

Lee, AK and Cockburn, A. 1985. *Evolutionary ecology of marsupials*. Cambridge University Press, Cambridge.

Lee, AK and Martin RW. 1988. *The Koala: A natural history*. New South Wales University Press, Sydney.

Lee, AK, Handasyde, KA and Sanson, GD (eds). 1991. *Biology of the Koala*. Surrey Beatty and Sons, Sydney.

Lindenmayer, D. 2002. *Gliders of Australia: A natural history*. University of New South Wales Press, Sydney.

Lindsey, T. 2000. *Green guide to mammals of Australia*. New Holland Publishers, Sydney.

Lunney, D, Urquhart, CA and Reed, P (eds). 1990. *Koala summit: Managing Koalas in New South Wales*. NSW National Parks and Wildlife Service, Sydney.

Lydekker, R. 1896. *A hand-book to the Marsupialia and Monotremata*. Edward Lloyd, London.

Mansergh, I and Broome, L. 1994. *The Mountain Pygmy-possum of the Australian Alps*. New South Wales University Press, Sydney.

Marlow, B. 1965. *Marsupials of Australia*. Hesperian Press, Perth.

Martin, R. 2005. *Tree-kangaroos of Australia and New Guinea*. CSIRO Publishing, Melbourne.

Martin, R and Handasyde, K. 1999. *The Koala: Natural history, conservation and management*. University of New South Wales Press, Sydney.

Maxwell, S, Burbidge, AA and Morris, K (eds). 1996. *The 1996 action plan for Australian marsupials and monotremes*. Wildlife Australia, Canberra.

Menkhorst, PW (ed). 1995. *Mammals of Victoria: Distribution, ecology and conservation*. Oxford University Press, Melbourne.

Menkhorst, P and Knight, F. 2004. *A field guide to the mammals of Australia*, 2nd ed. Oxford University Press, Melbourne.

Montague, TL (ed). 2000. *The Brushtail Possum: Biology, impact and management of an introduced marsupial*. Manaaki Whenua Press, Lincoln, New Zealand.

Nowak, RM (ed). 2005. *Walker's marsupials of the world*. The Johns Hopkins University Press, Baltimore.

Owen, D. 2003. *Thylacine: The tragic tale of the Tasmanian Tiger*. Allen and Unwin, Sydney.

Owen, D and Pemberton, D. 2005. *Tasmanian Devil: A unique and threatened animal*. Allen and Unwin, Sydney.

Paddle, R. 2000. *The last Tasmanian Tiger: The history and extinction of the Thylacine*. Cambridge University Press, Cambridge.

Ride, WDL. 1970. *A guide to the native mammals of Australia*. Oxford University Press, Melbourne.

Roberts, M, Carnio, J, Crawshaw, G and Hutchins, M (eds). 1993. *The biology and management of Australasian carnivorous marsupials*. Metropolitan Toronto Zoo, Toronto, and the American Association of Zoological Parks and Aquariums, Washington, DC.

Russell, R. 1980. *Spotlight on possums*. University of Queensland Press, Brisbane.

Saunders, NR and Hinds, LA (eds). 1997. *Marsupial biology: Recent research, new perspectives*. University of New South Wales Press, Sydney.

Seebeck, JH, Brown, PR, Wallis, RL and Kemper, CM (eds). 1990. *Bandicoots and bilbies*. Surrey Beatty and Sons, and the Australian Mammal Society, Sydney.

Serventy, V and Serventy, C. 2002. *Koalas*. New Holland Publishing, Sydney.

Smith, AP and Hume, ID (eds). 1984. *Possums and gliders*. Surrey Beatty and Sons, and the Australian Mammal Society, Sydney.

Smith, A and Winter, J. 1997. *A key and field guide to the possums, gliders and Koala*. Surrey Beatty and Sons, Sydney.

Smith, B. 1995. *Caring for possums*. Kangaroo Press, Sydney.

Smith, MJ. 1980. *Marsupials of Australia*. Volume 1: Possums, the Koala and wombats. Illustrated by R Woodford Ganf. Lansdowne Editions, Melbourne.

Stonehouse, B and Gilmore, D (eds). 1977. *The biology of marsupials*. Macmillan Press, London.

Strahan, R (ed). 1983. *The Australian Museum complete book of Australian mammals*. Angus and Robertson, Sydney.

Strahan, R (ed). 1995. *The mammals of Australia*, 2nd ed. Australian Museum and Reed Books, Sydney.

Strahan, R and Conder, P. 2007. *Dictionary of Australian and New Guinean mammals*. CSIRO Publishing, Melbourne.

Taylor, JM. 1984. *The Oxford guide to mammals of Australia*. Oxford University Press, Melbourne.

Triggs, B. 1996. *The wombat: Common Wombats in Australia*, revised ed. University of New South Wales Press, Sydney.

Triggs, B. 2004. *Tracks, scats and other traces: A field guide to Australian mammals*, revised ed. Oxford University Press, Melbourne.

Troughton, E. 1967. *Furred animals of Australia*, 9th ed. Angus and Robertson, Sydney.

Turner, JR. 2004. *Mammals of Australia: An introduction to their classification, biology and distribution*. Pensoft, Sofia, Bulgaria.

Tyndale-Biscoe, H. 1973. *Life of marsupials*. Edward Arnold, London.

Tyndale-Biscoe, H. 2005. *Life of marsupials.* CSIRO Publishing, Melbourne.

Tyndale-Biscoe, H and Renfree, M. 1987. *Reproductive physiology of marsupials.* Cambridge University Press, Cambridge.

Van Dyck, S and Strahan, R (eds). 2007. *The mammals of Australia,* 3rd ed. Reed New Holland, Sydney.

Waterhouse, GR. 1843. *The naturalist's library.* Volume VIII: *Mammalia. Marsupialia or pouched animals.* WH Lizars, Edinburgh.

Watts, D. 1993. *Tasmanian mammals: A field guide.* Peregrine Press, Kettering, Tasmania.

Wells, RT and Pridmore, PA (eds). 1999. *Wombats.* Surrey Beatty and Sons, Sydney.

Williams, A and Williams, R. 1999. *Caring for kangaroos and wallabies.* Kangaroo Press, Sydney.

Wilson, M and Croft, DB (eds). 2005. *Kangaroos: Myths and realities.* Australian Wildlife Protection Council, Melbourne.

Woodford, J. 2001. *The secret life of wombats.* Text Publishing, Melbourne.

Wood Jones, F. 1923–24. *The mammals of South Australia.* Part I: *The monotremes and carnivorous marsupials.* Part II: *The bandicoots and the herbivorous marsupials.* REE Rogers, Government Printer, Adelaide.

Specific publications on Australian marsupials

These works are mostly technical papers and publications in the primary scientific literature. They were consulted in compiling the text of this book and can be used by interested readers to follow up particular points of interest. They are available at university and other research libraries, or online (for a fee) from the publishers of the journals.

What is a marsupial?

Abbie, AA. 1939. A masticatory adaptation peculiar to some diprotodont marsupials. *Proceedings of the Zoological Society of London,* B, 109, 261–79.

Aplin, KP and Archer, M. 1987. Recent advances in marsupial systematics with a new syncretic classification. In M Archer (ed) *Possums and opossums: Studies in evolution.* Volume 1, pp xv–lxxii. Surrey Beatty and Sons, and Royal Zoological Society of New South Wales, Sydney.

Archer, M. 1982. Genesis: and in the beginning there was an incredible carnivorous mother. In M Archer (ed) *Carnivorous marsupials.* Volume 1, pp vii–x. Royal Zoological Society of New South Wales, Sydney.

Archer, M and Kirsch, JAW. 2006. The evolution and classification of marsupials. In PJ Armati, CR Dickman and ID Hume (eds) *Marsupials,* pp 1–21. Cambridge University Press, Cambridge.

Arrese, CA, Hart, NS, Thomas, N, Beazley, LD and Shand, J. 2002. Trichromacy in Australian marsupials. *Current Biology,* 12, 657–60.

Arrese, CA, Oddy, AY, Runham, PB, Hart, NS, Shand, J, Hunt, DM and Beazley, LD. 2005. Cone topography and spectral sensitivity in two potentially trichromatic marsupials, the Quokka (*Setonix brachyurus*) and Quenda (*Isoodon obesulus*). *Proceedings of the Royal Society,* B, 272, 791–96.

Byers, JA. 1999. The distribution of play behaviour among Australian marsupials. *Journal of Zoology,* London, 247, 349–56.

Calaby, JH. 1984. Foreword. In AP Smith and ID Hume (eds) *Possums and gliders,* pp iii–iv. Surrey Beatty and Sons, and the Australian Mammal Society, Sydney.

Cannon, JR, Bakker, HR, Bradshaw, SD and McDonald, IR. 1976. Gravity as the sole navigational aid to the newborn Quokka. *Nature,* 259, 42.

Croft, DB and Eisenberg, JF. 2006. Behaviour. In PJ Armati, CR Dickman and ID Hume (eds) *Marsupials,* pp 229–98. Cambridge University Press, Cambridge.

Dickman, CR. 2005. Marsupials of the world: An introduction. In RM Nowak (ed) *Walker's marsupials of the world,* pp 1–67. The Johns Hopkins University Press, Baltimore.

Flannery, TF. 1995. *Mammals of New Guinea,* revised ed. Reed Books, Sydney.

Flannery, TF. 1995. *Mammals of the south-west Pacific and Moluccan Islands.* Reed Books, Sydney.

Gemmell, RT and Rose, RW. 1989. The senses involved in movement of some newborn Macropodoidea and other marsupials from cloaca to pouch. In GC Grigg, P Jarman and ID Hume (eds) *Kangaroos, wallabies and rat-kangaroos.* Volume 1, pp 339–47. Surrey Beatty and Sons, Sydney.

Grant, T. 1995. *The Platypus: A unique mammal.* University of New South Wales Press, Sydney.

Grassé, P-P. 1955. Ordre des Marsupiaux. In P-P Grassé (ed) *Traité de zoologie.* Volume 17, No. 1, pp 93–185. Masson et Cie, Paris.

Green, B. 1984. Composition of milk and energetics of growth in marsupials. *Symposia of the Zoological Society of London,* 51, 369–87.

Green, B and Merchant, JC. 1988. The composition of marsupial milk. In CH Tyndale-Biscoe and PA Janssens (eds) *The developing marsupial: Models for biomedical research,* pp 41–54. Springer-Verlag, Berlin.

Griffin, AS, Blumstein, DT and Evans, CS. 2000. Training captive-bred or translocated animals to avoid predators. *Conservation Biology,* 14, 1317–26.

Hartman, CG. 1921. Traditional belief concerning the generation of the opossum (*Didelphis virginiana* L.). *Journal of American Folklore,* 34, 321–23.

Hughes, RL and Hall, LS. 1988. Structural adaptations of the newborn marsupial. In CH Tyndale-Biscoe and PA Janssens (eds) *The developing marsupial: Models for biomedical research,* pp 8–27. Springer-Verlag, Berlin.

Krockenberger, A. 2006. Lactation. In PJ Armati, CR Dickman and ID Hume (eds) *Marsupials,* pp 108–36. Cambridge University Press, Cambridge.

Lapointe, F-J and Legendre, P. 1996. Evolution of the marsupial brain: Does it reflect the evolution of behavior? In T Cabana (ed) *Animals in their environment: A tribute to Paul Pirlot,* pp 187–212. Éditions Orbis Publishing, Québec.

Linnaeus, C. 1758. *Systema naturae per regna tria naturae, secundum classes, ordines, genera, species, cum characteribus, differentiis, synonymis, locis.* Tomus I. Editio decima, reformata. Laurentii Salvi, Stockholm. Facsimile ed. 1956, British Museum (Natural History), London.

McKenna, MC and Bell, SK. 1997. *Classification of mammals above the species level.* Columbia University Press, New York.

McLean, IG, Schmitt, NT, Jarman, PJ, Duncan, C and Wynne, CDL. 2000. Learning for life: Training marsupials to recognise introduced predators. *Behaviour,* 137, 1361–76.

Reilly, SM and White, TD. 2003. Hypaxial motor patterns and the function of epipubic bones in primitive mammals. *Science,* 299, 400–02.

Renfree, MB. 1981. Embryonic diapause in marsupials. *Journal of Reproduction and Fertility,* Supplement, 29, 67–78.

Russell, EM. 1982. Patterns of parental care and parental investment in marsupials. *Biological Reviews,* 57, 423–86.

Russell, EM. 1985. The metatherians: Order Marsupialia. In RE Brown and DW Macdonald (eds) *Social odours in mammals,* pp 45–104. Clarendon Press, Oxford.

Sánchez-Villagra, MR. 2001. Ontogenetic and phylogenetic transformations of the vomeronasal complex and nasal floor elements in marsupial mammals. *Zoological Journal of the Linnean Society,* 131, 459–79.

Sánchez-Villagra, MR and Smith, KK. 1997. Diversity and evolution of the marsupial mandibular angle. *Journal of Mammalian Evolution,* 4, 119–44.

Shaw, G. 2006. Reproduction. In PJ Armati, CR Dickman and ID Hume (eds) *Marsupials,* pp 83–107. Cambridge University Press, Cambridge.

Shield, J. 1968. Reproduction of the Quokka, *Setonix brachyurus,* in captivity. *Journal of Zoology,* London, 155, 427–44.

Smith, KK. 1997. Comparative patterns of craniofacial development in eutherian and metatherian mammals. *Evolution,* 51, 1663–78.

Toftegaard, CL and Bradley, AJ. 2003. Chemical communication in dasyurid marsupials. In ME Jones, CR Dickman and M Archer (eds) *Predators with pouches: The biology of carnivorous marsupials,* pp 347–57. CSIRO Publishing, Melbourne.

Tyndale-Biscoe, CH. 1966. The marsupial birth canal. *Symposia of the Zoological Society of London,* 15, 233–50.

Tyndale-Biscoe, CH. 1984. Mammals: Marsupials. In GE Lamming (ed) *Marshall's physiology of reproduction,* 4th ed. Volume 1, pp 386–454. Churchill Livingstone, Edinburgh.

Tyndale-Biscoe, H and Renfree, M. 1987. *Reproductive physiology of marsupials.* Cambridge University Press, Cambridge.

Van Oosterzee, P. 2006. Drawing the Wallace Line. In JR Merrick, M Archer, GM Hickey and MSY Lee (eds) *Evolution and biogeography of Australasian vertebrates,* pp 95–110. Auscipub, Sydney.

Walton, DW and Richardson, BJ (eds). 1989. *Fauna of Australia.* Volume 1B: *Mammalia.* Australian Government Publishing Service, Canberra.

Westerman, M, Springer, MS, Dixon, J and Krajewski, C. 1999. Molecular relationships of the extinct Pig-footed Bandicoot *Chaeropus ecaudatus* (Marsupialia: Perameloidea) using 12S rRNA sequences. *Journal of Mammalian Evolution,* 6, 271–88.

Wilson, DE and Reeder, DM. 2005. *Mammal species of the world. A taxonomic and geographic reference,* 3rd ed. The Johns Hopkins University Press, Baltimore.

Woolley, P. 1974. The pouch of *Planigale subtilissima* and other dasyurid marsupials. *Journal of the Royal Society of Western Australia,* 57, 11–15.

Origins and evolution

Abbie, AA. 1941. Marsupials and the evolution of mammals. *Australian Journal of Science,* 4, 77–92.

Amrine-Madsen, H, Scally, M, Westerman, M, Stanhope, MJ, Krajewski, C and Springer, MS. 2003. Nuclear gene sequences provide evidence for the monophyly of Australidelphian marsupials. *Molecular Phylogenetics and Evolution,* 28, 186–96.

Aplin, KP, Baverstock, PR and Donnellan, SC. 1993. Albumin immunological evidence for the time and mode of origin of the New Guinean terrestrial mammal fauna. *Science in New Guinea,* 19, 131–45.

Archer, M. 1981. A review of the origins and radiations of Australian mammals. In A Keast (ed) *Ecological biogeography of Australia,* pp 1437–88. Junk, The Hague.

Archer, M. 1984. The Australian marsupial radiation. In M Archer and G Clayton (eds) *Vertebrate zoogeography and evolution in Australasia*, pp 633–808. Hesperian Press, Perth.

Archer, M and Hand, SJ. 2006. The Australian marsupial radiation. In JR Merrick, M Archer, GM Hickey and MSY Lee (eds) *Evolution and biogeography of Australasian vertebrates*, pp 575–646. Auscipub, Sydney.

Archer, M and Kirsch, J. 2006. The evolution and classification of marsupials. In PJ Armati, CR Dickman and M Archer (eds) *Marsupials*, pp 1–21. Cambridge University Press, Cambridge.

Archer, M, Arena, R, Bassarova, M, Black, K, Brammall, J, Cooke, B, Creaser, P, Crosby, K, Gillespie, A, Godthelp, H, Gott, M, Hand, SJ, Kear, B, Krikmann, A, Mackness, B, Muirhead, J, Musser, A, Myers, T, Pledge, N, Wang, Y and Wroe, S. 1999. The evolutionary history and diversity of Australian mammals. *Australian Mammalogy*, 21, 1–45.

Archer, M, Godthelp, HJ and Hand, SJ. 1993. Early Eocene marsupial from Australia. In F Schrenk and K Ernst (eds) *Kaupia: Darmstädter Beiträge zur Naturgeschichte. Monument Grube Messel – Perspectives and relationships*. Part 2, pp 193–200. Hessisches Landesmuseum Darmstadt, Darmstadt.

Asher, RJ, Horovitz, I and Sánchez-Villagra, MR. 2004. First combined cladistic analysis of marsupial mammal interrelationships. *Molecular Phylogenetics and Evolution*, 33, 240–50.

Averianov, AO, Archibald, JD and Martin, T. 2003. Placental nature of the alleged marsupial from the Cretaceous of Madagascar. *Acta Palaeontologica Polonica*, 48, 149–51.

Bensley, BA. 1903. On the evolution of the Australian Marsupialia; with remarks on the relationships of the marsupials in general. *Transactions of the Linnean Society of London, Zoology*, 9, 83–217.

Burk, A, Westerman, M, Kao, DJ, Kavanagh, JR and Springer, MS. 1999. An analysis of marsupial interordinal relationships based on 12S rRNA, tRNA, valine, 16S rRNA, and cytochrome *b* sequences. *Journal of Mammalian Evolution*, 6, 317–34.

Cardillo, M, Bindra-Emonds, ORP, Boakes, E and Purvis, A. 2004. A species-level phylogenetic supertree of marsupials. *Journal of Zoology, London*, 264, 11–31.

Cifelli, RL and Davis, BM. 2003. Marsupial origins. *Science*, 302, 1899–900.

Colgan, DJ. 1999. Phylogenetic studies of marsupials based on phosphoglycerate kinase DNA sequences. *Molecular Phylogenetics and Evolution*, 11, 13–26.

Darlington, PJ. 1957. *Zoogeography: The geographical distribution of animals*. Wiley, New York.

Darlington, PJ. 1965. *Biogeography of the southern end of the world*. Harvard University Press, Cambridge.

Dow, DB and Sukamto, R. 1984. Late Tertiary to Quaternary tectonics of Irian Jaya. *Episodes*, 7, 3–9.

Flannery, TF. 1989. Origins of the Australo-Pacific land mammal fauna. *Australian Zoological Reviews*, 1, 15–24.

Flannery, TF. 1994. The fossil land mammal record of New Guinea: A review. *Science in New Guinea*, 20, 39–48.

Flannery, TF. 1995. *Mammals of New Guinea*, revised ed. Reed Books, Sydney.

Flynn, JJ and Wyss, AR. 1999. New marsupials from the Eocene-Oligocene transition of the Andean Main Range, Chile. *Journal of Vertebrate Paleontology*, 19, 533–49.

Gemmell, NJ and Westerman, M. 1994. Phylogenetic relationships within the Class Mammalia: A study using mitochondrial 12S RNA sequences. *Journal of Mammalian Evolution*, 2, 3–23.

Godthelp, H, Wroe, S and Archer, M. 1999. A new marsupial from the early Eocene Tingamarra Local Fauna of Murgon, southeastern Queensland: A prototypical Australian marsupial? *Journal of Mammalian Evolution*, 6, 289–313.

Goin, FJ. 2003. Early marsupial radiations in South America. In ME Jones, CR Dickman and M Archer (eds) *Predators with pouches: The biology of carnivorous marsupials*, pp 30–42. CSIRO Publishing, Melbourne.

Goin, FJ, Case, J, Woodburne, MO, Vizcaíno, SF and Reguero M. 1999. New discoveries of 'opossum-like' marsupials from Antarctica (Seymour Island, medial Eocene). *Journal of Mammalian Evolution*, 6, 335–65.

Hershkovitz, P. 1999. *Dromiciops gliroides* Thomas, 1894, last of the Microbiotheria (Marsupialia), with a review of the Family Microbiotheriidae. *Fieldiana: Zoology*, New Series, 93, 1–60.

Horovitz, I and Sánchez-Villagra, MR. 2003. A morphological analysis of marsupial mammal higher-level phylogenetic relationships. *Cladistics*, 19, 181–212.

Ji, Q, Luo, Z-X, Yuan, C-X, Wible, JR, Zhang, J-P and Georgi, JA. 2002. The earliest known eutherian mammal. *Nature*, 416, 816–22.

Johnson, C. 2006. *Australia's mammal extinctions: A 50 000 year history*. Cambridge University Press, Cambridge.

Kemp, TS. 2005. *The origin and evolution of mammals*. Oxford University Press, Oxford.

Kirsch, JAW and Springer, MS. 1993. Timing of the molecular evolution of New Guinean marsupials. *Science in New Guinea*, 19, 147–56.

Kirsch, JAW, Dickerman, AW, Reig, OA and Springer, MS. 1991. DNA hybridization evidence for the Australasian affinity of the American marsupial *Dromiciops australis*. *Proceedings of the National Academy of Sciences USA*, 88, 10465–69.

Kirsch, JAW, Lapointe, F-J and Springer, MS. 1997. DNA-hybridisation studies of marsupials and their implications for metatherian classification. *Australian Journal of Zoology*, 45, 211–80.

Krajewski, C, Painter, J, Driskell, AC, Buckley, L and Westerman, M. 1993. Molecular systematics of New Guinean dasyurids (Marsupialia: Dasyuridae). *Science in New Guinea*, 19, 157–65.

Krause, DW. 2001. Fossil molar from a Madagascan marsupial. *Nature*, 412, 497–98.

Long, J, Archer, M, Flannery, T and Hand, S. 2002. *Prehistoric mammals of Australia and New Guinea: One hundred million years of evolution*. University of New South Wales Press, Sydney.

Luo, Z-X, Cifelli, RL and Kielan-Jaworowska, Z. 2001. Dual origin of tribosphenic mammals. *Nature*, 409, 53–57.

Luo, Z-X, Qiang, J, Wible, JR and Yuan C-X. 2003. An early Cretaceous tribosphenic mammal and metatherian evolution. *Science*, 302, 1934–40.

Marshall, LG. 1988. Land mammals and the great American interchange. *American Scientist*, 76, 380–88.

Martin, HA. 1998. Tertiary climatic evolution and the development of aridity in Australia. *Proceedings of the Linnean Society of New South Wales*, 119, 115–36.

McGowran, B and Li, Q. 1994. The Miocene oscillation in southern Australia. *Records of the South Australian Museum*, 27, 197–212.

Nilsson, M, Gullberg, A, Spotorno, AE, Arnasson, U and Janke A. 2003. Radiation of extant marsupials after the K/T boundary: Evidence from complete mitochondrial genomes. *Journal of Molecular Evolution*, 57, 1–10.

Palma, RE. 2003. Evolution of American marsupials and their phylogenetic relationships with Australian metatherians. In ME Jones, CR Dickman and M. Archer (eds) *Predators with pouches: The biology of carnivorous marsupials*, pp 21–29. CSIRO Publishing, Melbourne.

Patterson, D and Pascual, R. 1972. The fossil mammal fauna of South America. In A Keast, FC Erk and B Glass (eds) *Evolution, mammals, and southern continents*, pp 247–309. State University of New York Press, New York.

Phillips, MJ, Lin, Y-H, Harrison, GL and Penny, D. 2001. Mitochondrial genomes of a bandicoot and a brushtail possum confirm the monophyly of Australidelphian marsupials. *Proceedings of the Royal Society*, B, 268, 1533–38.

Rich, TH. 1991. Monotremes, placentals, and marsupials: Their record in Australia and its biases. In P Vickers-Rich, JM Monaghan, RF Baird and TH Rich (eds) *Vertebrate palaeontology of Australasia*, pp 893–1004. Pioneer Design Studio and Monash University Publications Committee, Melbourne.

Ride, WDL. 1962. On the evolution of Australian marsupials. In GW Leeper (ed) *The evolution of living organisms*, pp 281–306. Melbourne University Press, Melbourne.

Ride, WDL, Pridmore, PA, Barwick, RE, Wells, RT and Heady, RD. 1997. Towards a biology of *Propleopus oscillans* (Marsupialia: Propleopinae, Hypsiprymnodontidae). *Proceedings of the Linnean Society of New South Wales*, 117, 243–328.

Springer, MS, Kirsch, JAW and Case, JA. 1997. The chronicle of marsupial evolution. In TJ Givnish and KJ Sytsma (eds) *Molecular evolution and adaptive radiation*, pp 129–61. Cambridge University Press, Cambridge.

Springer, MS, Westerman, M, Kavanagh, JR, Burk, A, Woodburne, MO, Kao, DJ and Krajewski, C. 1998. The origin of the Australasian marsupial fauna and the affinities of the enigmatic Monito del Monte and Marsupial Mole. *Proceedings of the Royal Society*, B, 265, 2381–86.

Szalay, FS. 1982. A new appraisal of marsupial phylogeny and classification. In M Archer (ed) *Carnivorous marsupials*. Volume 2, pp 621–40. Royal Zoological Society of New South Wales, Sydney.

Szalay, FS. 1994. *Evolutionary history of the marsupials and an analysis of osteological characters*. Cambridge University Press, New York.

Troughton, E Le G. 1959. The marsupial fauna: Its origin and radiation. In A Keast, RL Crocker and CS Christian (eds) *Biogeography and ecology in Australia*, pp 69–88. Uitgeverij Dr W Junk, The Hague.

Woodburne, MO and Case, JA. 1996. Dispersal, vicariance, and the late Cretaceous to early Tertiary land mammal biogeography from South America to Australia. *Journal of Mammalian Evolution*, 3, 121–61.

Woodburne, MO and Zinsmeister, WJ. 1984. The first land mammal from Antarctica and its biogeographic implications. *Journal of Paleontology*, 58, 913–48.

Woodburne, MO, Rich, TH and Springer, MS. 2003. The evolution of tribosphery and the antiquity of mammalian clades. *Molecular Phylogenetics and Evolution*, 28, 360–85.

Worthy, TH, Tennyson, AJD, Archer, M, Musser, AM, Hand, SJ, Jones, C, Douglas, BJ, McNamara, JA and Beck, RMD. 2006. Miocene mammal reveals a Mesozoic ghost lineage on insular New Zealand, southwest Pacific. *Proceedings of the National Academy of Sciences USA*, 103, 19419–23.

Wroe, S. 2003. Australian marsupial carnivores: Recent advances in palaeontology. In ME Jones, CR Dickman and M Archer (eds) *Predators with pouches: The biology of carnivorous marsupials*, pp 102–23. CSIRO Publishing, Melbourne.

Wroe, S and Archer, M. 2006. Origins and early radiations of marsupials. In JR Merrick, M Archer, GM Hickey and MSY Lee (eds) *Evolution and biogeography of Australasian vertebrates*, pp 551–74. Auscipub, Sydney.

Recent history

Barrows, TT and Juggins, S. 2005. Sea-surface temperatures around the Australian margin and Indian Ocean during the Last Glacial Maximum. *Quaternary Science Reviews*, 24, 1017–47.

Bowler, JM and Magee, JW. 2000. Redating Australia's oldest human remains: A sceptic's view. *Journal of Human Evolution*, 38, 719–26.

Bowman, DMJS. 1998. Tansley Review no. 101: The impact of Aboriginal landscape burning on the Australian biota. *New Phytologist*, 140, 385–410.

Broom, R. 1898. On the affinities and habits of *Thylacoleo*. *Proceedings of the Linnean Society of New South Wales*, 22, 57–74.

Cupper, M. 2007. Ice age cold case. *Australasian Science*, 28, 16–17.

European Project for Ice Coring in Antarctica. 2004. Eight glacial cycles from an Antarctic ice core. *Nature*, 429, 623–28.

Flannery, TF. 1994. *The future eaters*. Reed Books, Melbourne.

Flannery, TF. 1995. *Mammals of New Guinea*, revised ed. Reed Books, Sydney.

Flood, J. 2004. *Archaeology of the dreamtime: The story of prehistoric Australia and its people*, revised ed. JB Publishing, Adelaide.

Gill, ED. 1954. Ecology and distribution of the extinct giant marsupial, *Thylacoleo*. *Victorian Naturalist*, 71, 18–35.

Hesse, PP, Magee, JW and van der Kaars, S. 2004. Late Quaternary climates of the Australian arid zone: A review. *Quaternary International*, 118–119: 87–102.

Johnson, C. 2006. *Australia's mammal extinctions: A 50 000 year history*. Cambridge University Press, Cambridge.

Johnson, CN and Wroe, S. 2003. Causes of extinction of vertebrates during the Holocene of mainland Australia: Arrival of the Dingo, or human impact? *The Holocene*, 13, 941–48.

Kershaw, AP, Clark, JS, Gill, AM and D'Costa, DM. 2002. A history of fire in Australia. In RA Bradstock, JE Williams and AM Gill (eds) *Flammable Australia: The fire regimes and biodiversity of a continent*, pp 3–25. Cambridge University Press, Cambridge.

Kohen, JL. 1995. *Aboriginal environmental impacts*. University of New South Wales Press, Sydney.

Long, J. 2007. Megafauna theory faces extinction. *Australasian Science*, 28, 18–19.

Long, J, Archer, M, Flannery, T and Hand, S. 2002. *Prehistoric mammals of Australia and New Guinea: One hundred million years of evolution*. University of New South Wales Press, Sydney.

Miller, GH, Fogel, ML, Magee, JW, Gagan, MK, Clarke, SJ and Johnson, BJ. 2005. Ecosystem collapse in Pleistocene Australia and a human role in megafaunal extinction. *Science*, 309, 287–90.

Mulvaney, J and Kamminga, J. 1999. *Prehistory of Australia*. Allen and Unwin, Sydney.

Murray, PF. 1984. Extinctions Downunder: A bestiary of extinct Australian late Pleistocene monotremes and marsupials. In PS Martin and RG Klein (eds) *Quaternary extinctions: A prehistoric revolution*, pp 600–28. University of Arizona Press, Tucson.

O'Connell, JF and Allen, J. 2004. Dating the colonization of Sahul (Pleistocene Australia – New Guinea): A review of recent research. *Journal of Archaeological Science*, 31, 835–53.

Owen, R. 1838. Fossil remains from Wellington Valley, Australia. Appendix in TL Mitchell *Three expeditions into the interior of eastern Australia, with descriptions of the recently explored region of Australia Felix, and of the present colony of New South Wales*. Volume 2, pp 359–69. T and W Boone, London.

Owen, R. 1859. On the fossil mammals of Australia. Part 1. Description of a mutilated skull of a large marsupial carnivore (*Thylacoleo carnifex* Owen), from a calcareous conglomerate stratum eighty miles SW of Melbourne, Victoria. *Philosophical Transactions of the Royal Society*, 149, 309–22.

Roberts, RG, Flannery, TF, Ayliffe, LK, Yoshida, H, Olley, JM, Prideaux, GJ, Laslett, GM, Baynes, A, Smith, MA, Jones, R and Smith, BL. 2001. New ages for the last Australian megafauna: Continent-wide extinction about 46,000 years ago. *Science*, 292, 1888–92.

Roberts, RG, Jones, R, Spooner, NA, Head, MJ, Murray, AS and Smith MA. 1994. The human colonization of Australia: Optical dates of 53,000 and 60,000 years bracket human arrival at Deaf Adder Gorge, Northern Territory. *Quaternary Science Reviews*, 13, 575–83.

Wroe, S, Crowther, M, Dortch, J and Chong, J. 2004. The size of the largest marsupial and why it matters. *Proceedings of the Royal Society*, B (Supplement), 271, S34–36.

Wroe, S, Field, J, Fullagar, R and. Jermiin, LS. 2004. Megafaunal extinction in the late Quaternary and the global overkill hypothesis. *Alcheringa*, 28, 291–331.

Wroe, S, McHenry, C and Thomason, J. 2005. Bite club: Comparative bite force in big biting mammals and the prediction of predatory behaviour in fossil taxa. *Proceedings of the Royal Society*, B, 272, 619–25

Natural history

Adams, J. 1985. The definition and interpretation of guild structure in ecological communities. *Journal of Animal Ecology*, 54, 43–59.

Aitken, PF. 1971. The distribution of the Hairy-nosed Wombat (*Lasiorhinus latifrons* (Owen)) part 1: Yorke Peninsula, Eyre Peninsula, the Gawler Ranges and Lake Harris. *South Australian Naturalist*, 45, 93–103.

Amico, G and Aizen, MA. 2000. Mistletoe seed dispersal by a marsupial. *Nature*, 408, 929–30.

Andrew, DL and Settle, GA. 1982. Observations on the behaviour of species of *Planigale* (Dasyuridae, Marsupialia) with particular reference to the Narrow-nosed Planigale (*Planigale tenuirostris*). In M Archer (ed) *Carnivorous marsupials*. Volume 1, pp 311–24. Royal Zoological Society of New South Wales, Sydney.

Andrewartha, HG and Birch, LC. 1954. *The distribution and abundance of animals*. University of Chicago Press, Chicago.

Arnold, J and Shield, J. 1970. Oxygen consumption and body temperature of the Chuditch (D*asyurus geoffroii*). *Journal of Zoology*, London, 160, 391–404.

Bailey, P and Best, L. 1992. A Red Kangaroo, *Macropus rufus*, recovered 25 years after marking in north-western New South Wales. *Australian Mammalogy*, 15, 141.

Banks, PB, Newsome, AE and Dickman, CR. 2000. Predation by Red Foxes limits recruitment in populations of Eastern Grey Kangaroos. *Austral Ecology*, 25, 283–91.

Barry, SJ. 1984. Small mammals in a south-east Queensland rainforest: The effects of soil fertility and past logging disturbance. *Australian Wildlife Research*, 11, 31–39.

Baynes, A and Johnson, KA. 1996. The contributions of the Horn Expedition and cave deposits to knowledge of the original mammal fauna of central Australia. In SR Morton and DJ Mulvaney (eds) *Exploring central Australia: Society, the environment and the 1894 Horn Expedition*, pp 168–86. Surrey Beatty and Sons, Sydney.

Beal, AM. 1989. Differences in salivary flow and composition among kangaroo species: Implications for digestive efficiency. In GC Grigg, P Jarman and ID Hume (eds) *Kangaroos, wallabies and rat-kangaroos*. Volume 1, pp 189–95. Surrey Beatty and Sons, Sydney.

Benshemesh, J and Johnson, K. 2003. Biology and conservation of marsupial moles (*Notoryctes*). In ME Jones, CR Dickman and M Archer (eds) *Predators with pouches: The biology of carnivorous marsupials*, pp 464–74. CSIRO Publishing, Melbourne.

Blackhall, S. 1980. Diet of the Eastern Native-cat, *Dasyurus viverrinus* (Shaw), in southern Tasmania. *Australian Wildlife Research*, 7, 191–97.

Bowman, DMJS and Woinarski, JCZ. 1994. Biogeography of Australian monsoon rainforest mammals: Implications for the conservation of rainforest mammals. *Pacific Conservation Biology*, 1, 98–106.

Bradley, AJ. 1997. Reproduction and life-history in the Red-tailed Phascogale, *Phascogale calura* (Marsupialia: Dasyuridae): The adaptive-stress senescence hypothesis. *Journal of Zoology*, London, 241, 739–55.

Bradley, AJ. 2003. Stress, hormones and mortality in small carnivorous marsupials. In ME Jones, CR Dickman and M Archer (eds) *Predators with pouches: The biology of carnivorous marsupials*, pp 254–67. CSIRO Publishing, Melbourne.

Bradshaw, FJ and Bradshaw, SD. 2001. Maintenance nitrogen requirement of an obligate nectarivore, the Honey Possum, *Tarsipes rostratus*. *Journal of Comparative Physiology*, B, 171, 59–67.

Braithwaite, LW, Dudziñski, ML and Turner, J. 1983. Studies on the arboreal marsupial fauna of eucalypt forests being harvested for woodpulp at Eden, NSW. II. Relationship between the fauna density, richness and diversity, and measured variables of the habitat. *Australian Wildlife Research*, 10, 231–47.

Braithwaite, LW, Turner, J and Kelly, J. 1984. Studies on the arboreal marsupial fauna of eucalypt forests being harvested for woodpulp at Eden, NSW. III. Relationships between faunal densities, eucalypt occurrence and foliage nutrients, and soil parent materials. *Australian Wildlife Research*, 11, 41–48.

Braithwaite, RW. 1990. Australia's unique biota: Implications for ecological processes. *Journal of Biogeography*, 17, 347–54.

Broome, LS. 2001. Density, home range, seasonal movements and habitat use of the Mountain Pygmy-possum *Burramys parvus* (Marsupialia: Burramyidae) at Mt Blue Cow, Kosciuszko National Park. *Austral Ecology*, 26, 275–92.

Brown, GD and Main, AR. 1967. Studies on marsupial nutrition. V. The nitrogen requirements of the Euro *Macropus robustus*. *Australian Journal of Zoology*, 15, 7–27.

Calaby, JH. 1960. Observations on the Banded Ant-eater *Myrmecobius f. fasciatus* Waterhouse (Marsupialia), with particular reference to its food habits. *Proceedings of the Zoological Society of London*, 135, 183–207.

Carthew, SM and Goldingay, RL. 1997. Non-flying mammals as pollinators. *Trends in Ecology and Evolution*, 12, 104–08.

Caughley, GJ. 1964. Density and dispersion of two species of kangaroo in relation to habitat. *Australian Journal of Zoology*, 12, 238–49.

Caughley, G, Grigg, GC, Caughley, J and Hill, GJE. 1980. Does Dingo predation control the densities of kangaroos and emus? *Australian Wildlife Research*, 7, 1–12.

Charles-Dominique, P, Atramentowicz, M, Charles-Dominique, M, Gérard, H, Hladik, CM and Prévost, MF. 1981. Les mammifères frugivores arboricoles nocturnes d'une forêt guyannaise: Inter-relations plantes–animaux. *Revue d'Ecologie* (*La Terre et la Vie*), 35, 341–435.

Chen, X, Dickman, CR and Thompson, MB. 1998. Diet of the Mulgara, *Dasycercus cristicauda* (Marsupialia: Dasyuridae), in the Simpson Desert, central Australia. *Wildlife Research*, 25, 233–42.

Chilcott, MJ. 1984. Coprophagy in the Common Ringtail Possum, *Pseudocheirus peregrinus* (Marsupialia: Petauridae). *Australian Mammalogy*, 7, 107–10.

Christensen, PES. 1980. The biology of *Bettongia penicillata* Gray 1837, and *Macropus eugenii* (Desmarest, 1817), in relation to fire. *Forestry Department of Western Australia Bulletin*, 91, 1–90.

Christensen, P, Maisey, K and Perry, DH. 1984. Radiotracking the Numbat, *Myrmecobius fasciatus*, in the Perup forest of Western Australia. *Australian Wildlife Research*, 11, 275–88.

Claridge, AW and Barry, SC. 2000. Factors influencing the distribution of medium-sized ground-dwelling mammals in southeastern mainland Australia. *Austral Ecology*, 25, 676–88.

Claridge, A, Seebeck, J and Rose, R. 2007. *Bettongs, potoroos and the Musky Rat-kangaroo*. CSIRO Publishing, Melbourne.

Claridge, AW, Tanton, MT and Cunningham, RB. 1993. Hypogeal fungi in the diet of the Long-nosed Potoroo (*Potorous tridactylus*) in mixed-species and regrowth eucalypt forest stands in south-eastern Australia. *Wildlife Research*, 20, 321–37.

Cockburn, A. 1990. Life history of the bandicoots: Developmental rigidity and phenotypic plasticity. In JH Seebeck, PR Brown, RL Wallis and CM Kemper (eds) *Bandicoots and bilbies*, pp 285–92. Surrey Beatty and Sons, and the Australian Mammal Society, Sydney.

Cockburn, A. 1997. Living slow and dying young: Senescence in marsupials. In NR Saunders and LA Hinds (eds) *Marsupial biology: Recent research, new perspectives*, pp 163–74. University of New South Wales Press, Sydney.

Copley, PB. 1983. Studies on the Yellow-footed Rock-wallaby, *Petrogale xanthopus* Gray (Marsupialia: Macropodidae). I. Distribution in South Australia. *Australian Wildlife Research*, 10, 47–61.

Cork, SJ. 1986. Foliage of *Eucalyptus punctata* and the maintenance nitrogen requirements of Koalas, *Phascolarctos cinereus*. *Australian Journal of Zoology*, 34, 17–24.

Cork, SJ and Foley, WJ. 1997. Digestive and metabolic adaptations of arboreal marsupials for dealing with plant antinutrients and toxins. In NR Saunders and LA Hinds (eds) *Marsupial biology: Recent research, new perspectives*, pp 204–26. University of New South Wales Press, Sydney.

Cork, SJ, Hume, ID and Dawson, TJ. 1983. Digestion and metabolism of a mature foliar diet (*Eucalyptus punctata*) by an arboreal marsupial, the Koala (*Phascolarctos cinereus*). *Journal of Comparative Physiology, B*, 153, 181–90.

Crowther, MS. 2002. Distributions of species of the *Antechinus stuartii–A. flavipes* complex as predicted by bioclimatic modelling. *Australian Journal of Zoology*, 50, 77–91.

Cunningham, SA. 1991. Experimental evidence for pollination of *Banksia* spp. by non-flying mammals. *Oecologia*, 87, 86–90.

Dawson, TJ. 1969. Temperature regulation and evaporative water loss in the Brush-tailed Possum *Trichosurus vulpecula*. *Comparative Biochemistry and Physiology*, 28, 401–07.

Dawson, TJ. 1989. Diets of macropodoid marsupials: General patterns and environmental influences. In GC Grigg, P Jarman and ID Hume (eds) *Kangaroos, wallabies and rat-kangaroos*. Volume 1, pp 129–42. Surrey Beatty and Sons, Sydney.

Dawson, TJ. 1995. *Kangaroos: Biology of the largest marsupials*. University of New South Wales Press, Sydney.

Dawson, TJ, Finch, E, Freedman, L, Hume, ID, Renfree, MB and Temple-Smith, PD. 1989. Morphology and physiology of the Metatheria. In DW Walton and BJ Richardson (eds) *Fauna of Australia*. Volume 1B: *Mammalia*, pp 451–504. Australian Government Publishing Service, Canberra.

Degabriele, R. 1983. Nitrogen and the Koala (*Phascolarctos cinereus*): Some indirect evidence. *Australian Journal of Ecology*, 8, 75–76.

Degabriele, R and Dawson, TJ. 1979. Metabolism and heat balance in an arboreal marsupial, the Koala *Phascolarctos cinereus*. *Journal of Comparative Physiology, B*, 134, 293–301.

Dennis, AJ. 2002. The diet of the Musky Rat-kangaroo, *Hypsiprymnodon moschatus*, a rainforest specialist. *Wildlife Research*, 29, 209–19.

Dennis, AJ. 2003. Scatter-hoarding by Musky Rat-kangaroos, *Hypsiprymnodon moschatus*, a tropical rain-forest marsupial from Australia: Implications for seed dispersal. *Journal of Tropical Ecology*, 19, 619–27.

Dennis, AJ and Marsh, H. 1997. Seasonal reproduction in Musky Rat-kangaroos, *Hypsiprymnodon moschatus*: A response to changes in resource availability. *Wildlife Research*, 24, 561–78.

Denny, MJS. 1975. The occurrence of the Eastern Grey Kangaroo (*Macropus giganteus*, Shaw) west of the Darling River. *Search*, 6, 89–90.

Denny, MJS. 1982. Adaptations of the Red Kangaroo and Euro (Macropodidae) to aridity. In WR Barker and PJM Greenslade (eds) *Evolution of the flora and fauna of arid Australia*, pp 179–83. Peacock Publications, Adelaide.

Dickman, CR. 1984. Competition and coexistence among the small marsupials of Australia and New Guinea. *Acta Zoologica Fennica*, 172, 27–31.

Dickman, CR. 1986. Niche compression: Two tests of an hypothesis using narrowly sympatric predator species. *Australian Journal of Ecology*, 11, 121–34.

Dickman, CR. 1988. Body size, prey size, and community structure in insectivorous mammals. *Ecology*, 69, 569–80.

Dickman, CR. 1989. Patterns in the structure and diversity of marsupial carnivore communities. In DW Morris, Z Abramsky, BJ Fox and MR Willig (eds) *Patterns in the structure of mammalian communities*, pp 241–51. Texas Tech University Press, Lubbock.

Dickman, CR. 1993. Evolution of semelparity in male dasyurid marsupials: A critique,

and an hypothesis of sperm competition. In M Roberts, J Carnio, G Crawshaw and M Hutchins (eds) *The biology and management of Australasian carnivorous marsupials*, pp 25–38. Metropolitan Toronto Zoo, Ontario, and American Association of Zoological Parks and Aquariums, Washington, DC.

Dickman, CR. 1994. Native mammals of western New South Wales: Past neglect, future rehabilitation? In D Lunney, S Hand, P Reed and D Butcher (eds) *Future of the fauna of western New South Wales*, pp 81–92. Royal Zoological Society of New South Wales, Sydney.

Dickman, CR. 1996. Vagrants in the desert. *Nature Australia*, 25, 54–62.

Dickman, CR. 2003. Distributional ecology of dasyurid marsupials. In ME Jones, CR Dickman and M Archer (eds) *Predators with pouches: The biology of carnivorous marsupials*, pp 318–31. CSIRO Publishing, Melbourne.

Dickman, CR. 2005. Marsupials of the world: An introduction. In RM Nowak (ed) *Walker's marsupials of the world*, pp 1–67. The Johns Hopkins University Press, Baltimore.

Dickman, CR. 2006. Species interactions: Indirect effects. In P Attiwill and BA Wilson (eds) *Ecology: An Australian perspective*, 2nd ed, pp 303–16. Oxford University Press, Oxford.

Dickman, CR and Braithwaite, RW. 1992. Postmating mortality of males in the dasyurid marsupials, *Dasyurus* and *Parantechinus*. *Journal of Mammalogy*, 73, 143–47.

Dickman, CR and Vieira, EM. 2006. Ecology and life histories. In PJ Armati, CR Dickman and ID Hume (eds) *Marsupials*, pp 199–228. Cambridge University Press, Cambridge.

Dickman, CR, Downey, FJ and Predavec, M. 1993. The Hairy-footed Dunnart *Sminthopsis hirtipes* (Marsupialia: Dasyuridae) in Queensland. *Australian Mammalogy*, 16, 69–72.

Dubost, G. 1979. The size of African forest artiodactyls as determined by the vegetation structure. *African Journal of Ecology*, 17, 1–17.

Duncan-Kemp, AM. 1933. *Our sandhill country: Nature and man in south-western Queensland*. Angus and Robertson, Sydney.

Dungan, RJ, O'Cain, MJ, Lopez, ML and Norton, DA. 2002. Contribution by possums to seed rain and subsequent seed germination in successional vegetation, Canterbury, New Zealand. *New Zealand Journal of Ecology*, 26, 121–28.

Ealey, EHM. 1967. Ecology of the Euro, *Macropus robustus* (Gould) in north-western Australia. *CSIRO Wildlife Research*, 12, 9–80.

Eberhard, IH. 1978. Ecology of the Koala, *Phascolarctos cinereus* (Goldfuss) Marsupialia: Phascolarctidae, in Australia. In GG Montgomery (ed) *The ecology of arboreal folivores*, pp 315–28. Smithsonian Institution Press, Washington, DC.

Efford, M. 2000. Possum density, population structure, and dynamics. In TL Montague (ed) *The Brushtail Possum: Biology, impact and management of an introduced marsupial*, pp 47–61. Manaaki Whenua Press, Lincoln.

Eldridge, MDB and Close, RL. 1992. Taxonomy of rock-wallabies, *Petrogale* (Marsupialia: Macropodidae). I. A revision of the eastern *Petrogale* with the description of three new species. *Australian Journal of Zoology*, 40, 605–25.

Eldridge, MDB, Johnston, PG and Lowry, PS. 1992. Chromosomal rearrangements in rock-wallabies, *Petrogale* (Marsupialia: Macropodidae). VII. G-banding analysis of *P. brachyotis* and *P. concinna*: Species with dramatically altered karyotypes. *Cytogenetics and Cell Genetics*, 61, 34–39.

Eldridge, MDB, Wilson, ACC, Metcalfe, CJ, Dollin, AE, Bell, JN, Johnson, PM, Johnston, PG and Close, RL. 2001. Taxonomy of rock-wallabies, *Petrogale* (Marsupialia: Macropodidae). III. Molecular data confirm the species status of the Purple-necked Rock-wallaby *Petrogale purpureicollis* Le Souef 1924. *Australian Journal of Zoology*, 49, 323–43.

Evans, MC. 1992. Diet of the Brushtail Possum *Trichosurus vulpecula* (Marsupialia: Phalangeridae) in central Australia. *Australian Mammalogy*, 15, 25–30.

Finlayson, HH. 1932. *Caloprymnus campestris*: Its recurrence and characters. *Transactions of the Royal Society of South Australia*, 56, 148–67.

Finlayson, HH. 1935. On mammals from the Lake Eyre Basin. Part II. The Peramelidae. *Transactions of the Royal Society of South Australia*, 59, 227–36.

Finlayson, HH. 1961. On central Australian mammals. Part IV. The distribution and status of central Australian species. *Records of the South Australian Museum*, 14, 141–91.

Fisher, DO and Dickman, CR 1993. The body size–prey size relationship in dasyurid marsupials: Tests of three hypotheses. *Ecology*, 74, 1871–83.

Fisher, DO, Owens, IPF and Johnson CN. 2001. The ecological basis of life history variation in marsupials. *Ecology*, 82, 3531–40.

Fox, BJ. 1989. Community ecology of macropodoids. In GC Grigg, P Jarman and ID Hume (eds) *Kangaroos, wallabies and rat-kangaroos*. Volume 1, pp 89–104. Surrey Beatty and Sons, Sydney.

Fox, BJ. 1999. The genesis and development of guild assembly rules. In E Weiher and P Keddy (eds) *Ecological assembly rules: Perspectives, advances, retreats*, pp 23–57. Cambridge University Press, Cambridge.

Friend, GR. 1990. Breeding and population dynamics of *Isoodon macrourus* (Marsupialia: Peramelidae): Studies from the wet-dry tropics of northern Australia. In JH Seebeck, PR Brown, RL Wallis and CM Kemper (eds) *Bandicoots and bilbies*, pp 357–65. Surrey Beatty and Sons, and the Australian Mammal Society, Sydney.

Friend, GR, Morris, KD and McKenzie, NL. 1991. The mammal fauna of Kimberley rainforests. In NL McKenzie, RB Johnston and PG Kendrick (eds) *Kimberley rainforests of Australia*, pp 393–412. Surrey Beatty and Sons, Sydney, Department of Conservation and Land Management, Perth, and Department of Arts, Heritage and Environment, Canberra.

Friend, JA. 1989. Myrmecobiidae. In DW Walton and BJ Richardson (eds) *Fauna of Australia*. Volume 1B, *Mammalia*, pp 583–90. Australian Government Publishing Service, Canberra.

Frith, HJ and Calaby, JH. 1969. *Kangaroos*. FW Cheshire, Melbourne.

Garkaklis, MJ, Bradley, JS and Wooller RD. 1998. The effects of Woylie (*Bettongia penicillata*) foraging on soil water repellency and water infiltration in heavy textured soils in southwestern Australia. *Australian Journal of Ecology*, 23, 492–96.

Garkaklis, MJ, Bradley, JS and Wooller, RD. 2000. Digging by vertebrates as an activity promoting the development of water-repellent patches in sub-surface soil. *Journal of Arid Environments*, 45, 35–42.

Garkaklis, MJ, Bradley, JS and Wooller, RD. 2003. The relationship between animal foraging and nutrient patchiness in south-west Australian woodland soils. *Australian Journal of Soil Research*, 41, 665–73.

Garkaklis, MJ, Bradley, JS and Wooller RD. 2004. Digging and soil turnover by a mycophagous marsupial. *Journal of Arid Environments*, 56, 569–78.

Gaughwin, MD, Judson, GJ, Macfarlane, MV and Siebert, BD. 1984. Effect of drought on the health of wild Hairy-nosed Wombats, *Lasiorhinus latifrons*. *Australian Wildlife Research*, 11, 455–63.

Gemmell, RT. 1990. The initiation of the breeding season of the Northern Brown Bandicoot *Isoodon macrourus* in captivity. In JH Seebeck, PR Brown, RL Wallis and CM Kemper (eds) *Bandicoots and bilbies*, pp 205–12. Surrey Beatty and Sons, and Australian Mammal Society, Sydney.

Gibson, LA. 2001. Seasonal changes in the diet, food availability and food preference of the Greater Bilby (*Macrotis lagotis*) in south-western Queensland. *Wildlife Research*, 28, 121–34.

Gilfillan, SL. 2001. An ecological study of a population of *Pseudantechinus macdonnellensis* (Marsupialia: Dasyuridae) in central Australia. I. Invertebrate food supply, diet and reproductive strategy. *Wildlife Research*, 28, 469–80.

Glen, AS and Dickman, CR. 2006. Diet of the Spotted-tailed Quoll (*Dasyurus maculatus*) in eastern Australia: Effects of season, sex and size. *Journal of Zoology*, 269, 241–48.

Godfrey, GK. 1969. The influence of increased photoperiod on reproduction in the dasyurid marsupial *Sminthopsis crassicaudata*. *Journal of Mammalogy*, 50, 132–33.

Godthelp, H, Archer, M, Cifelli, R, Hand, SJ and Gilkeson, CF. 1992. Earliest known Australian Tertiary mammal fauna. *Nature*, 356, 514–16.

Goldingay, RL. 2000. Small dasyurid marsupials – Are they effective pollinators? *Australian Journal of Zoology*, 48, 597–606.

Goldingay, RL and Jackson, SM. 2004. A review of the ecology of the Australian Petauridae. In RL Goldingay and SM Jackson (eds) *The biology of Australian possums and gliders*, pp 376–400. Surrey Beatty and Sons, Sydney.

Goldingay, RL, Carthew, SM and Whelan, RJ. 1991. The importance of non-flying mammals in pollination. *Oikos*, 61, 79–87.

Gordon, G. 1974. Movements and activity of the short-nosed bandicoot, *Isoodon macrourus* Gould (Marsupialia). *Mammalia* 38, 405–31.

Gordon, G and Lawrie, BC. 1978. The Rufescent Bandicoot, *Echymipera rufescens* (Peters & Doria) on Cape York Peninsula. *Australian Wildlife Research*, 5, 41–45.

Gordon, G, Hall, LS and Atherton, RG. 1990. Status of bandicoots in Queensland. In JH Seebeck, PR Brown, RL Wallis and CM Kemper (eds) *Bandicoots and bilbies*, pp 37–42. Surrey Beatty and Sons, and the Australian Mammal Society, Sydney.

Guiler, ER. 1958. Observations on a population of small marsupials in Tasmania. *Journal of Mammalogy*, 39, 44–58.

Guiler, ER. 1970. Observations on the Tasmanian Devil, *Sarcophilus harrisii* (Marsupialia: Dasyuridae). I. Numbers, home range, movements, and food in two populations. *Australian Journal of Zoology*, 18, 49–62.

Harley, DKP, Worley, MA and Harley, TK. 2004. The distribution and abundance of Leadbeater's Possum *Gymnobelideus leadbeateri* in lowland swamp forest at Yellingbo Nature Conservation Reserve. *Australian Mammalogy*, 26, 7–15.

Hawdon, J. 1952. *The journal of a journey from New South Wales to Adelaide*. Georgian House, Melbourne.

Hayward, MW, de Tores, PJ, Augee, ML, Fox, BJ and Banks, PB. 2004. Home range and movements of the Quokka *Setonix brachyurus* (Macropodidae: Marsupialia), and its impact on the viability of the metapopulation on the Australian mainland. *Journal of Zoology*, 263, 219–28.

Heinsohn, GE. 1966. Ecology and reproduction of the Tasmanian bandicoots (*Perameles gunnii* and *Isoodon obesulus*). *University of California Publications in Zoology*, 80, 1–96.

Heinze, D, Broome, L and Mansergh, I. 2004. A review of the ecology and conservation of the Mountain Pygmy-possum *Burramys parvus*. In RL Goldingay and SM Jackson (eds) *The biology of Australian possums and gliders*, pp 254–67. Surrey Beatty and Sons, Sydney.

Henry, SR, Lee, AK and Smith AP. 1989. The trophic structure and species richness of assemblages of arboreal mammals in Australian forests. In DW Morris, Z Abramsky, BJ Fox and MR Willig (eds) *Patterns in the structure of mammalian communities*, pp 229–40. Texas Tech University Press, Lubbock.

Holloway, JC and Geiser, F. 1996. Reproductive status and torpor of the marsupial *Sminthopsis crassicaudata*: Effect of photoperiod. *Journal of Thermal Biology*, 21, 373–80.

How, RA. 1978. Population strategies of four species of Australian 'possums'. In GG Montgomery (ed) *The ecology of arboreal folivores*, pp 305–13. Smithsonian Institution Press, Washington, DC.

How, RA. 1981. Population parameters of two congeneric possums, *Trichosurus* spp., in north-eastern New South Wales. *Australian Journal of Zoology*, 29, 205–15.

Howard, J. 1989. Diet of *Petaurus breviceps* (Marsupialia: Petauridae) in a mosaic of coastal woodland and heath. *Australian Mammalogy*, 12, 15–21.

Huang, C, Ward, S and Lee, AK. 1987. Comparison of the diets of the Feathertail Glider, *Acrobates pygmaeus*, and the Eastern Pygmy-possum, *Cercartetus nanus* (Marsupialia: Burramyidae) in sympatry. *Australian Mammalogy*, 10, 47–50.

Hulbert, AJ and Dawson, TJ. 1974. Standard metabolism and body temperature of perameloid marsupials from different environments. *Comparative Biochemistry and Physiology*, 47A, 583–90.

Hulbert, AJ and Dawson, TJ. 1974. Thermoregulation in perameloid marsupials from different environments. *Comparative Biochemistry and Physiology*, 47A, 591–616.

Hulbert, AJ and Dawson, TJ. 1974. Water metabolism in perameloid marsupials from different environments. *Comparative Biochemistry and Physiology*, 47A, 617–33.

Hume, ID. 1999. *Marsupial nutrition*. Cambridge University Press, Melbourne.

Hume, ID. 2006. Nutrition and digestion. In PJ Armati, CR Dickman and ID Hume (eds) *Marsupials*, pp 137–58. Cambridge University Press, Cambridge.

Irlbeck, NA and Hume, ID. 2003. The role of *Acacia* in the diets of Australian marsupials – A review. *Australian Mammalogy*, 25, 121–34.

Ivine, R and Bender, R. 1997. Introduction of the Sugar Glider *Petaurus breviceps* into re-established forest of the Organ Pipes National Park, Victoria. *Victorian Naturalist*, 114, 230–39.

Jarman, PJ. 1984. The dietary ecology of macropod marsupials. *Proceedings of the Nutrition Society of Australia*, 9, 82–87.

Jarman, PJ, Johnson, CN, Southwell, CJ and Stuart-Dick, R. 1987. Macropod studies at Wallaby Creek. I. The area and animals. *Australian Wildlife Research*, 14, 1–14.

Johnson, CN. 1996. Interactions between mammals and ectomycorrhizal fungi. *Trends in Ecology and Evolution*, 11, 503–07.

Johnson, CN, Delean, S and Balmford, A. 2002. Phylogeny and the selectivity of extinction in Australian marsupials. *Animal Conservation*, 5, 135–42.

Johnson, P. 2003. *Kangaroos of Queensland*. Queensland Museum, Brisbane.

Jones, CG, Lawton, JH and Shachak, M. 1997. Positive and negative effects of organisms as physical ecosystem engineers. *Ecology*, 78, 1946–57.

Jones, KMW, Maclagan, SJ and Krockenberger, AK. 2006. Diet selection in the Green Ringtail Possum (*Pseudochirops archeri*): A specialist folivore in a diverse forest. *Austral Ecology*, 31, 799–807.

Jones, ME and Barmuta, LA. 1998. Diet overlap and relative abundance of sympatric dasyurid carnivores: A hypothesis of competition. *Journal of Animal Ecology*, 67, 410–21.

Jones, ME and Barmuta, LA. 2000. Niche differentiation among sympatric Australian dasyurid carnivores. *Journal of Mammalogy*, 81, 434–47.

Kaufmann, JH. 1974. Habitat use and social organization of nine sympatric species of macropod marsupials. *Journal of Mammalogy*, 55, 66–80.

Kavanagh, RP and Lambert, MJ. 1990. Food selection by the Greater Glider, *Petauroides volans*: Is foliar nitrogen a determinant of habitat quality? *Australian Wildlife Research*, 17, 285–99.

Kennedy, PM and Heinsohn, GE. 1974. Water metabolism of two marsupials – the Brush-tailed Possum, *Trichosurus vulpecula* and the rock-wallaby, *Petrogale inornata* in the wild. *Comparative Biochemistry and Physiology*, 47A, 829–34.

Kerle, JA. 1984. Variation in the ecology of *Trichosurus*: Its adaptive significance. In AP Smith and ID Hume (eds) *Possums and gliders*, pp 115–28. Surrey Beatty and Sons, and the Australian Mammal Society, Sydney.

Kerle, JA. 2001. *Possums: The brushtails, ringtails and Greater Glider*. University of New South Wales Press, Sydney.

Kitchener, DJ. 1972. The importance of shelter to the Quokka, *Setonix brachyurus* (Marsupialia) on Rottnest Island. *Australian Journal of Zoology*, 20, 281–99.

Kitchener, DJ, Stoddart, J and Henry J. 1984. A taxonomic revision of the *Sminthopsis murina* complex (Marsupialia, Dasyuridae) in Australia, including descriptions of four new species. *Records of the Western Australian Museum*, 11, 201–48.

Kraaijeveld, K, Kraaijeveld-Smit, FJL and Adcock, G. 2003. Does female mortality drive male semelparity in dasyurid marsupials? *Proceedings of the Royal Society*, B, 270, S251–53.

Lee, AK and McDonald, IR. 1985. Stress and population regulation in small mammals. *Oxford Reviews in Reproductive Biology*, 7, 261–304.

Lee, AK and Ward, SJ. 1989. Life histories of macropodoid marsupials. In GC Grigg, P Jarman and ID Hume (eds) *Kangaroos, wallabies and rat-kangaroos*. Volume 1, pp 105–15. Surrey Beatty and Sons, Sydney.

Lee, AK, Woolley, P and Braithwaite, RW. 1982. Life history strategies of dasyurid marsupials. In M Archer (ed) *Carnivorous marsupials*. Volume 1, pp 1–11. Royal Zoological Society of New South Wales, Sydney.

Lim, L. 1992. *Recovery plan for the Kowari Dasyuroides byrnei Spencer, 1896* (Marsupialia: Dasyuridae). Australian National Parks and Wildlife Service, Canberra.

Lindenmayer, DB. 1997. Differences in the biology and ecology of arboreal marsupials in forests of southeastern Australia. *Journal of Mammalogy*, 78, 1117-27.

Lindenmayer, DB. 2000. Factors at multiple scales affecting distribution patterns and their implications for animal conservation - Leadbeater's Possum as a case study. *Biodiversity and Conservation*, 9, 15-35.

Lindenmayer, D. 2002. *Gliders of Australia: A natural history*. University of New South Wales Press, Sydney.

Lindenmayer, DB, Boyle, S, Burgman, MA, McDonald, D and Tomkins, B. 1994. The sugar and nitrogen content of the gums of *Acacia* species in the Mountain Ash and Alpine Ash forests of central Victoria and its potential implications for exudivorous marsupials. *Australian Journal of Ecology*, 19, 169-77.

Lindenmayer, DB, Dubach, J and Viggers, KL. 2002. Geographic dimorphism in the Mountain Brushtail Possum (*Trichosurus caninus*): The case for a new species. *Australian Journal of Zoology*, 50, 369-93.

Logan, M. 2003. Effect of tooth wear on the rumination-like behavior, or merycism, of free-ranging Koalas (*Phascolarctos cinereus*). *Journal of Mammalogy*, 84, 897-902.

Lorini, ML, De Oliveira, JA and Persson, VG. 1994. Annual age structure and reproductive patterns in *Marmosa incana* (Lund, 1841) (Didelphidae, Marsupialia). *Zeitschrift für Säugetierkunde*, 59, 65-73.

Lyne, AG. 1964. Observations on the breeding and growth of the marsupial *Perameles nasuta* Geoffroy, with notes on other bandicoots. *Australian Journal of Zoology*, 12, 322-39.

Lyne, AG. 1976. Observations on oestrus and the oestrous cycle in the marsupials *Isoodon macrourus* and *Perameles nasuta*. *Australian Journal of Zoology*, 24, 513-21.

Mansergh, I and Broome, L. 1994. *The Mountain Pygmy-possum of the Australian Alps*. New South Wales University Press, Sydney.

Marsh, KJ, Foley, WJ, Cowling, A and Wallis, IR. 2003. Differential susceptibility to *Eucalyptus* secondary compounds explains feeding by the Common Ringtail (*Pseudocheirus peregrinus*) and Common Brushtail Possum (*Trichosurus vulpecula*). *Journal of Comparative Physiology*, B, 173, 69-78.

Martin, G. 2003. The role of small ground-foraging mammals in topsoil health and biodiversity: Implications to management and restoration. *Ecological Management and Restoration*, 4, 114-19.

Martin, RW. 1981. Age-specific fertility in three populations of the Koala, *Phascolarctos cinereus* Goldfuss, in Victoria. *Australian Wildlife Research*, 8, 275-83.

Martin, RW and Handasyde, KA. 1991. Population dynamics of the Koala (*Phascolarctos cinereus*) in south-eastern Australia. In AK Lee, KA Handasyde and GD Sanson (eds) *Biology of the Koala*, pp 75-84. Surrey Beatty and Sons, Sydney.

McAllan, B. 2003. Timing of reproduction in carnivorous marsupials. In ME Jones, CR Dickman and M Archer (eds) *Predators with pouches: The biology of carnivorous marsupials*, pp 147-68. CSIRO Publishing, Melbourne.

McAllan, BM and Dickman, CR. 1986. The role of photoperiod in the timing of reproduction in the dasyurid marsupial *Antechinus stuartii*. *Oecologia*, 68, 259-64.

McAllan, BM, Dickman, CR and Crowther, MS. 2006. Photoperiod as a reproductive cue in the marsupial genus *Antechinus*: Ecological and evolutionary consequences. *Biological Journal of the Linnean Society*, 87, 365-79.

McIlroy, JC. 1976. Aspects of the ecology of the Common Wombat, *Vombatus ursinus*. I. Capture, handling, marking and radio-tracking techniques. *Australian Wildlife Research*, 3, 105-16.

McNamara, JA. 1997. Some smaller macropod fossils of South Australia. *Proceedings of the Linnean Society of New South Wales*, 117, 97-105.

Merchant, JC. 1976. Breeding biology of the Agile Wallaby, *Macropus agilis* (Gould) (Marsupialia: Macropodidae), in captivity. *Australian Wildlife Research*, 3, 93-103.

Merchant, JC and Calaby, JH. 1981. Reproductive biology of the Red-necked Wallaby (*Macropus rufogriseus banksianus*) and Bennett's Wallaby (*M. r. rufogriseus*) in captivity. *Journal of Zoology, London*, 194, 203-17.

Minta, SC, Clark, TW and Goldstraw, P. 1990. Population estimates and characteristics of the Eastern Barred Bandicoot in Victoria, with recommendations for population monitoring. In TW Clark and JH Seebeck (eds) *Management and conservation of small populations*, pp 47-75. Chicago Zoological Society, Chicago.

Mitchell, TL. 1838. *Three expeditions into the interior of eastern Australia, with descriptions of the recently explored region of Australia Felix, and of the present colony of New South Wales*, 2 volumes. T and W Boone, London.

Moir, RJ, Somers, M and Waring, H. 1956. Studies on marsupial nutrition. I. Ruminant-like digestion in a herbivorous marsupial (*Setonix brachyurus* Quoy and Gaimard). *Australian Journal of Biological Science*, 9, 293-304.

Morton, SR. 1978. An ecological study of *Sminthopsis crassicaudata* (Marsupialia: Dasyuridae). *Australian Wildlife Research*, 5, 151-211.

Murphy, MT, Garkaklis, MJ and St J Hardy, GE. 2005. Seed caching by Woylies *Bettongia penicillata* can increase Sandalwood *Santalum spicatum* regeneration in Western Australia. *Austral Ecology*, 30, 747-55.

Newsome, AE. 1965. The distribution of Red Kangaroos, *Megaleia rufa* (Desmarest), about sources of persistent food and water in central Australia. *Australian Journal of Zoology*, 13, 289-99.

Newsome, AE. 1965. Reproduction in natural populations of the Red Kangaroo, *Megaleia rufa* (Desmarest), in central Australia. *Australian Journal of Zoology*, 13, 735-59.

Newsome, AE. 1966. The influence of food on breeding in the Red Kangaroo in central Australia. *CSIRO Wildlife Research*, 11, 187-96.

Newsome, AE. 1997. Reproductive anomalies in the Red Kangaroo in central Australia explained by Aboriginal traditional knowledge and ecology. In NR Saunders and LA Hinds (eds) *Marsupial biology: Recent research, new perspectives*, pp 227-36. University of New South Wales Press, Sydney.

Newsome, AE, Catling, PC, Cooke, BD and Smyth, R. 2001. Two ecological universes separated by the dingo barrier fence in semi-arid Australia: Interactions between landscapes, herbivory and carnivory, with and without Dingoes. *Rangeland Journal*, 23, 71-98.

Nguyen, VP, Needham, AD and Friend, JA. 2005. A quantitative dietary study of the 'critically endangered' Gilbert's Potoroo *Potorous gilbertii*. *Australian Mammalogy*, 27, 1-6.

Noble, JC. 1993. Relict surface-soil features in semi-arid mulga (*Acacia aneura*) woodlands. *Rangeland Journal*, 15, 48-70.

Noble, JC. 1999. Fossil features of Mulga *Acacia aneura* landscapes: Possible imprinting by extinct Pleistocene fauna. *Australian Zoologist*, 31, 396-402.

Oakwood, M, Bradley, AJ and Cockburn, A. 2001. Semelparity in a large marsupial. *Proceedings of the Royal Society*, B, 268, 407-11.

Oliver, AJ. 1966. Grey kangaroos near Wiluna. *Western Australian Naturalist*, 10, 74-75.

Pine, RH, Dalby, PL and Matson, JO. 1985. Ecology, postnatal development, morphometrics, and taxonomic status of the short-tailed opossum, *Monodelphis dimidiata*, an apparently semelparous annual marsupial. *Annals of the Carnegie Museum*, 54, 195-231.

Pizzuto, TA, Finlayson, GR, Crowther, MS and Dickman, CR. 2007. Microhabitat use of the Brush-tailed Bettong (*Bettongia penicillata*) and Burrowing Bettong (*B. lesueur*) in semi-arid New South Wales: Implications for reintroduction programs. *Wildlife Research*, 34, 271-79.

Procter-Gray, E. 1984. Dietary ecology of the Coppery Brushtail Possum, Green Ringtail Possum and Lumholtz's Tree-kangaroo in north Queensland. In AP Smith and ID Hume (eds) *Possums and gliders*, pp 129-35. Surrey Beatty and Sons, and the Australian Mammal Society, Sydney.

Rand, AL. 1937. Some original observations on the habits of *Dactylopsila trivirgata* Gray. *American Museum Novitates*, 957, 1-7.

Redford, KH and Eisenberg, JF. 1992. Mammals of the Neotropics. Volume 2: *The southern cone: Chile, Argentina, Uruguay, Paraguay*. University of Chicago Press, Chicago.

Rhind, SG and Bradley, JS. 2002. The effect of drought on body size, growth and abundance of wild Brush-tailed Phascogales (*Phascogale tapoatafa*) in south-western Australia. *Wildlife Research*, 29, 235-45.

Rhind, SG, Bradley, JS and Cooper, NK. 2001. Morphometric variation and taxonomic status of Brush-tailed Phascogales, *Phascogale tapoatafa* (Meyer, 1793) (Marsupialia: Dasyuridae). *Australian Journal of Zoology*, 49, 345-68.

Rich, TH, Flannery, TF, Trusler, P, Kool, L, van Klaveren, N and Vickers-Rich, P. 2001. A second tribosphenic mammal from the Mesozoic of Australia. *Records of the Queen Victoria Museum*, 110, 1-9.

Rich, TH, Vickers-Rich, P, Constantine, A, Flannery, TF, Kool, L and van Klaveren, N. 1997. A tribosphenic mammal from the Mesozoic of Australia. *Science*, 278, 1438-42.

Richardson, KC, Wooller, RD and Collins, BG. 1986. Adaptations to a diet of nectar and pollen in the marsupial *Tarsipes rostratus* (Marsupialia: Tarsipedidae). *Journal of Zoology, London*, 208, 285-97.

Root, RB. 1967. The niche exploitation pattern of the Blue-gray Gnatcatcher. *Ecological Monographs*, 37, 317-49.

Rose, RW. 1989. Embryonic growth rates of marsupials with a note on monotremes. *Journal of Zoology, London*, 218, 11-16.

Rounsevell, DE, Taylor, RJ and Hocking, GJ. 1991. Distribution records of native terrestrial mammals in Tasmania. *Wildlife Research*, 18, 699-717.

Rowston, C and Catterall, CP. 2004. Habitat segregation, competition and selective deforestation: Effects on the conservation status of two similar *Petaurus* gliders. In D Lunney (ed) *Conservation of Australia's forest fauna*, 2nd ed, pp 741-47. Royal Zoological Society of New South Wales, Sydney.

Runcie, MJ. 1999. Movements, dens and feeding behaviour of the tropical Scaly-tailed Possum (*Wyulda squamicaudata*). *Wildlife Research*, 26, 367-73.

Russell, EM. 1974. Recent ecological studies on Australian marsupials. *Australian Mammalogy*, 1, 189-211.

Russell, R. 1980. *Spotlight on possums*. University of Queensland Press, Brisbane.

Sanson, GD. 1978. The evolution and significance of mastication in the Macropodidae. *Australian Mammalogy*, 2, 23-28.

Sanson, GD. 1980. The morphology and occlusion of the molariform cheek teeth in some Macropodinae (Marsupialia: Macropodidae). *Australian Journal of Zoology*, 28, 341-65.

Scott, LK, Hume, ID and Dickman, CR. 1999. Ecology and population biology of Long-nosed Bandicoots (*Perameles nasuta*) at North Head, Sydney Harbour National Park. *Wildlife Research*, 26, 805-21.

Seebeck, JH and Rose, RW. 1989. Potoroidae. In DW Walton and BJ Richardson (eds) *Fauna of Australia*. Volume 1B, *Mammalia*, pp 716–39. Australian Government Publishing Service, Canberra.

Seebeck, JH, Bennett, AF and Scotts, DJ. 1989. Ecology of the Potoroidae – A review. In GC Grigg, P Jarman and ID Hume (eds) *Kangaroos, wallabies and rat-kangaroos*. Volume 1, pp 67–88. Surrey Beatty and Sons, Sydney.

Seebeck, JH, Warneke, RM and Baxter, BJ. 1984. Diet of the Bobuck, *Trichosurus caninus* (Ogilby) (Marsupialia: Phalangeridae) in a mountain forest in Victoria. In AP Smith and ID Hume (eds) *Possums and gliders*, pp 145–54. Surrey Beatty and Sons, and the Australian Mammal Society, Sydney.

Sharman, GB. 1955. Studies on marsupial reproduction. III. Normal and delayed pregnancy in *Setonix brachyurus*. *Australian Journal of Zoology*, 3, 56–70.

Shield, J. 1968. Reproduction of the Quokka, *Setonix brachyurus*, in captivity. *Journal of Zoology, London*, 155, 427–44.

Shimmin, GA, Taggart, DA and Temple-Smith, PD. 2002. Mating behaviour in the Agile Antechinus *Antechinus agilis* (Marsupialia: Dasyuridae). *Journal of Zoology, London*, 258, 39–48.

Short, J and Milkovits, G. 1990. Distribution and status of the Brush-tailed Rock-wallaby in south-eastern Australia. *Australian Wildlife Research*, 17, 169–79.

Smith, A. 1982. Is the Striped Possum (*Dactylopsila trivirgata*; Marsupialia, Petauridae) an arboreal anteater? *Australian Mammalogy*, 5, 229–34.

Smith, AP. 1982. Diet and feeding strategies of the Sugar Glider in temperate Australia. *Journal of Animal Ecology*, 51, 149–66.

Smith, AP. 1982. Leadbeater's Possum and its management. In RH Groves and WDL Ride (eds) *Species at risk: Research in Australia*, pp 129–47. Australian Academy of Science, Canberra.

Southgate, RI. 1990. Distribution and abundance of the Greater Bilby *Macrotis lagotis* Reid (Marsupialia: Peramelidae). In JH Seebeck, PR Brown, RL Wallis and CM Kemper (eds) *Bandicoots and bilbies*, pp 293–302. Surrey Beatty and Sons, and the Australian Mammal Society, Sydney.

Spencer, PBS, Rhind, SG and Eldridge, MDB. 2001. Phylogeographic structure within *Phascogale* (Marsupialia: Dasyuridae) based on partial cytochrome *b* sequence. *Australian Journal of Zoology*, 49, 369–77.

Stoddart, DM and Braithwaite, RW. 1979. A strategy for utilization of regenerating heathland habitat by the Brown Bandicoot (*Isoodon obesulus*; Marsupialia, Peramelidae). *Journal of Animal Ecology*, 48, 165–79.

Sturt, C. 1833. *Two expeditions into the interior of southern Australia, during the years 1828, 1829, 1830, and 1831: With observations on the soil, climate, and general resources of the colony of New South Wales*, 2 volumes. Smith, Elder and Co., London.

Suckling, GC and Goldstraw, P. 1989. Progress of Sugar Glider, *Petaurus breviceps*, establishment at the Tower Hill State Game Reserve, Victoria. *Victorian Naturalist*, 106, 179–83.

Sumner, J and Dickman, CR. 1998. Distribution and identity of species in the *Antechinus stuartii–A. flavipes* group (Marsupialia: Dasyuridae) in south-eastern Australia. *Australian Journal of Zoology*, 46, 27–41.

Taylor, JM, Calaby, JH and Redhead, TD. 1982. Breeding in wild populations of the marsupial-mouse *Planigale maculata sinualis* (Dasyuridae, Marsupialia). In M Archer (ed) *Carnivorous marsupials*. Volume 1, pp 83–87. Royal Zoological Society of New South Wales, Sydney.

Telfer, WR and Bowman, DMJS. 2006. Diet of four rock-dwelling macropods in the Australian monsoon tropics. *Austral Ecology*, 31, 817–27.

Thompson, SD. 1987. Body size, duration of parental care, and the intrinsic rate of natural increase in eutherian and metatherian mammals. *Oecologia*, 71, 201–09.

Thomson, JA and Owen, WH. 1964. A field study of the Australian Ringtail Possum *Pseudocheirus peregrinus* (Marsupialia: Phalangeridae). *Ecological Monographs*, 34, 27–52.

Tulloch, A. 2004. The importance of food and shelter for habitat use and conservation of the burramyids in Australia. In RL Goldingay and SM Jackson (eds) *The biology of Australian possums and gliders*, pp 268–84. Surrey Beatty and Sons, Sydney.

Turner, V. 1982. Marsupials as pollinators in Australia. In JA Armstrong, JM Powell and AJ Richards (eds) *Pollination and evolution*, pp 55–66. Royal Botanic Gardens, Sydney.

Tyndale-Biscoe, CH. 1965. The female urogenital system and reproduction of the marsupial *Lagostrophus fasciatus*. *Australian Journal of Zoology*, 13, 255–67.

Tyndale-Biscoe, CH. 1989. The adaptiveness of reproductive processes. In GC Grigg, P Jarman and ID Hume (eds) *Kangaroos, wallabies and rat-kangaroos*. Volume 1, pp 277–85. Surrey Beatty and Sons, Sydney.

Tyndale-Biscoe, H. 2005. *Life of marsupials*. CSIRO Publishing, Melbourne.

Tyndale-Biscoe, H and Renfree, M. 1987. *Reproductive physiology of marsupials*. Cambridge University Press, Cambridge.

Van Dyck, S. 1982. The relationships of *Antechinus stuartii* and *A. flavipes* (Dasyuridae, Marsupialia) with special reference to Queensland. In M Archer (ed) *Carnivorous marsupials*. Volume 2, pp 723–66. Royal Zoological Society of New South Wales, Sydney.

Van Dyck, S. 1987. The Bronze Quoll, *Dasyurus spartacus* (Marsupialia: Dasyuridae), a new

species from the savannahs of Papua New Guinea. *Australian Mammalogy*, 11, 145–56.

Van Dyck, S and Crowther, MS. 2000. Reassessment of northern representatives of the *Antechinus stuartii* complex (Marsupialia: Dasyuridae): *A. subtropicus* sp. nov., and *A. adustus* new status. *Memoirs of the Queensland Museum*, 45, 611–35.

van Tets, IG. 1998. Can flower-feeding marsupials meet their nitrogen requirements on pollen in the field? *Australian Mammalogy*, 20, 383–90.

Vernes, K. 2000. Immediate effects of fire on survivorship of the Northern Bettong (*Bettongia tropica*): An endangered Australian marsupial. *Biological Conservation*, 96, 305–09.

Vernes, K and Pope, LC. 2006. Population density of the Northern Bettong *Bettongia tropica* in northeastern Queensland. *Australian Mammalogy*, 28, 87–92.

Ward, SJ. 1990. Life history of the Eastern Pygmy-possum, *Cercartetus nanus* (Burramyidae: Marsupialia), in south-eastern Australia. *Australian Journal of Zoology*, 38, 287–304.

Ward, SJ. 1998. Numbers of teats and pre- and post-natal litter sizes in small diprotodont marsupials. *Journal of Mammalogy*, 79, 999–1008.

Wells, RT. 1978. Thermoregulation and activity rhythms in the Hairy-nosed Wombat, *Lasiorhinus latifrons* (Owen), (Vombatidae). *Australian Journal of Zoology*, 26, 639–51.

White, TCR. 1978. The importance of a relative shortage of food in animal ecology. *Oecologia*, 33, 71–86.

White, TCR. 2005. *Why does the world stay green? Nutrition and survival of plant-eaters*. CSIRO Publishing, Melbourne.

Williams, PA. 2003. Are possums important dispersers of large-seeded fruit? *New Zealand Journal of Ecology*, 27, 221–23.

Williams, SE, Pearson, RG and Walsh PJ. 1996. Distributions and biodiversity of the terrestrial vertebrates of Australia's Wet Tropics: A review of current knowledge. *Pacific Conservation Biology*, 2, 327–62.

Winter, JW. 1988. Ecological specialization of mammals in Australian tropical and sub-tropical rainforest: Refugial or ecological determinism? *Proceedings of the Ecological Society of Australia*, 15, 127–38.

Winter, JW. 1997. Responses of non-volant mammals to late Quaternary climatic changes in the Wet Tropics region of north-eastern Australia. *Wildlife Research*, 24, 493–511.

Winter, JW. 2006. Historical records of the Eastern Grey Kangaroo *Macropus giganteus* on Cape York Peninsula: Clues to a postulated range extension. *Australian Mammalogy*, 28, 249–52.

Wolfe, KM, Mills, HR, Garkaklis, MJ and Bencini, R. 2004. Post-mating survival in a small marsupial is associated with nutrient inputs from seabirds. *Ecology*, 85, 1740–46.

Wooller, RD, Richardson, KC, Saffer, VM, Garavanter, CAM, Bryant, KA, Everaardt, AN and Wooller, SJ. 2004. The Honey Possum *Tarsipes rostratus*: An update. In RL Goldingay and SM Jackson (eds) *The biology of Australian possums and gliders*, pp 312–17. Surrey Beatty and Sons, Sydney.

Woolley, P. 1966. Reproduction in *Antechinus* spp. and other dasyurid marsupials. *Symposia of the Zoological Society of London*, 15, 281–94.

Woolley, PA. 2003. Reproductive biology of some dasyurid marsupials of New Guinea. In ME Jones, CR Dickman and M Archer (eds) *Predators with pouches: The biology of carnivorous marsupials*, pp 169–82. CSIRO Publishing, Melbourne.

Woolley, PA. 2005. The species of *Dasycercus* Peters, 1875 (Marsupialia: Dasyuridae). *Memoirs of Museum Victoria*, 62, 213–21.

Wright, W, Sanson, GD and McArthur, C. 1991. The diet of the extinct bandicoot *Chaeropus ecaudatus*. In P Vickers-Rich, JM Monaghan, RF Baird and TH Rich (eds) *Vertebrate Palaeontology of Australasia*, pp 229–45. Pioneer Design Studio, Melbourne.

Yom-Tov, Y. 1985. The reproductive rates of Australian rodents. *Oecologia*, 66, 250–55.

Cultural history

Archer, M and Beale, B. 2004. *Going native: Living in the Australian environment*. Hodder Headline, Sydney.

Bennett, G. 1860. *Gatherings of a naturalist in Australasia: Being observations principally on the animal and vegetable productions of New South Wales, New Zealand, and some of the austral islands*. John van Voorst, London.

Bonyhady, T. 2000. *The colonial earth*. Melbourne University Press, Melbourne.

Bowen, J. 1987. *Kidman: The forgotten king*. Angus and Robertson, Sydney.

Breeden, S and Wright, B. 2001. *Kakadu: Looking after country – the Gagudju way*. JB Books, Adelaide.

Burbidge, AA and McKenzie, NL. 1989. Patterns in the modern decline of Western Australia's vertebrate fauna: Causes and conservation implications. *Biological Conservation*, 50, 143–98.

Burbidge, AA, Johnson, KA, Fuller, PJ and Southgate, RI. 1988. Aboriginal knowledge of the mammals of the central deserts of Australia. *Australian Wildlife Research*, 15, 9–39.

Calaby, JH. 1965. Early European description of an Australian mammal. *Nature*, 205, 516–17.

Carnegie, DW. 1898. *Spinifex and sand: A narrative of five years' pioneering and exploration in Western Australia*. C Arthur Pearson, London.

Clune, F. 1945. *Captain Starlight: Reckless rascal of 'robbery under arms'*. Hawthorn Press, Melbourne.

Darwin, C. 1845. *The voyage of the 'Beagle'*. Facsimile ed. 1968, Heron Books, Geneva.

Davies, RG, Webber, LM and Barnes, GS. 2004. Urban wildlife management – It's as much about people! In D Lunney and S Burgin (eds) *Urban wildlife: More than meets the eye*, pp 38–43. Royal Zoological Society of New South Wales, Sydney.

de Blainville, HMD. 1816. Prodrome d'une nouvelle distribution systématique du règne animal. *Bulletin des Sciences par la Société philomatique de Paris*, 8, 105–24.

Denny, M. 1992. *Historical and ecological study of the effects of European settlement on inland NSW*. Nature Conservation Council of New South Wales, Sydney.

Dickman, C, Hutchings, P and Lunney, D. 2004. Threatened species legislation: Just one act in the play. In P Hutchings, D Lunney and C Dickman (eds) *Threatened species legislation: Is it just an Act?*, pp 180–92. Royal Zoological Society of New South Wales, Sydney.

Donovan, V and Wall, C (eds). 2004. *Making connections: A journey along central Australian Aboriginal trading routes*. Arts Queensland, Brisbane.

Finlayson, HH. 1935. *The red centre: Man and beast in the heart of Australia*. Angus and Robertson, Sydney.

Finney, CM. 1984. *To sail beyond the sunset: Natural history in Australia 1699–1829*. Rigby, Adelaide.

Fleay, D. 1947. *Gliders of the gum trees*. Bread and Cheese Club, Melbourne.

Gammage, B. 2003. *Australia under Aboriginal management: 15th Barry Andrews Memorial Lecture*. University of New South Wales – Australian Defence Force Academy, Canberra.

Garran, JC and White, L. 1985. *Merinos, myths and Macarthurs: Australian graziers and their sheep, 1788–1900*. Australian National University Press, Canberra.

Gould, J. 1845–63. *The mammals of Australia*, 3 volumes. The Author, London.

Hrdina, F. 1997. Marsupial destruction in Queensland 1877–1930. *Australian Zoologist*, 30, 272–86.

Jarman, PJ and Brock, MA. 2004. The evolving intent and coverage of legislation to protect biodiversity in New South Wales. In P Hutchings, D Lunney and C Dickman (eds) *Threatened species legislation: Is it just an Act?*, pp 1–19. Royal Zoological Society of New South Wales, Sydney.

Jarman, PJ and Johnson, KA. 1977. Exotic mammals, indigenous mammals, and land-use. *Proceedings of the Ecological Society of Australia*, 10, 146–63.

Kocan, P. 1975. *Paperback poets second series 11: The other side of the fence*. University of Queensland Press, Brisbane.

Krefft, G. 1871. *The mammals of Australia, illustrated by Miss Harriett Scott, and Mrs Helena Forde, for the Council of Education; with a short account of the species hitherto described*. Thomas Richards, Government Printer, Sydney.

Letnic, M. 2007. The impacts of pastoralism on the fauna of arid Australia. In C Dickman, D Lunney and S Burgin (eds) *Animals of arid Australia: Out on their own?*, pp 65–75. Royal Zoological Society of New South Wales, Sydney.

Lloyd, N and Mulcock, J. 2006. Human–animal studies in Australia: Perspectives from the arts, humanities and social sciences. *Australian Zoologist*, 33, 290–94.

Lunney, D. 2004. A test of our civilisation: Conserving Australia's forest fauna across a cultural landscape. In D Lunney (ed) *Conservation of Australia's forest fauna*, 2nd ed, pp 1–22. Royal Zoological Society of New South Wales, Sydney.

Mahoney, JA. 1982. Identities of the rodents (Muridae) listed in T. L. Mitchell's 'Three expeditions into the interior of eastern Australia, with descriptions of the recently explored region of Australia Felix, and of the present colony of New South Wales' (1st ed, 1838; 2nd ed, 1839). *Australian Mammalogy*, 5, 15–36.

Marshall, AJ. 1966. On the disadvantages of wearing fur. In AJ Marshall (ed) *The great extermination: A guide to Anglo-Australian cupidity wickedness and waste*, pp 9–42. William Heinemann, Melbourne.

Rolls, EC. 1969. *They all ran wild*. Angus and Robertson, Sydney.

Saunders, G, Lane, C, Harris, S and Dickman, CR. 2006. *Foxes in Tasmania: A report of an incursion by an invasive species*. Invasive Animals Cooperative Research Centre, Canberra.

Smith, N. 2006. Thank your mother for the rabbits: Bilbies, bunnies and redemptive ecology. *Australian Zoologist*, 33, 369–78.

Splatt, W and Bruce, S. 1978. *100 masterpieces of Australian landscape painting*. Rigby, Adelaide.

Stubbs, BJ. 2001. From 'useless brutes' to national treasures: A century of evolving attitudes towards native fauna in New South Wales, 1860s to 1960s. *Environment and History*, 7, 23–56.

Ward, R. 1958. *The Australian legend*. Oxford University Press, Melbourne.

Waterhouse, GR. 1843. *The naturalist's library*. Volume VIII: *Mammalia. Marsupialia or pouched animals*. WH Lizars, Edinburgh.

Conservation

Abbott, I. 2002. Origin and spread of the Cat, *Felis catus*, on mainland Australia, with a discussion of the magnitude of its early impact on native fauna. *Wildlife Research*, 29, 51–74.

Abbott, I. 2006. Mammalian faunal collapse in Western Australia, 1875–1925: The hypothesised role of epizootic disease and a conceptual model of its origin, introduction, transmission, and spread. *Australian Zoologist*, 33, 530–61.

Allen, H. 1980. Aborigines of the western plains of New South Wales. *Parks and Wildlife*, 2, 33–43.

Anon. 1901. *Royal Commission to inquire into the condition of the Crown Tenants of the Western Division of New South Wales*. Government Printer, Sydney.

Armstrong, R. 2004. Baiting operations: Western Shield review – February 2003. *Conservation Science Western Australia*, 5, 31–50.

Bainbridge, DRJ and Jabbour, HN. 1998. Potential of assisted breeding techniques for the conservation of endangered mammalian species in captivity: A review. *The Veterinary Record*, 143, 159–68.

Banks, SC, Hoyle, SD, Horsup, A, Sunnucks, P and Taylor, AC. 2003. Demographic monitoring of an entire species (the Northern Hairy-nosed Wombat, *Lasiorhinus krefftii*) by genetic analysis of non-invasively collected material. *Animal Conservation*, 6, 101–07.

Banks, SC, Piggott, MP, Hansen, BD, Robinson, NA and Taylor, AC. 2002. Wombat coprogenetics: Enumerating a Common Wombat population by microsatellite analysis of faecal DNA. *Australian Journal of Zoology*, 50, 193–204.

Bradshaw, CJA and Brook, BW. 2005. Disease and the devil: Density-dependent epidemiological processes explain historical population fluctuations in the Tasmanian Devil. *Ecography*, 28, 181–90.

Brereton, R, Bennett, S and Mansergh, I. 1995. Enhanced greenhouse climate change and its potential effect on selected fauna of south-eastern Australia: A trend analysis. *Biological Conservation*, 72, 339–54.

Burbidge, AA and Eisenberg, JF. 2006. Conservation and management. In PJ Armati, CR Dickman and ID Hume (eds) *Marsupials*, pp 299–330. Cambridge University Press, Cambridge.

Burbidge, AA and McKenzie, NL. 1989. Patterns in the modern decline of Western Australia's vertebrate fauna: Causes and conservation implications. *Biological Conservation*, 50, 143–98.

Burbidge, AA, Johnson, KA, Fuller, PJ and Southgate, RI. 1988. Aboriginal knowledge of the mammals of the central deserts of Australia. *Australian Wildlife Research*, 15, 9–39.

Caughley, J. 1980. Native quolls and Tiger Quolls. In C Haigh (ed) *Endangered animals of New South Wales*, pp 45–48. NSW National Parks and Wildlife Service, Sydney.

Cogger, HG, Ford, HA, Johnson, CN, Holman, J and Butler, D. 2003. *Impacts of land clearing on Australian wildlife in Queensland*. WWF-Australia, Brisbane.

Courtenay, J and Friend T. 2003. *Gilbert's Potoroo recovery plan: July 2003–June 2008*. Department of Conservation and Land Management, Wanneroo.

Dickman, CR. 1991. Use of trees by ground-dwelling mammals: Implications for management. In D Lunney (ed) *Conservation of Australia's forest fauna*, pp 125–36. Royal Zoological Society of New South Wales, Sydney.

Dickman, CR. 1994. Native mammals of western New South Wales: Past neglect, future rehabilitation? In D Lunney, S Hand, P Reed and D Butcher (eds) *Future of the fauna of western New South Wales*, pp 81–92. Royal Zoological Society of New South Wales, Sydney.

Dickman, CR. 1994. Mammals of New South Wales: Past, present and future. *Australian Zoologist*, 29, 158–65.

Dickman, CR. 1996. *Overview of the impacts of feral cats on Australian native fauna*. Australian Nature Conservation Agency, Canberra.

Dickman, CR and Read, DG. 1992. The biology and management of dasyurids of the arid zone in New South Wales. *New South Wales National Parks and Wildlife Service Species Management Report*, 11, 1–112.

Dickman, CR, Pressey, RL, Lim, L and Parnaby, HE. 1993. Mammals of particular conservation concern in the Western Division of New South Wales. *Biological Conservation*, 65, 219–48.

Grigg, G. 2002. Conservation benefit from harvesting kangaroos: Status report at the start of a new millennium: A paper to stimulate discussion and research. In D Lunney and CR Dickman (eds) *A zoological revolution: Using native fauna to assist in its own survival*, pp 53–76. Royal Zoological Society of New South Wales, and Australian Museum, Sydney.

Eldridge, MDB, King, JM, Loupis, AK, Spencer, PBS, Taylor, AC, Pope, LC and Hall, GP. 1999. Unprecedented low levels of genetic variation and inbreeding depression in an island population of the Black-footed Rock-wallaby. *Conservation Biology*, 13, 531–41.

Fickel, J, Wagener, A and Ludwig, A. 2007. Semen cryopreservation and the conservation of endangered species. *European Journal of Wildlife Research*, 53, 81–88.

Finlayson, HH. 1932. *Caloprymnus campestris*: Its recurrence and characters. *Transactions of the Royal Society of South Australia*, 56, 148–67.

Fisher, DO, Blomberg, SP and Owens, IPF. 2003. Extrinsic versus intrinsic factors in the decline and extinction of Australian marsupials. *Proceedings of the Royal Society*, B, 270, 1801-08.

Fleming, PJS. 2001. Legislative issues relating to control of Dingoes and other wild dogs in New South Wales. II. Historical and technical justifications for current policy. In CR Dickman and D Lunney (eds) *A symposium on the Dingo*, pp 42-48. Royal Zoological Society of New South Wales, Sydney.

Forsyth, DM, Duncan, RP, Bomford, M and Moore, G. 2004. Climatic suitability, life-history traits, introduction effort, and the establishment and spread of introduced mammals in Australia. *Conservation Biology*, 18, 557-69.

Frankham, R. 1998. Inbreeding and extinction: Island populations. *Conservation Biology*, 12, 665-75.

Friend, JA and Thomas, ND. 2003. Conservation of the Numbat (*Myrmecobius fasciatus*). In ME Jones, CR Dickman and M Archer (eds) *Predators with pouches: The biology of carnivorous marsupials*, pp 452-63. CSIRO Publishing, Melbourne.

Glen, AS and Dickman, CR. 2005. Complex interactions among mammalian carnivores in Australia, and their implications for wildlife management. *Biological Reviews*, 80, 387-401.

Gordon, G and Hrdina, F. 2005. Koala and possum populations in Queensland during the harvest period, 1906-1936. *Australian Zoologist*, 33, 69-99.

Gordon, G, Hrdina, F and Patterson, R. 2006. Decline in the distribution of the Koala *Phascolarctos cinereus* in Queensland. *Australian Zoologist*, 33, 345-58.

Herbert, CA. 2007. From the urban fringe to the Abrolhos Islands: Management challenges of burgeoning marsupial populations. In D Lunney, P Eby, P Hutchings and S Burgin (eds) *Pest or guest: The zoology of overabundance*, pp 129-41. Royal Zoological Society of New South Wales, Sydney.

Horsup, A. 2004. *Recovery plan for the Northern Hairy-nosed Wombat* Lasiorhinus krefftii *2004-2008*. Environmental Protection Agency, Brisbane.

Hundloe, T and Hamilton, C. 1997. *Koalas and tourism: An economic evaluation*. Discussion Paper Number 13. The Australia Institute, Sydney.

IUCN. 2001. *IUCN red list categories and criteria: Version 3.1*. IUCN Species Survival Commission, Gland.

IUCN. 2006. *2006 IUCN red list of threatened species* at www.iucnredlist.org/

Jarman, PJ and Johnson, KA. 1977. Exotic mammals, indigenous mammals, and land-use. *Proceedings of the Ecological Society of Australia*, 10, 146-63.

Jeans, DN. 1972. *An historical geography of New South Wales to 1901*. Reed Education, Sydney.

Jenkins, CFH. 1974. The decline of the Dalgite (*Macrotis lagotis*) and other wild life in the Avon Valley. *Western Australian Naturalist*, 12, 169-72.

Johnson, C, Cogger, H, Dickman, C and Ford, H. 2007. *The impacts of the approved clearing of native vegetation on Australian wildlife in New South Wales*. WWF-Australia, Sydney.

Johnson, CN, Delean, S and Balmford, A. 2002. Phylogeny and the selectivity of extinction in Australian marsupials. *Animal Conservation*, 5, 135-42.

Kemp, LF, Johnston, G and Carthew, SM. 2007. An experimental reintroduction of the previously extinct South Australian mainland Tammar Wallaby (*Macropus eugenii eugenii*). In RT Kingsford (ed) *The biodiversity extinction crisis: An Australasian and Pacific response - Final program*, p 28. ICMS, Sydney.

Kinnear, JE, Onus, ML and Sumner, NR. 1998. Fox control and rock-wallaby population dynamics. II. An update. *Wildlife Research*, 25, 81-88.

Kinnear, JE, Sumner, NR and Onus, ML. 2002. The Red Fox in Australia - An exotic predator turned biocontrol agent. *Biological Conservation*, 108, 335-59.

Krefft, G. 1866. On the vertebrated animals of the lower Murray and Darling, their habits, economy, and geographical distribution. *Transactions of the Philosophical Society of New South Wales*, 1862-1865, 1-33.

Kutt, AS, Van Dyck, S and Christie, SJ. 2005. A significant range extension for the Chestnut Dunnart *Sminthopsis archeri* (Marsupialia: Dasyuridae) in north Queensland. *Australian Zoologist*, 33, 265-68.

Langford, D. 1999. *The Mala* (Lagorchestes hirsutus) *recovery plan*. Parks and Wildlife Commission of the Northern Territory, Alice Springs.

Lim, L. 1992. *Recovery plan for the Kowari* Dasyuroides byrnei *Spencer, 1896* (*Marsupialia: Dasyuridae*). Australian National Parks and Wildlife Service, Canberra.

Lunney, D. 1994. Royal Commission of 1901 on the western lands of New South Wales - An ecologist's summary. In D Lunney, S Hand, P Reed and D Butcher (eds) *Future of the fauna of western New South Wales*, pp 221-40. Royal Zoological Society of New South Wales, Sydney.

Lunney, D. 2001. Causes of the extinction of native mammals of the Western Division of New South Wales: An ecological interpretation of the nineteenth century historical record. *Rangeland Journal*, 23, 44-70.

Lunney, D, Curtin, AL, Ayers, D, Cogger, HG, Dickman, CR, Maitz, W, Law, B and Fisher, D. 2000. The threatened and non-threatened native vertebrate fauna of New South Wales:

Status and ecological attributes. *New South Wales National Parks and Wildlife Service Environmental and Heritage Monograph*, series 4, 1-134.

Lunney, D, O'Connell, M, Sanders, J and Forbes, S. 1989. Habitat of the White-footed Dunnart *Sminthopsis leucopus* (Gray) (Dasyuridae: Marsupialia) in a logged, burnt forest near Bega, New South Wales. *Australian Journal of Ecology*, 14, 335-44.

Maxwell, S, Burbidge, AA and Morris, K (eds). 1996. *The 1996 action plan for Australian marsupials and monotremes*. Wildlife Australia, Canberra.

McKenzie, NL, Burbidge, AA, Baynes, A, Brereton, RN, Dickman, CR, Gordon, G, Gibson, LA, Menkhorst, PW, Robinson, AC, Williams, MR and Woinarski, JCZ. 2007. Analysis of factors implicated in the recent decline of Australia's mammal fauna. *Journal of Biogeography*, 34, 597-611.

Morris, K, Johnson, B, Orell, P, Gaikhorst, G, Wayne, A and Moro, D. 2003. Recovery of the threatened Chuditch (*Dasyurus geoffroii*): A case study. In ME Jones, CR Dickman and M Archer (eds) *Predators with pouches: The biology of carnivorous marsupials*, pp 435-51. CSIRO Publishing, Melbourne.

Murray, AJ, Poore, RN and Dexter, N. 2006. *Project deliverance: The response of 'critical-weight-range' mammals to effective Fox control in mesic forest habitats in far East Gippsland, Victoria*. Department of Sustainability and Environment, Melbourne.

Newsome, AE, Pech, R, Smyth, R, Banks, P and Dickman, CR. 1997. *Potential impacts on Australian native fauna of Rabbit calicivirus disease*. Australian Nature Conservation Agency, Canberra.

Paddle, R. 2000. *The last Tasmanian Tiger: The history and extinction of the Thylacine*. Cambridge University Press, Cambridge.

Richards, JD and Short, J. 2003. Reintroduction of the Western Barred Bandicoot to mainland Australia. *Biological Conservation*, 109, 181-95.

Robinson, AC and Young, MC. 1983. The Toolache Wallaby (*Macropus greyi* Waterhouse). *Department of Environment and Planning South Australia Special Publication*, 2, 1-54.

Rolls, EC. 1969. *They all ran wild*. Angus and Robertson, Sydney.

Salo, P, Korpimäki, E, Banks, PB, Nordström, M and Dickman, CR. 2007. Alien predators are more dangerous than native predators to prey populations. *Proceedings of the Royal Society*, B, 274, 1237-43.

Saunders, G, Coman, B, Kinnear, J and Braysher, M. 1995. *Managing vertebrate pests: Foxes*. Australian Government Publishing Service, Canberra.

Selwood, L and Coulson, G. 2006. Marsupials as models for research. *Australian Journal of Zoology*, 54, 137-38.

Sharland, M. 1962. *Tasmanian wildlife: A popular account of the furred land mammals, snakes and introduced mammals of Tasmania*. Melbourne University Press, Melbourne.

Short, J, Richards, JD and Turner, B. 1998. Ecology of the Western Barred Bandicoot (*Perameles bougainville*) (Marsupialia: Peramelidae) on Dorre and Bernier islands, Western Australia. *Wildlife Research*, 25, 567-86.

Smith, MJ. 1998. Establishment of a captive colony of *Bettongia tropica* (Marsupialia: Potoroidae) by cross-fostering; and observations on reproduction. *Journal of Zoology*, London, 244, 43-50.

Stead, DG. 1935. *The Rabbit in Australia: History, life story, habits, effect upon Australian primary production and best means of extermination*. Winn and Co., Sydney.

Strahan, R (ed). 1995. *The mammals of Australia*, 2nd ed. Australian Museum and Reed Books, Sydney.

Troughton, E Le G. 1932. Australian furred animals, their past, present, and future. *Australian Zoologist*, 7, 173-93.

Troughton, E Le G. 1944. The imperative need for federal control of post-war protection of nature. *Proceedings of the Linnean Society of New South Wales*, 69, iv-xv.

US Fish and Wildlife Service. 2007. *Draft recovery plan for the Ivory-billed Woodpecker* (Campephilus principalis). US Fish and Wildlife Service, Atlanta.

Van Dyck, S and Strahan, R (eds). 2007. *The mammals of Australia*, 3rd ed. Reed New Holland, Sydney.

Wilson, BA and Aberton, JG. 2006. Effects of landscape, habitat and fire on the distribution of the White-footed Dunnart *Sminthopsis leucopus* (Marsupialia: Dasyuridae) in the eastern Otways, Victoria. *Australian Mammalogy*, 28, 27-38.

Wilson, BA, Dickman, CR and Fletcher, TP. 2003. Dasyurid dilemmas: Problems and solutions for conserving Australia's small carnivorous marsupials. In ME Jones, CR Dickman and M Archer (eds) *Predators with pouches: The biology of carnivorous marsupials*, pp 407-21. CSIRO Publishing, Melbourne.

Woinarski, JCZ. 2004. In a land with few possums, even the common are rare: Ecology, conservation and management of possums in the Northern Territory. In RL Goldingay and SM Jackson (eds) *The biology of Australian possums and gliders*, pp 51-62. Surrey Beatty and Sons, Sydney.

Wood Jones, F. 1925. *The mammals of South Australia*. Part III. *The Monodelphia*. REE Rogers, Government Printer, Adelaide.

ACKNOWLEDGEMENTS

Author's acknowledgements

I am most grateful to David Happold for giving me the chance to begin my formal studies on marsupials, and to many friends and colleagues for showing me how to find and identify different species in the bush. In Canberra, Pippa Carron, Ken Green, Will Osborne, Viv Read, Tony Stewart, Chris Tidemann and Dedee Woodside were particularly helpful to the novice in their midst. Barry Fox, Tony Lee and Steve Van Dyck also provided much appreciated advice and encouragement at this time. Adrian Bradley, Don Bradshaw, Bert Main, Norm McKenzie, Keith Morris, Wayne Packer and Phil Withers helped me to learn about the magnificent marsupials of Western Australia, and Patsy Armati, Tony English, Ian Hume, Dan Lunney, Alan Newsome and Ron Strahan have been just as generous in sharing their experiences with the spectacular fauna of central and eastern Australia. My students, research assistants and other collaborators over the years have kept me on my toes, sharing their insights and discoveries, and providing great camaraderie in the lab and the field. I have learned much from the written works of others, too numerous to list here but must single out for special mention the pioneering and visionary studies in recent years of Mike Archer, Andrew Burbidge, Tim Flannery, Steve Morton, Hugh Tyndale-Biscoe and Pat Woolley.

Most of the funding for my research has been provided by the Australian Research Council, with additional support from Australian Geographic and CSIRO-University awards. Institutional support for my frequent absences and dented field vehicles has come from the Australian National University, the University of Western Australia and, for the last 19 years, the University of Sydney. My field work on the marsupials of central Australia has taken place on properties owned by David and Paula Smith, Howard and Shirley Jukes, Gordon and Colleen McDonald, Tom and Leanne Churches and, most recently, Bush Heritage Australia. I am most grateful to all of them for their interest and hospitality in hosting our field teams for so long.

This book owes much to the efforts of many people. In the first place I am extremely grateful to Nola Mallon for encouraging me to write this account, and for providing the opportunity to work with Rosemary Woodford Ganf. Nola suggested many excellent ideas for the content, design and direction of the book, and gentle but necessary reminders when deadlines loomed. Rosemary contributed the beautiful paintings of marsupials that grace the pages of this work and bring their subjects so vividly to life. I thank Joe Benshemesh, Diana Fisher, Menna Jones, David Lindenmayer, Dan Lunney and Peter McRae for providing first hand accounts of the marsupials that they know better than anyone else, Kado Muir for providing an Indigenous perspective on marsupials, Tim Flannery for writing the Foreword, and Adele Haythornthwaite for generously agreeing to compile accounts of all the currently accepted species. Gus Bernardi, Mathew Crowther and Carol McKechnie gave advice on various aspects on the manuscript, and Janice Chanson and Mike Hoffmann of the IUCN kindly checked forthcoming changes in the status of various marsupial species. Mathew Crowther, Liz Denny, Graeme Finlayson, Kris Helgen, Clare McArthur and the late John Kirsch read one or more chapters, and Hugh Tyndale-Biscoe generously reviewed the entire manuscript. Their comments helped immeasurably. Kathie Stove reviewed and edited all parts of the book, providing thoughtful and constructive comments throughout, and Lynn Twelftree has designed the book beautifully. I greatly appreciate their contributions.

Finally, I owe a great debt to my family. My parents always encouraged their wayward sons to follow their passion for natural history, and my brother Keith and I have been fortunate to be able to do this and so never really grow up. My wife Carol and daughter Alice have shared in the whole enterprise, nurturing marsupials and many other creatures in our home over the years, holding the fort during my prolonged absences in the field and at my desk, providing excellent advice and comments on the book as it progressed, and sustaining me at all times. This book is for them.

CRD

Artist's acknowledgements

The initial planning and gratefully received advice for this collection of paintings came from Dr Robert Warneke and the late Peter Aitken.

Many institutions throughout Australia provided me with valuable information. They include CSIRO, Division of Wildlife Research; South Australian Museum and its Evolutionary Biology Unit; Queensland National Parks and Wildlife; Museum Victoria; Perth Zoo; The University of Adelaide, Waite Arboretum; Adelaide Botanic Gardens; Queen Victoria Museum Launceston and Punch Bowl Zoo Launceston.

I am grateful to the many zoologists and botanists working in many fields of natural science who freely shared their knowledge with me. Without their help it would have been impossible for me to put together such a collection of work.

For the preparation of the paintings of the possums, koala and wombats I owe thanks to Dr John Winter, Robert Atherton, Rupert Russell, Dr R Fairfax, Joan Dixon, the late Professor WD Williams, Dr S Barker, Norman Parish, Dr AC Robinson, Dr RT Wells, Dr A Wells, Dr ME Christian, Dr G Gordon, Dr PA Woolley, Robert Hawkes, Dr Judy West, Joy Fox, Dr Margaret Lawrence, Dr M Andrew, Joan Powling and David Symons.

Similarly I would like to thank Dr Catherine Kemper, Dr Chris Watts, Dr Heather Aslin, Dr PA Woolley, Mr WK Youngson, Mr RH Green, Montague and Glen Turner, the late Dr Meredith Smith, Dr ME Christian, Jiri Lochman, Dr Tim Flannery and, most especially, the wonderful late Dr John Calaby for their help in preparing the paintings of the carnivorous marsupials, bandicoots, kangaroos, wallabies and rat-kangaroos.

My sincere thanks go to Kathie Stove, the editor, and Lynn Twelftree, the designer, for their contributions.

Finally my deepest thanks must go to the publisher, Nola Mallon, and the author, Christopher Dickman, who against all odds have brought this book to fruition.

I dedicate my contribution to this book to my inspirational family, Dr Michael Woodford, June Woodford, Dr George Ganf and Ross Ganf. I will forever be grateful to you all.

RWG

INDEX